GENOME EVOLUTION

Based on the Proceedings of an International Symposium

GENOME
EVOLUTION

PHENOTYPIC
VARIATION

held in Cambridge, England

THE SYSTEMATICS ASSOCIATION
SPECIAL VOLUME NO. 20

GENOME EVOLUTION

Edited by

G. A. DOVER

*Department of Genetics
and Fellow, King's College
University of Cambridge, U.K.*

R. B. FLAVELL

*Plant Breeding Institute
Trumpington, Cambridge, U.K.*

1982

Published for the
SYSTEMATICS ASSOCIATION

by
ACADEMIC PRESS
LONDON NEW YORK
PARIS SAN DIEGO SAN FRANCISCO SAO PAULO
SYDNEY TOKYO TORONTO

ACADEMIC PRESS INC. (LONDON) LTD.
24/28 Oval Road
London NW1

United States Edition published by
ACADEMIC PRESS INC.
111 Fifth Avenue
New York, New York 10003

British Library Cataloguing in Publication Data
Genome evolution—(The Systematics Association
   special volume, ISSN 0309-2593; no. 20)
   1. Chromosomes—Congresses
   I. Dover, G.A.   II. Flavell, R.B.
   III. Series
   574.87'322   QH600

   ISBN 0-12-221380-7
   ISBN 0-12-221382-3 Pbk

Printed in Great Britain

PREFACE

Over the past two decades it has become clear that the nuclear genomes of eukaryotes contain amounts of DNA that are far in excess of that required to code for the proteins of the organisms. The nature and unexpected behaviour of this DNA has been examined in detail in a wide variety of species with a battery of increasingly sophisticated techniques. Speculation on the biological functions of such sequences has run through the gamut of possibilities from useless junk to reserves of "slave" genes. More recently it was suggested that the accumulation of excess DNA represents nothing more than the selfish propagation of selectively neutral molecules. This idea challenged some cherished concepts concerning biological adaptation and evolutionary processes. In addition it rekindled some age-old debates about the nature of genomic mutations and their phenotypic consequences.

It was against this background that we considered the time appropriate for a meeting to discuss some concepts of evolutionary biology from the perspective of the nuclear genome. Unfortunately, neither time nor money were sufficient to debate all aspects of the genome. Accordingly, we invited topics for discussion which would serve as a framework for general issues in genome biology. The majority of the issues raised at the meeting is represented in the essays presented in this book.

Initially our invitations to the contributors included some general guidelines as to the questions which might be addressed in the oral and written presentations. There is nothing new under the sun, however, and we discovered later that Richard Goldschmidt had asked the same basic questions in his last book, Theoretical Genetics (1955). We took his formulations as the theme of the meeting:

(1) What is the nature of the genetic material?
(2) How does the genetic material act in controlling specific development?
(3) How do the nature and action of the genetic material account for evolution?

It is appropriate, 26 years later, to attempt to clarify
current problems associated with the same questions. In accordance with this, the meeting was designed to assess the types,
rates and magnitudes of changes in multigene and non-coding
families, and to discuss their possible significance for the
ontogeny and phylogeny of organisms. Each author has been
free to develop these questions as he or she saw fit and we
have not exercised editorial control over their contributions.
      The meeting was divided into four sections dealing with
(a) modes of genome evolution (b) evolution of gene families
(c) nuclear organisation and DNA content and (d) genome
evolution and speciation. It is in the nature of biology that
the lines between such sections can become blurred and it is
entirely appropriate that authors range freely from one to
another in their contributions.
      The meeting that finally took place had an atmosphere that
reflected the growing interest of molecular and evolutionary
biologists in the implications of the newer data and concepts
emerging from genome studies. It is difficult to assess what
makes for a successful discussion meeting. Nevertheless,
there is no doubt that much of the credit goes to the five
chairmen who were able to guide lengthy and stimulating discussions involving a large number of participants. We are
very grateful for the experience and wisdom of Sydney Brenner,
Walter Bodmer, Bryan Clarke, John Fincham and John Thoday.
Although their words are not recorded here their influence on
the meeting was substantial. In addition we are indebted to
John Maynard Smith for his entertaining and clear overview of
the problems facing biologists of all description.
      Finally, we are appreciative of the oral presentations of
Michael Ashburner (Cambridge), Georgio Bernardi (Paris) and
Allan Wilson (Berkeley) which are not reproduced here. A
written account of Georgio Bernardi's contribution can be
found in his recent paper in Cold Spring Harbor Symp. Quant.
Biol. (1982) **46.**
      We always suspected that in entering unfamiliar territory
between molecular and evolutionary biology, we were likely to
tread on some conceptual landmines. If some of the forays
have proved to be hazardous at least they have helped explore
the territory. Once the dust has settled then it might be
valuable (if our spouses are willing!) to launch another
Cambridge Genome Meeting.
      We are happy to acknowledge the financial support of the
Royal Society; King's College Research Centre and Bedford
Fund; the Systematics Association and the British Council.

Gabriel Dover and Richard Flavell
January, 1982.

CONTENTS

LIST OF CONTRIBUTORS

BAYEV, A.A., *Department of Molecular Biology, King's Buildings, Mayfield Road, University of Edinburgh, Scotland, EH9 3JR (present address: Institute for Molecular Biology, Academy of Science of USSR, Vavilov Str., Moscow B-334, USSR)*

BENNETT, M.D., *Plant Breeding Institute, Cambridge CB2 2LQ, England*

BENTLEY, D.L., *Medical Research Council Laboratory of Molecular Biology, Hills Road, Cambridge CB2 2QH, England*

BOSTOCK, C.J., *Medical Research Council, Mammalian Genome Unit, King's Buildings, West Mains Road, Edinburgh, EH9 3JT, Scotland*

BOWCOCK, A.M., *Department of Molecular Biology, King's Buildings, Mayfield Road, University of Edinburgh, EH9 3JR, Scotland*

BRITTEN, R.J., *Division of Biology, California Institute of Technology, Pasadena, Calif. 91125, U.S.A.*

BROWN, L., *Department of Molecular Biology, King's Buildings, Mayfield Road, University of Edinburgh, EH9 3JR, Scotland*

BROWN, S., *Department of Genetics, University of Cambridge, Cambridge CB2 3EH, England*

COEN, E., *Department of Genetics, University of Cambridge, Cambridge CB2 3EH, England*

DALLAS, J., *Department of Genetics, University of Cambridge, Cambridge CB2 3EH, England*

DAVIDSON, E.H., *Division of Biology, California Institute of Technology, Pasadena, Calif. 91125, U.S.A.*

DONEHOWER, L., *Laboratory of Tumor Virus Genetics, National Cancer Institute, Bethesda, MD 20050, U.S.A.*

DOOLITTLE, W.F., *Department of Biochemistry, Dalhousie University, Halifax, Nova Scotia, Canada B3H 4H7*

DOVER, G.A., *Department of Genetics, University of Cambridge, Cambridge CB2 3EH, England*

FINNEGAN, D.J., *Department of Molecular Biology, King's Buildings, Mayfield Road, University of Edinburgh EH9 3JR, Scotland*

FLAVELL, R.B., *Plant Breeding Institute, Cambridge CB2 2LQ, England*

FORSTER, A., *Medical Research Council, Laboratory of Molecular*

Biology, Hills Road, Cambridge, CB2 2QH, England
GILLESPIE, D., *Barry Ashbee Laboratories, Orlowitz Cancer
Institute, Hahnemann Medical College, Philadelphia,
PA 19102, U.S.A.*
HÖCHTL, J., *Institut für Physiologische Chemie, Physikalische
Biochemie und Zellbiologie der Universität München, Goethe-
strasse 33, 8000 München 2, Federal Republic of
Germany*
HUTCHINSON, J., *Plant Breeding Institute, Cambridge CB2 2LQ,
England*
JEFFREYS, A.J., *Genetics Department, University of Leicester,
Leicester LE1 7RH, England*
JENKINS, G., *Department of Agricultural Botany, U.C.W.,
Aberystwyth, Wales, U.K.*
JONES, K.W., *Department of Genetics, University of Edinburgh,
King's Buildings, West Mains Road, Edinburgh EH9 3JN,
Scotland*
LEIPOLDT, M., *Abteilung Humangenetik, Zentrum Hygiene und
Humangenetik der Universität, Nikolausberger Weg 5a,
D-3400 Göttingen, Federal Republic of Germany*
MACGREGOR, H.C., *Department of Zoology, University of Leices-
ter, Leicester LE1 7RH, England*
MANUELIDIS, L., *Section of Neuropathology, Yale University
School of Medicine, 333 Cedar Street, New Haven,
Conn. 06510, U.S.A.*
MATTHYSSENS, G., *Medical Research Council, Laboratory of
Molecular Biology, Hills Road, Cambridge CB2 2QH, England
(present address: Free University of Brussels, Institute
of Molecular Biology, B-1640 St Genesius-Rode, Belgium)*
MIKLOS, G.L.G., *Research School of Biological Sciences,
Australian National University, Canberra, A.C.T., Australia*
MILSTEIN, C.P., *Medical Research Council, Laboratory of
Molecular Biology, Hills Road, Cambridge CB2 2QH, England
(permanent address: Institute of Animal Physiology,
Babraham, Cambridge, England)*
NEUMAIER, P.S., *Institut für Physiologische Chemie, Physikal-
ische Biochemie und Zellbiologie der Universität München,
Goethestrasse 33, 8000 München 2, Federal Republic of
Germany*
PAGÈS, M., *Equipe de Recherche de Biophysique 140 CNRS,
Bâtiment Chimie Extension, Université des Sciences et
Techniques du Langudoc, Place Eugène Bataillon -34060
Montpelier Cedex, France*
PECH, M., *Institut für Physiologische Chemie, Physikalische
Biochemie und Zellbiologie der Universität München,
Goethestrasse 33, 8000 München 2, Federal Republic of
Germany*
RABBITTS, T.H., *Medical Research Council, Laboratory of
Molecular Biology, Hills Road, Cambridge CB2 2QH, England*

REES, H., *Department of Agricultural Botany, U.C.W.,
Aberystwyth, Wales, U.K.*

ROIZÈS, G.P., *Equipe de Recherche de Biophysique 140 CNRS,
Bâtiment Chimie Extension, Université des Sciences et
Techniques du Languedoc, Place Eugène Bataillon -34060
Montpelier Cedex, France*

SCHELLER, R.H., *Division of Biology, California Institute of
Technology, Pasadena, Calif. 91125, U.S.A. (present
address: Institute of Cancer Research, 701 West 168th
Street, New York, N.Y. 10068, U.S.A.)*

SCHMIDTKE, J., *Abteilung Humangenetik, Zentrum Hygiene und
Humangenetik der Universität, Nikolausberger Weg 5a,
D-3400 Göttingen, Federal Republic of Germany*

SCHNELL, H., *Institut für Physiologische Chemie, Physikalische
Biochemie und Zellbiologie der Universität München, Goethe-
strasse 33, 8000 München 2, 'Federal Republic of
Germany*

SEAL, A.G., *Plant Breeding Institute, Cambridge CB2 2LQ,
England*

SINGH, L., *Department of Genetics, University of Edinburgh,
King's Buildings, West Mains Road, Edinburgh EH9 3JN,
Scotland*

SMITH, J.M., *School of Biological Sciences, University of
Sussex, Brighton BN1 9QG, England*

STRACHAN, T., *Department of Genetics, University of Cambridge,
Cambridge CB2 3EH, England*

STRAYER, D., *Barry Ashbee Laboratories, Orlowitz Cancer Institute, Hahnemann Medical College, Philadelpha, PA 19102,
U.S.A.*

THOMAS, T.L., *Division of Biology, California Institute of
Technology, Pasadena, Calif. 91125, U.S.A.*

TRICK, M., *Department of Genetics, University of Cambridge,
Cambridge CB2 3EH, England*

TYLER-SMITH, C., *Medical Research Council, Mammalian Genome
Unit, King's Buildings, West Mains Road, Edinburgh EH9 3JT,
Scotland*

WILL, B.M., *Department of Molecular Biology, King's Buildings,
Mayfield Road, University of Edinburgh EH9 3JR, Scotland*

ZACHAU, H.G., *Institut für Physiologische Chemie, Physikal-
ische Biochemie und Zellbiologie der Universität München,
Goethestrasse 33, 8000 München 2, Federal Republic of
Germany*

SYSTEMATICS ASSOCIATION PUBLICATIONS

1. BIBLIOGRAPHY OF KEY WORKS FOR THE IDENTIFICATION OF THE
   BRITISH FAUNA AND FLORA
   3rd edition (1967)
   Edited by G.J. Kerrich, R.D. Meikle and N. Tebble
2. FUNCTION AND TAXONOMIC IMPORTANCE (1959)
   Edited by A.J. Cain
3. THE SPECIES CONCEPT IN PALAEONTOLOGY (1956)
   Edited by P.S. Sylvester-Bradley
4. TAXONOMY AND GEOGRAPHY (1962)
   Edited by D. Nichols
5. SPECIATION IN THE SEA (1963)
   Edited by J.P. Harding and N. Tebble
6. PHENETIC AND PHYLOGENETIC CLASSIFICATION (1964)
   Edited by V.H. Heywood and J. McNeill
7. ASPECTS OF TETHYAN BIOGEOGRAPHY (1967)
   Edited by C.G. Adams and D.V. Ager
8. THE SOIL ECOSYSTEM (1969)
   Edited by H. Sheals
9. ORGANISMS AND CONTINENTS THROUGH TIME (1973)†
   Edited by N.F. Hughes

LONDON. Published by the Association

SYSTEMATICS ASSOCIATION SPECIAL VOLUMES

1. THE NEW SYSTEMATICS (1940)
   Edited by Julian Huxley (Reprinted 1971)
2. CHEMOTAXONOMY AND SEROTAXONOMY (1968)*
   Edited by J.G. Hawkes
3. DATA PROCESSING IN BIOLOGY AND GEOLOGY (1971)*
   Edited by J.L. Cutbill
4. SCANNING ELECTRON MICROSCOPY (1971)*
   Edited by V.H. Heywood
5. TAXONOMY AND ECONOMY (1973)*
   Edited by V.H. Heywood

---

* Published by Academic Press for the Systematics Association

† Published by the Palaeontological Association in conjunction
  with the Systematics Association

---

* Published by Academic Press for the Systematics Association

PART I
MODES OF GENOME EVOLUTION

Selfish DNA after Fourteen Months

W. FORD DOOLITTLE

Department of Biochemistry, Dalhousie University, Nova Scotia

"But under us all moved, and moved us, gently, up and down, and from side to side." ... Samuel Beckett

Against the sound advice of Dr Brenner, I have felt obliged to defend the ideas on "selfish DNA" presented by us and Drs Orgel and Crick numerous times during the 14 months since the simultaneous appearance of our articles in Nature (Doolittle and Sapienza, 1980; Orgel and Crick, 1980). My faith remains unshaken and the ideas remain essentially unchanged, but I have learned some useful things about the nature of the opposition to them. After reminding you of what it is that we and Orgel and Crick actually said and discussing the barriers to its acceptance (some rational, some not), I will attempt to relate the selfish DNA notion to the theme of this conference.

THE NOTION OF SELFISH DNA

Pieces of genomic DNA exist in environments which promote their replication and allow the accumulation by them of heritable mutations. Pieces of DNA which code for nothing useful run the risk of deletion. If there are sequence-specific strategies by which deletion can be avoided, natural selection operating within genomes, independently of organismal phenotype or population adaptability, will give rise to and maintain DNAs which adopt such sequence-specific strategies (Doolittle and Sapienza, 1980; Orgel and Crick, 1980; Dover and Doolittle, 1980; Orgel *et al.*, 1980; Sapienza and Doolittle, 1981; Doolittle, 1981a). This argument and its corollary, that we need seek no other explanation for the existence of genomic DNAs whose behaviours ensure survival, seem logically unassailable. Both are sterile, however, if no sequence-specific survival strategies are possible, or if no known genomic DNAs can be shown to have adopted them.

1. *Demonstrably transposable elements*

Transposition which produces faithful and equally transposable
copies of the transposed element elsewhere in the genome,
while preserving the integrity of the original copy at its
original site, is a sequence-specific survival strategy.
Elements adopting it can be eliminated from the genome only by
highly improbable, simultaneous, multiple deletion events. All
adequately characterized bacterial insertion sequences and
transposons (with the possible exception of IS4) appear to
have adopted this strategy (Shapiro, 1979, 1981; Arthur and
Sheratt, 1979; Kopecko, 1980; Calos and Miller, 1980; Harshey
and Bukhari, 1981).

All well-characterized eukaryotic transposable elements
(that is, the mobile genetic elements of yeast and *Drosophila*
and the proviral genomes of vertebrate retroviruses) show
astonishing similarities in structure to bacterial transposons
and, like them, generate short flanking repeats of genomic
sequences initially present only once at the site of insertion
(e.g. see Farabaugh and Fink, 1980; Gafner and Phillipsen,
1980; Dunsmuir *et al.*, 1980; Shimotohno *et al.*, 1980; Majors
and Varmus, 1981; Hishinuma *et al.*, 1981). This analogy in
structure, as well as certain sequence homologies, has been
taken as evidence for evolutionary homology, that is, common
descent (Temin, 1980; Levis *et al.*, 1980; Ju and Skalka,
1980). They are not necessarily that (see Doolittle, 1980,
1981a) and there is as yet no reason to suppose that structural analogies reflect an underlying similarity in molecular
mechanisms of transposition. One can imagine other mechanisms.

For instance, it has been suggested that genomic DNA copies
of retroviral genomes can transpose themselves through transcription, reverse transcription, circularization and reintegration, a process similar in its consequences (but not its
molecular details) to bacterial transposition, and one which
formally ensures self-preservation in the same way (Temin,
1980; Shoemaker *et al.*, 1980; Hishinuma *et al.*, 1981).

It has yet to be shown that this really happens with integrated retrovirus proviral genomes, but this is probably what
does happen with *Drosophila* nomadic DNA elements. These are
heavily transcribed, the transcripts are predominantly nuclear, some are large enough (as with retroviruses) to include
all element-specific information, and circular double-
stranded DNAs containing this information are (again as with
retroviruses) found in the nucleus (Stanfield and Helinski,
1976; Young, 1979; Stanfield and Lengyel, 1979, 1980; Levis
et al., 1980; Young and Schwartz, 1981). I would like to
predict (in the absence of any evidence) that similar

mechanisms will apply to yeast Ty-1-like elements, which are also abundantly transcribed (Cameron *et al.*, 1979).

For both *Drosophila* and yeast mobile elements, we can *at least* eliminate transposition mechanisms which involve excision and insertion of individual copies without *some* sort of replication event; such non-conservative mechanisms could not easily account for the observed behaviours. Copy number could increase in a haphazard way *via* recombination between chromosomes bearing elements at different sites, but the constancy in copy number and variability in position observed for nomadic DNAs in different *Drosophila melanogaster* strains would not then be expected, nor would the increase in copy number observed in cultured cells (Young, 1979; Young and Schwartz, 1981). Furthermore, elements transposing by excision and reinsertion without duplication, and amplified by recombination in a purely random way, should all be equally "old", having enjoyed the same number of rounds of DNA replication regardless of the number of times they have suffered transposition. Thus they should be equally divergent in sequence and should not be divisible into subfamilies whose members show greater homology to each other than to members of other subfamilies of the same superfamily. It now seems clear that yeast Ty-1 comprises a superfamily divisible into subfamilies in just this way (Williamson *et al.*, 1981; Kingsman *et al.*, 1981). Finally, at least one apparent case of insertion of a new Ty-1 copy without destruction of any pre-existing copies has been documented (Cameron *et al.*, 1979).

Excision and reintegration without duplication is not a strategy for self-preservation, although it would probably still be considered "transposition", since this word seems only loosely defined. Certain recombinational switch mechanisms (Silverman and Simon, 1980; Howe, 1980) and perhaps controlling elements in maize (Peterson, 1981; Burr and Burr, 1981) may behave in this way. I would not call the DNAs involved in such events selfish; genetic variability of a predictable and specific kind induced by their movements during the life-time of a single organism or within a few generations of a population could easily be maintained by natural selection operating through phenotype.

Because self-preservation *is* ensured by strategies which are identical in their consequences (if not in their mechanisms) to bacterial transposition, we have placed most emphasis on genetic elements employing such strategies (Doolittle and Sapienza, 1980; Sapienza and Doolittle, 1981). (Ohta and Kimura (1981) have recently presented a mathematical treatment of the evolutionary behaviour of such elements.) However, similar arguments can be applied, with varying degrees of force, to other components of the eukaryotic genome.

2. *Middle-repetitive DNAs as a class*

The transposable elements of yeast and *Drosophila* make up most
or all of the middle-repetitive DNA components of these organisms (excluding multigene families coding for ribosomal RNAs,
histones and the like). They are generally rather long (several thousand base-pairs) and, in general, dispersed throughout the genome. Physically similar DNAs are found in many
eukaryotes (Adams *et al.*, 1980; Anderson *et al.*, 1981; Sapienza and Doolittle, 1981), although few have been well-
characterized. However, in species showing the *Xenopus*, or
short-period interspersion pattern, most middle-repetitive
elements are much smaller (300 to 500 base-pairs). Are these
too transposable elements to which we need assign no other
function than self-preservation?

The best known of these short dispersed repetitive DNAs is
a 300 base-pair sequence repeated some 300,000 times in the
human genome. Members of this family, the *Alu*I family, represent more than half of all human middle-repetitive DNA, and
their relatives are identifiable in other mammals (Rubin *et
al.*, 1980; Jelinek *et al.*, 1980). Cloned family members show
75 to 85% sequence homology. All lack the terminal inverted
or direct repeats characteristic of bacterial transposable
elements and the longer mobile DNAs of yeast and *Drosophila*
(Bell *et al.*, 1980; Pan *et al.*, 1981). However, all cloned
copies appear to be flanked by direct repeats which lie outside the boundaries of the element and which are of *different*
sequence, just as is found with integrated copies of bacterial, yeast and *Drosophila* transposable elements and retroviral proviruses (Bell *et al.*, 1980; Pan *et al.*, 1981).
Elder *et al.* (1981) find that a majority of cloned *Alu*I DNAs
can be transcribed *in vitro* by RNA polymerase III, and that
transcription begins very near one end of the element and
terminates at variable lengths beyond the other end. They
suggest that the RNA polymerase III recognition site involves
a sequence similar to that found in RNA polymerase III-trans-
cribed adenovirus and transfer RNA genes and located some 70
nucleotides downstream from the initiation site. I have suggested that a site quite homologous to the intragenic RNA
polymerase III recognition site of vertebrate 5 S ribosomal
RNA genes, and located near the centre of the element, is involved (Doolittle, 1981b). Although the abundant 7 S (ca.
300 nucleotide) cytoplasmic RNAs which form hybrids with
*Alu*I DNA do not appear to be direct transcripts of most members of this family (Weiner, 1980), and although many nuclear
transcripts of *Alu*I sequences may be initiated by RNA

polymerase II at sites outside the element (Jelinek *et al.*,
1980), the findings of Pan *et al.* (1981) and Elder *et al.*
(1981) are tantalizing. As the latter authors suggest,

> "a transposition mechanism (*which I would further suggest operates via an RNA polymerase III-generated intermediate*) seems a plausible if unproved way to understand the interspersion of these elements throughout the genome. Moreover, the fact that these sequences retain extensive homology within and between mammalian species argues that they are functional in some way - even if their only function is self preservation".

It is unlikely that selection for "function" operating
through phenotype on each member of any large family of
middle-repetitive DNAs *could* maintain the observed degrees of
sequence homogeneity (Salser and Isaacson, 1976; Doolittle and
Sapienza, 1980) or explain the usual observation of greater
within-species than between-species sequence conservation.
As Klein *et al.* (1978) and Anderson *et al.* (1981) admit, the
data on sea urchin middle-repetitive DNAs are most easily
accommodated by some mechanism formally analogous to transposition. Excision-reintegration mechanisms of transposition
will probably not easily explain the data and do not ensure
genomic survival. Transposition of either the bacterial or
retroviral type will do both, as will some other mechanisms
for saltatory replication (Britten and Davidson, 1971; Anderson *et al.*, 1981), for whose operation on dispersed repetitive
DNAs there is no evidence, even circumstantial.
    We had predicted earlier, on the basis of such arguments,
that most or all eukaryotic middle-repetitive DNAs (except
for members of multigene families) will prove to be transposable elements (or the degenerate descendants thereof) and
thus potentially selfish, no function being *required* to explain their origin or maintenance (Doolittle and Sapienza,
1980; Sapienza and Doolittle, 1981). Data on the *Alu*I family
suggesting its involvement in a variety of cellular functions
were thus a source of some discomfort (Rubin *et al.*, 1980;
Jelinek *et al.*, 1980). I am delighted to see evidence for
the transposability of the members of this family accumulate,
and would like to see more extensive study of specific small
middle-repetitive DNAs in genetically manipulatable organisms
showing the *Xenopus* pattern of interspersion. Perhaps *Caenorhabditis elegans*, which has a variety of such DNAs (Emmons
et al., 1981) and a genome even smaller than *Drosophila*'s, is
the organism of choice.

8   W.F. DOOLITTLE

3. *Highly-repetitive DNAs*

As far as I'm aware, most of the data bearing on the evolution of tandemly repeated (highly-repetitive or "satellite")
DNAs can be explained by the unequal crossing-over model art-
iculated mostly clearly by Smith (1976, 1978; and see reviews
by John and Miklos, 1977; Walker, 1978; Bostock, 1980; Brutlag,
1980). Tandem arrays of repeated elements, once generated,
are in a sense self-perpetuating, although the lengths of such
arrays can be varied by the same mechanism which creates them.
Unequal crossing-over provides a mechanism for conserving
within-species homogeneity, even of sequences under no functional constraints. No selection of any sort need be invoked
unless there are certain sequences, or repeat lengths, which
are favoured by unequal crossing-over mechanisms. *If* there
are favoured sequences or lengths, DNAs which have acquired
them can be considered selfish.

    Data clearly supporting this possibility are limited and
difficult for me to assess. Brown and Dover (1979,1980) provide some evidence for preferential amplification of specific
sequence variants in mouse satellite DNAs. Christie and Skinner (1980) invoke "selective amplification super-imposed on a
biased pattern of mutation" to explain the evolutionary behaviour of a crab satellite DNA. Donehower *et al*. (1980) suggest
that the behaviour of conserved and variable regions in the
sequence of the 342-343 base-pair repeat unit of primate
highly repetitive DNAs indicates sequence and size constraints
imposed by recombination mechanisms and nucleosomal packaging.
Hsieh and Brutlag (1979a) find that sequence divergence in the
1.688g cm^{-3} *D. melanogaster* satellite is largely limited to
nine sites within the 359 base-pair monomer sequence, and
Brutlag (1980) suggests that the "limitation of divergence to
a few prominent locations may be a result of selective amplification of repeat units containing the particular changes".
Similarly, the significant homology in sequence between this
invertebrate satellite and the major tandemly repetitive DNAs
of primates may indicate some role of sequence in promoting
amplification, although other explanations are possible (Brutlag, 1980). The original suggestion by Musich, Brown and co-
workers (Musich *et al*., 1977; Brown *et al*., 1979), that repeats
of nucleosomal length are favoured, can be questioned (Fittler
and Zachau, 1979; Singer, 1979; Wu and Manuelidis, 1980).
However, the finding by Hsieh and Brutlag (1979b) of a protein
which specifically binds to a highly symmetric region of the
1.688 g cm^{-3} *D. melanogaster* satellite has, as Brutlag (1980)
stresses,

    "evolutionary relevance since they [such binding proteins]

give rise to functional significance to the repeated
sequence itself, which would introduce selectable properties on both the amounts and sequence of satellite
DNAs".

The question here of course is what one means by functional
significance. It is possible that DNA-binding proteins can
favour the amplification of sequences they recognize even
though the sequence (or indeed even the binding) may be of no
other significance.
This is all pretty shaky ground, and it is rendered no less
treacherous by the uncertainty about what roles, if any,
highly repetitive DNAs play in chromosome behaviour and evolution (John and Miklos, 1977; Bostock, 1980; Brutlag, 1980;
Dover *et al*., 1982) and about whether, if they do play such
roles, they do so "accidentally" (Dover, 1978) or because
they have been selected to do so. The distinction between
"ignorant" (Dover, 1980) and selfish highly repetitive DNAs,
although meaningful, is not easy to make (Dover and Doolittle,
1980). I would not wish to base my defence of our ideas on
the properties of this class of DNAs.

4. *Unique-sequence non-coding DNA*

The functions of much of the non-coding unique-sequence DNA
of eukaryotic genomes, including many "spacers" (Federoff,
1979) and all but the 5' and 3' termini of intervening sequences, remain undefined (Crick, 1980). We cannot claim such DNA
as selfish, since its sequence is not generally well-conserved
and it has no obvious protection against elimination. Clearly,
some of it arises by gene duplication and subsequent drift
(Ohno, 1980; Proudfoot and Maniatis, 1980). But some of it
may represent degenerate or no longer transposable copies of
once-transposable elements, selfish in origin but now defence-
less and moribund. Some notion of the fraction of genomic DNA
which is of this sort might be derived by extrapolation from
hybridization experiments performed under conditions of decreasing stringency (Burr and Schimke, 1980; Murray *et al*.,
1981), but exact determination of that fraction of truly
unique sequence DNA which is of selfish origin may be, by
definition, impossible. I do not want to argue strongly for
this position either. I am sure it is partly correct, but
the main arguments about whether DNA can properly be called
selfish hinge on interpretations, not data, and I think it
best to concentrate on DNAs such as transposable elements
where at least the data begin to look clear.

BARRIERS TO ACCEPTANCE

1. Irrational barriers

Both we and Drs Orgel and Crick were surprised at the sometimes rather violent negative reactions provoked by our explicit articulation of ideas which we thought intuitively obvious.
Only some of the more moderate and scientifically defensible
objections have appeared in the pages of Nature or elsewhere
(Dover, 1980; Cavalier-Smith, 1980; Jain, 1980; Shapiro, 1981;
Dover et al., 1981). Apparently, what we did was point out a
serious disharmony between the ways in which many molecular
biologists who believe they understand the evolutionary process and most contemporary population geneticists (who probably do understand it) think about how natural selection
really works. This disharmony is reflected in the fact that
many of the criticisms we have heard are directed at the words
we have used, at our alliance with sociobiologists whose ideas,
if carried to extremes which most of them (Dawkins, 1976; Wilson, 1978) would reject, have politically distasteful implications, or at our cavalier disregard of earlier articulations
of similar ideas which had escaped our own notice and made no
significant impact on molecular biological thinking. I will
not address these criticisms.

Others seem to have mistaken the selfish DNA notion for a
simple revival of the suggestion that much genomic DNA is
"junk". For instance, Manuelidis (1980) complains that "the
lack of knowledge about the function of various repeated DNAs
may prematurely relegate some of them to essentially meaningless categories". Ohno (1981) is even more outspoken; saying
"not surprisingly, the concept of "junk" DNA was revived, this
time disguised as "selfish" DNA".

That is not what we meant at all. Junk (as usually
defined) is indeed meaningless and uninteresting (Orgel et
al., 1980). A piece of DNA which has been selected only for
intragenomic survival is, on the other hand, a fascinating
thing: a subcellular "organism" whose biology and evolution
are so simple we may eventually be able to understand them
completely, but so subtle that this understanding will not
come without effort or excitement.

2. Rational barriers

(a) The "testability" problem. Respectable scientific hypotheses are supposed to be testable. I part slightly with Drs
Orgel and Crick (1980) on the ease with which we could be proven right. The argument that selfish DNAs are an inevitable
product of selection operating within the genome is derived

from what we hoped were logical considerations. We could
never prove that any given piece of DNA we think selfish did
not arise through individual or group selection to play some
role so subtle that it may never be identified (Gould and
Lewontin, 1979). If someone could show (1) that a family or
families of transposable elements contributed nothing to individual fitness and (2) that the activities of the family or
families established within the population a level of genetic
variability which was in fact greater than the optimum needed
to allow evolutionary adaptation to plausible environmental
challenge (and hence was detrimental), we could rest our case.
I think Michael Young (Young, 1979; Young and Schwartz, 1981)
may have already accomplished the first feat, with *D. melanogaster* and *D. simulans*. But in what conceivable experimental
or natural system could one accomplish the second? The situation *can* be dealt with theoretically (see, for example, Ohta
and Kimura (1981)) and Donal Hickey (personal communication)
has calculated that selfish elements transposing in the bacterial way can be driven to fixation in a sexually-reproducing
population *even if their effect on individual fitness is significantly negative!*

Creationists have claimed, and not without at least some
philosophical support (Little, 1980; but see Rosenberg, 1978),
that the theory of evolution by natural selection is not a
falsifiable hypothesis, but one would not wish to side with
them in denying its utility in understanding what biology is
all about. (I must quickly add that neither we nor Drs Orgel
and Crick for a minute pretend that the selfish DNA notion
partakes of the grandeur of the theory of natural selection.
That is a big thing, and what we have proposed is but a little
thing, a minor and as I see it perfectly orthodox articulation
of the bigger thing.) However, the selfish DNA idea shares
with the Darwinian paradigm the awkward feature of being in-
accessible to unequivocal experimental test.

(b) The everyday utility of transposable elements. Genetic
determinants for antibiotic and heavy-metal resistance, toxin
production, catabolism of novel substrates and doubtless many
other traits are frequently transposon-borne (Bukhari *et al.*,
1977; Stuttard and Rozee, 1980; Kopecko, 1980; Calos and
Miller, 1980). The acquisition of such determinants is
clearly advantageous both to the transposon and its host, but
this need not mean that the former was not selfish in origin
before such acquisition. The conversion of parasites into
symbionts is a recurring theme in evolutionary biology (Price,
1980), and can even be demonstrated in the laboratory (Jeon
and Jeon, 1976).

Some genetic rearrangements are known to play roles in the
regulation of gene expression. As recently reviewed by Shapiro (1981), those which are at least partially understood at
the molecular level include the systems responsible for (1)
phase variation in *Salmonella*, (2) host range determination in
phages Pl and Mu, (3) sex determination in yeast, (4) anti-
genic variation in trypanosomes, and (5) the generation of
antigen-specific immunoglobulins. These genetic rearrangements are frequently discussed with transposition or amplification events of the sort we consider selfish but, as I understand them, they are not necessarily the same kind of thing.
The first two involve simple inversions, not sequence duplication (Silverman and Simon, 1980; Howe, 1980). The third probably proceeds by "gene conversion" (Klar *et al.*, 1980; Haber
et al., 1980). The fourth may resemble transposition, but its
mechanism remains unclear, and the last involves deletion of
DNA between the segments to be joined (Davis *et al.*, 1980).
As Shapiro (1981) says,

> "there is no net synthesis of phosphodiester bonds in
> these recombination events, although it is possible that
> some single strand degradation and resynthesis does
> occur. Studies of transposable elements in bacteria
> have revealed a very different situation ...".

No matter what model for bacterial transposition one favours
(Shapiro, 1979; Arthur and Sheratt, 1979; Harshey and Bukhari,
1981), replication and retention of the original copy of the
transposable element at its original site is required, and I
suspect this will prove to be true of most eukaryotic transposable elements present in multiple copies.

Genetic rearrangements involved in differentiation may
have originated from transposition mechanisms employed by
originally selfish elements, but there is no compelling reason
to believe that they have, and no reason whatever to assume by
extrapolation from these systems that all genetic rearrangements play meaningful selected roles in the regular determination of phenotype. I do not regard the fact that transcription of middle-repetitive DNAs of yeast, *Drosophila* and man
is both frequent and, at least in the former case, controlled
by other genes (Errede *et al.*, 1980) to be compelling evidence
for function. In both *Drosophila* and man, such transcripts
are not predominantly associated with polysomes (Weiner,
1980; Young and Schwartz, 1981) and in the latter are heterogeneous in sequence. Nor do I regard the fact that some prokaryotic and eukaryotic transposable elements contain transcription start or stop sites which can turn on or off genes
downstream from the point of insertion (Chaleff and Fink,

1980; Calos and Miller, 1980; Errede *et al.*, 1980) as evidence
for function selected through phenotype. In most cases, the
genes affected are essentially randomly chosen, not clearly
the target of a specific regulatory system. (This may not be
true of maize controlling elements (Peterson, 1981) but the
physical basis of their "transposition" remains unclear.)
Retroviral genomes lacking their own *onc* genes appear to be
able to induce tumours by insertion near, and activation of,
chromosomal genes, but surely one would not consider this the
product of natural selection acting on either host or virus
(Hayward *et al.*, 1981; Neel *et al.*, 1981; Payne *et al.*, 1981).

*(c) The long-term evolutionary advantage of genomic rearrange-
ments.* Transposable elements promote genetic rearrangements,
and the kinds of rearrangements (transpositions, deletions and
inversions) seem similar in both prokaryotes and eukaryotes
(Calos and Miller, 1980; Roeder and Fink, 1980; Shapiro, 1981).
This (and the occasional turning on and off of genes adjacent
to the site of insertion) appears to be all that many, perhaps
most, transposable elements actually do for the organism which
bears them and it does not seem to be a good thing. Selection
operating on individuals should eliminate such elements. Thus
many have claimed that transposable elements are maintained
because they play important "evolutionary roles". This is
not a straw man which Carmen Sapienza and I set up in order to
have a hypothesis against which to pit the notion of selfish
DNA. I can only document this with quotations not, I hope,
taken out of context:

> "Whether they (insertion sequences) exert control func-
> tions at these positions or are simply kept in reserve
> as prefabricated units for the evolution of new control
> circuits remains unclear." (Nevers and Saedler, 1977).

> "It is possible that the sole function of these elements
> is to promote genetic variability ..." (Strobel *et al.*,
> 1979)

> "A tenable hypothesis regarding the function of trans-
> position is that it allows adaptation of a particular
> cell to a new environment." (Cameron *et al.*, 1979)

> "All these alterations could lead to changes in struc-
> tural gene function and in the control of gene expres-
> sion and could provide the organism with a means of
> rapid adaptation to environmental change." (Roeder and
> Fink, 1980)

Evolutionary roles have similarly been invoked for hetero-

chromatic highly repetitive DNAs, whose presence *does* affect
recombination in neighbouring and distant regions and whose
characteristics *may* (although the experimental evidence is
not strong) affect chromosome pairing (John and Miklos, 1977;
Bostock, 1980; Brutlag, 1980).

Neither we nor Drs Orgel and Crick denied that transposable elements or heterochromatic highly repetitive DNAs have
such evolutionary effects, nor that these effects might not
be important, perhaps even the basis for macroevolutionary
change. *What we were arguing against was the assumption that
these elements arose through and are maintained by natural
selection because of these effects.*

This assumption is often only implicit in the writings of
many who suggest that the only roles of mobile dispersed and
tandemly reiterated DNAs are evolutionary ones. Thus we have
been accused by some of these of misrepresenting their positions and thus indeed of attacking straw men after all. I
apologize to those who feel we have put words in their mouths.
But I do not see how statements that the only "functions" of
transposable elements or highly repetitive DNAs are to generate or modulate genetic variability can mean anything other
than that natural selection maintains, and probably even gave
rise to, such elements through selection for such "functions".
Shapiro (1981) has been brave enough to articulate this view
outright:

"Why, then, are insertion elements not removed from the
genome? I think the answer must be that there is a selective advantage in the ability to generate new chromosome primary structure."

Those who speculate on the function of excess DNA have formu-
lated this position in a more extreme way. For instance,
Jain (1980) states

"at a given point of time there will always be large
amounts of non-specific DNA. This fraction is best
described as "incidental" rather than "selfish" DNA.
We may call it incidental because it is a byproduct of
the inherent property of mutability of the genome, a
characteristic to which natural selection attaches
great importance even if it leads to the production
of repeated sequences and a wasteful deployment of
energy. Viewed in this light, non-functional DNA is
very much a product of natural selection - a selection
operating for mutability *per se*."

The question of whether natural selection operates in this

way, that is of whether the evolutionary process itself
evolves under the direct influence of natural selection, lies
at the root of the *real* controversy over whether self-
maintaining, structured, genomic components without phenotypic function can properly be called "selfish". This may
seem like a small and almost metascientific quibble. In fact
it is not; it is one of the most troublesome questions in evolutionary biology today. It manifests itself in debates over
the origin and maintenance of mechanisms involved in the
optimization of mutation rates (if these exist; Cox, 1976),
recombination, sexual reproduction, altruistic behaviours of
all sorts and even speciation (Hamilton, 1964; Wilson, 1975;
Dawkins, 1976; Alexander and Borgia, 1978; Maynard Smith,
1978; Stanley, 1979; Gould, 1980). Such mechanisms are not
clearly advantageous to, and can be detrimental to, the fitness of the individual. Yet they may increase the long-term
survival properties of the group to which the individual
belongs, thus seeming to be the product of what has been
called "group selection".

All invocations by molecular biologists of strictly evolutionary roles for transposable elements, highly repetitive
DNAs or even intervening sequences (Gilbert, 1978) are
strongly group selectionist, although this is not generally
appreciated. Nor is it generally appreciated by molecular
biologists that evolutionary biologists have become during the
last 15 years (Hamilton, 1964) increasingly restrictive in
defining the constraints under which group selection can
operate and have in general sought, with increasing urgency,
ways in which traits which appear only advantageous to the
group can be seen as advantageous to the individual (Wilson,
1975), or, in the final analysis, the trait-determining gene
(Dawkins, 1976). The debate remains active, the proponents
of more holistic views have not yet been defeated, and the
arguments are too complex for me to follow. However, I'm safe
in saying that whenever one *can* rationalize the origin and
maintenance of traits which generate genetic variability (variability upon which selection *may* act at the level of the
group) in terms of selection operating at lower levels (the
individual or the gene), one *should*. That is what we were
trying to do, and in our attempts went a bit further than Dawkins (1976), in defining a unit of selection lower than that
of the expressed gene, which spreads through a population by
virtue of its self-serving phenotypic expression. That unit
is the self-serving sequence, which spreads itself through the
genome without supervision by natural selection operating on
phenotype, and which can be spread between genomes in
sexually reproducing organisms more rapidly than other sequences, again without phenotypic supervision (Orgel and Crick,
1980; D. Hickey, unpublished results).

SELFISH DNA AND MACROEVOLUTION

Three separate threads of evidence seem to be converging to
produce an increasingly popular if not uncontested view of
macroevolutionary events and how they differ from microevolutionary events. The first is the punctuationalist (as opposed
to gradualist) interpretation of the fossil record championed
by Eldredge and Gould, Stanley and others (see Stanley, 1979;
Gould, 1980; Schopf, 1980). These authors argue that macro-
evolutionary events (the kind which are involved in speciation; or cladogenesis) occur over short periods of time (as
geologists measure time) and are different in kind from the
slow fixation of selectively favourable or neutral mutations
which form the basis of the "molecular clock" and which are
responsible for phyletic evolution (anagenesis). The second
is the evidence (Wilson *et al.*, 1974; Bush *et al.*, 1977; Stanley, 1979) that high rates of speciation are correlated (in
some causal way) with rates of chromosomal rearrangement (as
measured by karyotypic diversity). The third is the notion
that alternations in regulatory and developmental patterns,
rather than in the primary sequences of structural genes, are
responsible for most of what we observe as gross morphological
change in the evolution of higher organisms (Wilson *et al.*,
1974,1977; King and Wilson, 1975).

There are at least two possible ways to weave these threads
together. One is to assume that genomic rearrangements lead
to reproductive isolation within a deme (only those pairings
between individuals with similarly altered chromosomes being
fertile) and thus directly to (sympatric) speciation. Alterations in form and function then ensue by drift and normal
micro-evolutionary processes. In this view, the initial event
leading to speciation is not necessarily adaptive, and may
even temporarily lower fitness (Gould, 1980). Alternatively,
saltatory genome rearrangements occurring over a short period
could produce, through their effects on developmental and
regulatory patterns, dramatic but adaptive alterations in
organismal form and function ("systemtic mutations", Goldschmidt, 1940) and thus lead indirectly to reproductive isolation and speciation (Wilson *et al.*, 1974; Stanley, 1979;
Dover, 1980).

I find this latter, "hopeful monster" model for speciation
a bit fuzzy. It depends rather heavily on the assumption
that saltatory genome rearrangements *can* lead to orchestrated
rearrangements of patterns of gene expression. In the words
of Wilson *et al.* (1974), "adaptation is probably a complex
process requiring new interactions among many genes. The
reshuffling of genes may be an important mechanism by which
new interactions can occur". Unitary models for the

regulation of eukaryotic gene expression, like that of Davidson and Britten (1979), seem to encourage those who still hold
such views, although I am not sure why. Appealing as these
models are, there remains little direct evidence for them
(Davidson and Britten, 1979; Brown, 1981). I see no reason to
believe that karyotypic changes of the type shown by Bush *et
al*. (1977) to be associated with high rates of evolution have
anything to do with any meaningful reassortment of DNA sequences involved in regulation.

If the kinds of DNA we call selfish are involved in all
this, I suspect their role is "accidental" (Dover, 1978).
Selfish elements can in principle bring about chromosome rearrangements of sufficient magnitude to establish reproductive
isolation. This is not, however, why the elements are there,
but an incidental byproduct of the mechanisms by which they
maintain themselves. Such evolutionarily significant rearrangements may in fact be quite infrequent. Young (1979)
described two strains of *D. melanogaster* in which the number
and kinds of nomadic middle-repetitive DNAs were the same,
but in which the chromsomal locations of the elements were
almost entirely different. These strains interbreed freely.

There are, however, mobile repetitive genetic elements
which do have profound effects on interstrain fertility.
Included among these are P and I factors determining hybrid
dysgenesis in *Drosophila* (Bregliano *et al*., 1980; Engels,
1981). It's not clear to me why hybrid dysgenesis is a good
thing. It surely promotes reproductive isolation and hence
speciation, but it is not easy to argue that P and I elements
were selected for this purpose, nor in fact that the ability
to speciate has been "selected for" in the way we normally
use this term. Speciose taxa may be more likely to survive
random extinction, but this does not mean that natural selection has operated on mechanisms by which speciation is effected, at least not in the direct way it can be envisioned to
have operated on microevolutionary phyletic changes which
track environmental change (Stanley, 1979; Gould, 1980).
From my limited knowledge of dysgenesis, I suspect it could
be interpreted as the result of rampant mobilization, in a
novel cytoplasm, of transposable elements usually held in
check by other chromosomal or cytoplasmic genes in the strain
in which they normally reside (Engels, 1981). This behaviour
is not adaptive for the elements themselves; it limits "host
range". Furthermore, factors promoting dysgenesis are most
often lost in an apparently random way during laboratory maintenance, an observation most consistent with the assumption
that these elements move by excision and reinsertion without
replication (Engels, 1981). I am not sure I would call such
elements selfish, and do not think that those who wish to

attribute selected evolutionary roles to all transposable
middle-repetitive DNAs (Young and Schwartz, 1981; Shapiro,
1981) should derive as much comfort from the behaviour of
factors promoting dysgenesis as they seem to do.

GENOMIC PLASTICITY

I do not think I am alone in suspecting that the genomes of
eukaryotes are in general more "plastic" than those of prokaryotes, although it would be hard to justify this suspicion
quantitatively. Certainly those eukaryotes we know well have
more mobile elements, of more different kinds, than do any of
the prokaryotes we know well. They also usually have more
"excess" DNA of all kinds, which may provide the raw material
for the formation of new genes (Ohno, 1970; Hinegardner, 1977),
and the added advantage of being able to shuffle modules of
protein-coding sequences around in the manner envisioned by
Gilbert (1978) and indirectly confirmed by a variety of observations (summarized most recently by Artymiuk *et al.*, 1981).
Probably all of these things do help speed the evolutionary
process and perhaps that is why eukaryotes seem (anthropocen-
trism aside) to have developed a much greater range of morphological and behavioural (but not biochemical) adaptations than
have prokaryotes (Carlile, 1980).

If we accept all this, then it is reasonable to ask whether
the greater evolutionary potential of the eukaryotic genome
is a recent development and whether plasticity itself arose
through natural selection operating at some level higher than
that of the genome or the individual, to favour evolutionary
potential. I have already expressed my views on the latter
point. There is no need for us to assume that any of the
components of the eukaryotic genome which impart fluidity
(middle-repetitive DNAs, highly repetitive DNAs, spacers,
introns) arose through selection operating at levels higher
than that of the genome. Prudence dictates that we should
not.

There is also no reason to assume that agents promoting
fluidity arose in the eukaryotic nuclear genomic lineage only
after its enclosure within the nuclear membrane; that is,
only after the appearance of the first eukaryotic cells, an
event usually considered to have occurred between 1.2 and 1.5
thousand million years ago (Schopf, 1977). Woese and collab-
orators (Woese and Fox, 1977a,b), Darnell (1978) and I (Doolittle, 1978,1980) have argued elsewhere that the qualitative
differences in the ways in which eukaryotic nuclear and prokaryotic genomes are organized, expressed and regulated, as
well as the quantifiable differences in rRNA sequences, imply
that the eukaryotic nuclear genomic lineage diverged from

prokaryotic (or at least eubacterial) genomic lineages very
early in cellular evolution. So early, we believe, that these
qualitative and quantitative differences reflect separately
achieved solutions to problems of how genes should be organized, expressed and regulated, problems not yet solved in that
ancient last common ancestor.

I personally view many of the plasticity-promoting elements
of the modern eukaryotic nuclear genome as likely to have
characterized the genome of that common ancestor. Most people
other than Sidney Fox (Fox and Dose, 1977) like to think that
genotype evolved before phenotype so that nucleic acids which
by virtue of their structure alone were more likely to be
replicated (somehow) were the very first products of natural
selection. Such molecules are by definition selfish, and
presumably were important components of the first cellular
genomes (Eigen *et al.*, 1981). Darnell (1978) and I (Doolittle,
1978) pointed out the difficulty of envisioning a mechanism by
which, or a selection pressure because of which, introns could
be inserted between the domain-coding regions of already functioning intact genes. Thus we suggested that these too were
primitive structures, structures which have been *lost* in prokaryotes, not *gained* in eukaryotes (see also Sharp, 1981;
Chambon, 1981). Reanney (1974,1979) has argued, I believe
convincingly, that the earliest cells must have had large
genomes, in which sequences potentially expressible in selectable phenotype were multiply represented. One can see some
advantage to redundancy, plasticity and reassortment at the
RNA level (splicing) in systems which could neither replicate,
transcribe or translate genetic information with great accuracy. One can also see greater evolutionary flexibility.

Once replication, transcription and translation became
accurate, redundancy and DNAs not expressed in useful phenotype became a burden. Selection at the level of the individual should favour their loss. Flexibility and further evolutionary potential are lost at the same time, but selection
for their retention is less immediate and, I presume, less
intense.

But when radically new adaptations are called for, it is
those entities which by chance have not succumbed to the
enticement of more rapid growth through genomic streamlining
which are more likely to be able to rise to the call. Thus I
see the plasticity of the eukaryotic genome as a primitive
trait, a trait maintained only by chance and against selection at the level of the individual. Its preservation in us
more complex beings is the product of something more like
species selection (Stanley, 1979; Gould, 1980). Those who
have lost plasticity (prokaryotes, some "lower" eukaryotes)
are many in number but fewer in kind and more limited in

potential. It is not the deliberately meek but the accidentally primitive who have, at least temporarily, inherited the
earth.

ACKNOWLEDGEMENTS

I thank Carmen Sapienza for his active participation and
moral support in preparing what I hope will be the last paper
on selfish DNA I ever write; the Medical Research Council and
Natural Sciences and Engineering Research Council of Canada
for supporting research which has stimulated my thinking about
evolution but has nothing whatever to do with the eukaryotic
genome; and Jim Shapiro for strengthening my belief in natural
selection as a major determinant in the evolutionary history
of transposable elements. This review is dedicated to Carl
Woese on the occasion of his 41st birthday.

REFERENCES

Adams, J.W., Kaufman, R.E., Kretschmer, P.J., Harrison, M.
    and Nienhuis, A.W. (1980). A family of long reiterated
    DNA sequences, one copy of which is next to the human beta
    globin gene. *Nucl. Acids Res.* 8, 6113–6128.
Alexander, R.D. and Borgia, G. (1978). Group selection, alt-
    ruism, and the levels of organization of life. *Annu. Rev.
    Ecol. Syst.* **9**, 449–474.
Anderson, D.M., Scheller, R.H., Posakony, J.W., McAllister,
    L.B., Trabert, S.G., Beall, C., Britten, R.J. and Davidson,
    E.H. (1981). Repetitive sequences of the sea urchin gen-
    ome. Distribution of members of specific repetitive fami-
    lies. *J. Mol. Biol.* 145, 5–28.
Arthur, A. and Sheratt, D.J. (1979). Dissection of the trans-
    position process: a transposon-encoded site-specific recom-
    bination system. *Mol. Gen. Genet.* 175, 267–274.
Artymiuk, P.J., Blake, C.C.F. and Sippel, A.E. (1981). Genes
    pieced together – exons delineate homologous structures of
    diverged lysozymes. *Nature (London)* 290, 287–288.
Bell, G.I., Pictet, R. and Rutter, W.J. (1980). Analysis of
    the regions flanking the human insulin gene and sequence
    of an Alu-family member. *Nucl. Acids Res.* 8, 4091–4109.
Bostock, C. (1980). A function for satellite DNA? *Trends
    Biochem. Sci.* 5, 117–119.
Bregliano, J.C., Picard, G., Bucheton, A., Pelisson, A.,
    Lavige, J.M. and L'Heritier, P.L. (1980). Hybrid dys-
    genesis in *Drosophila melanogaster*. *Science* 207, 606–611.
Britten, R.J. and Davidson, E.H. (1971). Repetitive and non-
    repetitive DNA sequences and a speculation on the origins
    of evolutionary novelty. *Quart. Rev. Biol.* 46, 111–137.

Brown, D.D. (1981). Gene expression in eukaryotes. *Science* 211, 667–674.

Brown, F.L., Musich, P.R. and Maio, J.J. (1979). The repetitive sequence structure of component alpha DNA and its relationship to the nucleosomes of African green monkey. *J. Mol. Biol.* 131, 777–799.

Brown, S.D.M. and Dover, G.A. (1979). Conservation of sequences in related genomes of *Apodemus*: constraints on the maintenance of satellite DNA sequences. *Nucl. Acids Res.* 6, 2423–2434.

Brown, S.D.M. and Dover, G.A. (1980). Conservation of segmental variants of satellite DNA of *Mus musculus* in a related species: *Mus spretus*. *Nature (London)* 285, 47–49.

Brutlag, D.L. (1980). Molecular arrangement and evolution of heterochromatic DNA. *Annu. Rev. Genet.* 14, 121–144.

Bukhari, A.I., Shapiro, J.A. and Adhya, S.L. (1977). Editors of: "DNA insertion elements, plasmids, and episomes". Cold Spring Harbor Laboratory Press, Cold Spring Harbor.

Burr, B. and Burr, F.A. (1981). Detection of changes in maize DNA at the *Shrunken* locus due to the intervention of *Ds* elements. *Cold Spring Harbor Symp. Quant. Biol.* 45, 463–466.

Burr, H.E. and Schimke, R.T. (1980). Intragenomic DNA sequence homologies in the chicken and other members of the class *Aves*: DNA re-association under reduced stringency conditions. *J. Mol. Evol.* 15, 291–307.

Bush, G.L., Case, S.M., Wilson, A.C. and Patton, J.L. (1977). Rapid speciation and chromosomal evolution in mammals. *Proc. Nat. Acad. Sci., U.S.A.* 74, 3942–3946.

Calos, M.P. and Miller, J.H. (1980). Transposable elements. *Cell* 20, 579–595.

Cameron, J.R., Loh, E.Y. and Davis, R.W. (1979). Evidence for transposition of dispersed repetitive DNA families in yeast. *Cell* 16, 739–751.

Carlile, M.J. (1980). From prokaryote to eukaryote: gains and losses. *Symp. Soc. Gen. Microbiol.* 30, 1–40.

Cavalier-Smith, T. (1980). How selfish is DNA? *Nature (London)* 285, 617–618.

Chaleff, D.T. and Fink, G.R. (1980). Genetic events associated with an insertion mutation in yeast. *Cell* 21, 227–237.

Chambon, P. (1981). Split genes. *Sci. Amer.* 244, 60–71.

Christie, N.T. and Skinner, D.M. (1980). Evidence for non-random alterations in a fraction of the highly repetitive DNA of a eukaryote. *Nucl. Acids Res.* 8, 279–298.

Cox, E.C. (1976). Bacterial mutator genes and the control of spontaneous mutation. *Annu. Rev. Genet.* 10, 135–156.

Crick, F.H.C. (1979), Split genes and RNA splicing. *Science* 204, 264–271.

Darnell, J.E. Jr (1978).   Implications of RNA-RNA splicing in
    evolution of eukaryotic cells. *Science* **202**, 1257-1260.
Davidson, E.H. and Britten, R.J. (1979).   Regulation of gene
    expression: possible role of repetitive sequences. *Science*
    202, 1257-1260.
Davis, M.M., Kim, S.K. and Hood, L. (1980).   Immunoglobulin
    class switching: developmentally regulated DNA rearrange-
    ments during differentiation. *Cell* **22**, 1-2.
Dawkins, R. (1976).   "The Selfish Gene". Oxford University
    Press, New York.
Donehower, L., Furlong, C., Gillespie, D. and Kurnit, D.
    (1980).   DNA sequence of baboon highly repeated DNA: evi-
    dence for evolution by nonrandom unequal crossovers.
    Proc. Nat. Acad. Sci., U.S.A. **77**, 2129-2133.
Doolittle, W.F. (1978).   Genes-in-pieces: were they ever
    together? *Nature (London)* **272**, 581-582.
Doolittle, W.F. (1980).   Revolutionary concepts in evolution-
    ary cell biology. *Trends Biochem. Sci.* **5**, 146-149.
Doolittle, W.F. (1981a).   Prejudices and preconceptions about
    genome evolution. *In* "Evolution today: proceedings of the
    second international congress of systematic and evolution-
    ary biology: (Scudder, C.G.E. and Reveal, J., eds), Univer-
    sity Park Press, Baltimore, *in the press.*
Doolittle, W.F. (1981b).   5 S ribosomal RNA genes and the
    *Alu*I family: evolutionary and functional significance of a
    region of strong homology. *FEBS Letters* **126**, 147-149.
Doolittle, W.F. and Sapienza, C. (1980).   Selfish genes, the
    phenotype paradigm and genome evolution. *Nature (London)*
    284, 601-603.
Dover, G.A. (1978).   DNA conservation and speciation: adapt-
    ive or accidental? *Nature (London)* **272**, 123-124.
Dover, G. (1980).   Ignorant DNA? *Nature (London)* **285**, 618-
    620.
Dover, G. and Doolittle, W.F. (1980).   Modes of genome evolu-
    tion. *Nature (London)* **288**, 646-647.
Dover, G.A., Brown, S.D.M., Coen, E.S., Dallas, J., Strachan,
    T. and Trick, M. (1982).   The dynamics of genome evolution
    and species differentiation. *In* "Genome Evolution" (Dover,
    G.A. and Flavell, R.B. Eds). Academic Press, London.
Dunsmuir, P., Brorein, W.J., Simon, M.A. and Rubin, G.M.
    (1980).   Insertion of the *Drosophila* transposable element
    copia generates a 5 base-pair duplication. *Cell* **21**, 575-
    579.
Eigen, M., Gardiner, W., Schuster, P. and Winkler-Oswalisch,
    R. (1981).   The origin of genetic information. *Sci. Amer.*
    244, 88-118.

Elder, J.T., Pan, J., Duncan, C.H. and Weissman, S.M. (1981).
    Transcriptional analysis of interspersed repetitive poly-
    merase III transcription units in human DNA. *Nucl. Acids
    Res.* **9**, 1171-1189.
Emmons, S.W., Rosenzweig, B. and Hirsh, D. (1981). Arrange-
    ment of repeated sequences in the DNA of the nematode
    Caenorhabditis elegans. *J. Mol. Biol.* **144**, 481-500.
Engels, W.R. (1981). Hybrid dysgenesis in *Drosophila* and the
    stochastic loss hypothesis. *Cold Spring Harbor Symp. Quant.
    Biol.* **45**, 561-566.
Errede, B., Cardillo, T.S., Sherman, F., Dubois, E., Des-
    champs, J. and Wiane, J.-M. (1980). Mating signals control
    expression of mutations resulting from insertion of a trans-
    posable repetitive element adjacent to diverse yeast genes.
    Cell **22**, 427-436.
Farabaugh, P.J. and Fink, G.R. (1980). Insertion of the euk-
    aryotic transposable element Ty-1 creates a 5 base-pair
    duplication. *Nature (London)* **286**, 352-356.
Federoff, N. (1979). On spacers. *Cell* **16**, 697-710.
Fittler, F. and Zachau, H.G. (1979). Subunit structure of
    alpha-satellite DNA containing chromatin from African
    green monkey cells. *Nucl. Acids Res.* **7**, 1-14.
Fox, S.W. and Dose, K. (1977). "Molecular Evolution and the
    Origin of Life". Marcel Dekker, New York.
Gafner, J. and Phillipsen, P. (1980). Common features of
    transposition - a yeast transposon also generates duplica-
    tions of the target sequence. *Nature (London)* **286**, 414-
    418.
Gilbert, W. (1978). Why genes in pieces? *Nature (London)*
    271, 501.
Goldschmidt, R. (1940). "The Material Basis of Evolution".
    Yale University Press, New Haven.
Gould, S.J. (1980). Is a new and general theory of evolution
    emerging? *Paleobiology* **6**, 119-130.
Gould, S.J. and Lewontin, R.C. (1979). The spandrels of San
    Marco and the Panglossian paradigm: a critique of the
    adaptationist programme. *Proc. Roy. Soc. ser. B* **205**,
    581-598.
Haber, J.E., Rogers, D.T. and McCusker, J.H. (1980). Homo-
    thallic conversions of yeast mating-type genes occur by
    intrachromosomal recombination. *Cell* **22**, 277-289.
Hamilton, W.D. (1964). The genetical evolution of social
    behaviour, I, II. *J. Theor. Biol.* **7**, 1-52.
Harshey, R.M. and Bukhari, A.I. (1981). A mechanism of DNA
    transposition. *Proc. Nat. Acad. Sci., U.S.A.* **78**, 1090-
    1094.
Hayward, W.S., Neel, B.G. and Astrin, S.M. (1981). Activa-
    tion of a cellular *onc* gene by promoter insertion in ALV-

induced lymphoid leukosis. *Nature (London)* **290**, 475–480.

Hinegardner, R. (1977). Evolution of genome size. *In* "Molecular Evolution" (Ayala, F.J., ed.) pp. 179–199, Sinauer Associates, Sunderland, Mass.

Hishinuma, F., DeBona, P.J. Astrin, S. and Skalka, A.M. (1981). Nucleotide sequence of acceptor site and termini of integrated avian endogenous provirus *ev* 1: Integration creates a 6 bp repeat of host DNA. *Cell* **23**, 155–164.

Howe, M.M. (1980). The invertible G segment of phage Mu. *Cell* **21**, 605–606.

Hsieh, T. and Brutlag, D.L. (1979a). Sequence and sequence variation within the 1.688 g cm^{-3} satellite DNA of *Drosophila melanogaster*. *J. Mol. Biol.* 135, 465–481.

Hsieh, T. and Brutlag, D.L. (1979b). A protein that preferentially binds *Drosophila* satellite DNA. *Proc. Nat. Acad. Sci., U.S.A.* **76**, 726–730.

Jain, H.K. (1980). Incidental DNA. *Nature (London)* **288**, 647–648.

Jelinek, W.R., Toomey, T.P., Leinwand, L., Duncan, C.H., Biro, P.A., Choudary, P.V., Weissman, S.M., Rubin, C.M., Houck, C.M., Deininger, P.L. and Schmid, C.W. (1980). Ubiquitous interspersed repeated sequences in mammalian genomes. *Proc. Nat. Acad. Sci., U.S.A.* 77, 1398–1402.

Jeon, K.W. and Jeon, M.S. (1976). Endosymbiosis in amoebae: recently established endosymbionts have become required cytoplasmic components. *J. Cell Physiol.* 89, 337–344.

John, B. and Miklos, G.L.G. (1977). Functional aspects of satellite DNA and heterochromatin. *Int. Rev. Cytol.* 58, 1–113.

Ju, G. and Skalka, A.M. (1980). Nucleotide sequence analysis of the long terminal repeat (LTR) of avian retroviruses: structural similarities with transposable elements. *Cell* 22, 379–386.

King, M.C. and Wilson, A.C. (1975). Evolution at two levels in humans and chimpanzees. *Science* 188, 107–116.

Kingsman, A.J., Gimlich, R.L., Clarke, L., Chinault, A.C. and Carbon, J. (1981). Sequence variation in dispersed repetitive sequences in *Saccharaomyces cerevisiae*. *J. Mol. Biol.* 146, 619–632.

Klar, A.J.S., McIndoo, J., Strathern, J.N. and Hicks, J.B. (1980). Evidence for a physical interaction between the transposed and the substituted sequences during mating type gene transposition in yeast. *Cell* 22, 291–298.

Klein, W.H., Thomas, T.L., Lai, C., Scheller, R.H., Britten, R.J. and Davidson, E.H. (1978). Characteristics of individual repetitive sequence families in the sea urchin genome studied with cloned repeats. *Cell* 14, 889–900.

Kopecko, D. (1980). Involvement of specialized recombination

in the evolution and expression of bacterial genomes. *In*
"Plasmids and Transposons: Environmental Effects and Maintenance Mechanisms" (Stuttard, C. and Rozee, K.R., eds)
pp. 165-205, Academic Press, New York.

Levis, R., Dunsmuir, P. and Rubin, G.M. (1980). Terminal
repeats of the *Drosophila* transposable element *copia*:
nucleotide sequence and genomic organization. *Cell* **21**,
581-588.

Little, J. (1980). Evolution: myth, metaphysics, or science?
New Sci. **87**, 708-709.

Majors, J.E. and Varmus, H.E. (1981). Nucleotide sequences
at host-proviral junctions for mouse mammary tumour virus.
Nature (London) **289**, 253-25.

Manuelidis, L. (1980). Novel class of mouse repeated DNAs.
Nucl. Acids Res. **8**, 3247-3258.

Maynard Smith, J. (1978). "The Evolution of Sex". Cambridge
University Press, Cambridge.

Murray, M.G., Peters, D.L. and Thompson, W.F. (1981).
Ancient repeated sequences in the pea and mung bean genomes and implications for genome evolution. *J. Mol. Evol.*
17, 31-42.

Musich, P.R., Brown, F.L. and Maio, J.J. (1977). Mammalian
repetitive DNA and the subunit structure of chromatin.
Cold Spring Harbor Symp. Quant. Biol. **42**, 1147-1160.

Neel, B.G., Hayward, W.S., Robinson, H.L., Fang, J. and
Astrin, S.M. (1981). Avian leukosis virus-induced tumors
have common proviral integration sites and synthesize discrete new RNAs: oncogenesis by promoter insertion. *Cell*
23, 323-334.

Nevers, P. and Saedler, H. (1977). Transposable genetic elements as agents of gene instability and chromosomal rearrangements. *Nature (London)* **268**, 109-115.

Ohno, S. (1970). "Evolution by Gene Duplication". Springer
Verlag, Berlin.

Ohno, S. (1981). (AGCTG)(AGCTG)(AGCTG)(GGGTG) as the primor-
dial sequence of intergenic spacers: the roles in immunoglobulin class switch. *Differentiation* **18**, 65-74.

Ohta, T. and Kimura, M. (1981). Some calculations on the
amount of selfish DNA. *Proc. Nat. Acad. Sci., U.S.A.* **78**,
1129-1132.

Orgel, L.E. and Crick, F.H.C. (1980). Selfish DNA: the ultimate parasite. *Nature (London)* **284**, 604-607.

Orgel, L.E., Crick, F.H.C. and Sapienza, C. (1980). Selfish
DNA. *Nature (London)* **288**, 645-646.

Pan, J., Elder, J.T., Duncan, C.H. and Weissman, S.M. (1981).
Structural analysis of interspersed repetitive polymerase
III transcription units in human DNA. *Nucl. Acids Res.*
9, 1151-1170.

Payne, G.S., Courtneidge, S.A., Crittenden, L.B., Fadly, A.M., Bishop, J.M. and Varmus, H.E. (1981). Analysis of avian leukosis virus DNA and RNA in bursal tumors: viral gene expression is not required for maintenance of the tumor state. *Cell* **23**, 311-322.

Peterson, P.A. (1981). Diverse expression of controlling element components in maize: test of a model. *Cold Spring Harbor Symp. Quant. Biol.* **45**, 447-456.

Price, P.W. (1980). "Evolutionary Biology of Parasites". Princeton University Press, Princeton.

Proudfoot, N.J. and Maniatis, T. (1980). The structure of a human alpha-globin pseudogene and its relationship to α-globin gene duplication. *Cell* **21**, 537-544.

Reanney, D.C. (1974). On the origin of prokaryotes. *J. Theoret. Biol.* **48**, 243-251.

Reanney, D. (1979). RNA splicing and polynucleotide evolution. *Nature (London)* **277**, 598-600.

Roeder, G.S. and Fink, G.R. (1980). DNA rearrangements associated with a transposable element in yeast. *Cell* **21**, 239-249.

Rosenberg, A. (1978). The supervenience of biological concepts. *Phil. Sci.* **45**, 368-386.

Rubin, C.M., Houck, C.M., Deininger, P.L., Friedmann, T. and Schmid, C.W. (1980). Partial nucleotide sequence of the 300-nucleotide interspersed repeated human DNA sequences. *Nature (London)* **284**, 372-374.

Salser, W. and Isaacson, J.S. (1976). Mutation rates in globin genes: the genetic load and Haldane's dilemma. *Progr. Nucl. Acids. Res. Mol. Biol.* **19**, 205-220.

Sapienza, C. and Doolittle, W.F. (1981). Genes are things you have whether you want them or not. *Cold Spring Harbor Symp. Quant. Biol.* **45**, 177-182.

Schopf, J.W. (1977). Earliest evidence of fossil eukaryotes. *In* "Chemical evolution of the Early Precambrian" (Ponnanperuma, C., ed.), pp. 107-109, Academic Press, New York.

Schopf, T.J.M. (1980). *In* "Paleobotany, Paleoecology and Evolution" (Niklas, K.J., ed.), Praeger, New York, *in the press*.

Shapiro, J.A. (1979). Molecular model for the transposition and replication of bacteriophage Mu and other transposable elements. *Proc. Nat. Acad. Sci., U.S.A.* **76**, 1933-1937.

Shapiro, J.A. (1981). Changes in gene order and gene expression. *Nat. Cancer Inst. Monographs, in the press*.

Sharp, P.A. (1981). Speculations on RNA splicing. *Cell* **23**, 643-646.

Schimotohno, K., Mizutani, S. and Temin, M. (1980). Sequence of retrovirus provirus resembles that of bacterial transposable elements. *Nature (London)* **285**, 550-554.

Shoemaker, C., Goff, S., Gilboa, E., Paskind, M., Mitra, S. and Baltimore, D. (1980). Structure of a cloned circular Moloney murine leukema virus DNA molecule containing an inverted segment: implications for retrovirus integration. *Proc. Nat. Acad. Sci., U.S.A.* **77**, 3932–3936.

Silverman, M. and Simon, M. (1980). Phase variation: genetic analysis of switching mutants. *Cell* **19**, 845–854.

Singer, D.S. (1979). Arrangement of a highly repeated DNA sequence in the genome and chromatin of the African green monkey. *J. Biol. Chem.* **254**, 5506–5514.

Smith, G.P. (1976). Evolution of repeated DNA sequences by unequal cross-over. *Science* **191**, 528–535.

Smith, G.P. (1978). What is the origin and evolution of repetitive DNAs? *Trends Biochem. Sci.* **3**, N34–N36.

Stanfield, S. and Helinski, D, (1976). Small circular DNA in *Drosophila melanogaster*. *Cell* **9**, 333–345.

Stanfield, S.W. and Legnyel, J.A. (1979). Small circular DNA of *Drosophila melanogaster:* chromosomal homology and kinetic complexity. *Proc. Nat. Acad. Sci., U.S.A.* **76**, 6142–6146.

Stanfield, S.W. and Lengyel, J.A. (1980). Small circular deoxyribonucleic acid of *Drosophila melanogaster:* homologous transcripts in the nucleus and cytoplasm. *Biochemistry* **19**, 3873–3877.

Stanley, S.M. (1979). "Macroevolution: Pattern and Process." Freeman, San Francisco.

Strobel, E., Dunsmuir, P. and Rubin, G.M. (1979). Polymorphisms in the chromosomal locations of elements of the *412*, *copia* and *297* dispersed repeated gene families in Drosophila. *Cell* **17**, 429–439.

Stuttard, C. and Rozee, K.R. (1980). "Plasmids and Transposons: Environmental Effects and Maintenance Mechanisms." Academic Press, New York.

Temin, H.M. (1980). Origin of retroviruses from cellular moveable genetic elements. *Cell* **21**, 599–600.

Walker, P.M.B. (1978). Genes and non-coding DNA sequences. *Ciba Found. Symp.* **66**, 25–38.

Weiner, A.M. (1980). An abundant cytoplasmic 7 S RNA is complementary to the dominant interspersed middle repetitive DNA sequence family in the human genome. *Cell* **22**, 209–218.

Williamson, V.M., Young, E.T. and Ciriacy, M. (1981). Transposable elements associated with constitutive expression of yeast alcohol dehydrogenase II. *Cell* **23**, 605–614.

Wilson, A.C., Sarich, V.M. and Maxson, L.R. (1974). The importance of gene rearrangement in evolution: evidence from studies on rates of chromosomal, protein, and anatomical evolution. *Proc. Nat. Acad. Sci., U.S.A.* **71**, 3028–3030.

Wilson, A.C., Carlson, S.S. and White, T.T. (1977). Biochemical evolution. *Annu. Rev. Biochem.* **46**, 573-639.

Wilson, E.O. (1975). "Sociobiology: The New Synthesis". The Belknap Press of Harvard University Press, Cambridge, Mass.

Wilson, E.O. (1978). "On Human Nature". Harvard University Press, Cambridge, Mass.

Woese, C.R. and Fox, G.E. (1977a). The concept of cellular evolution. *J. Mol. Evol.* **10**, 1-6.

Woese, C.R. and Fox, G.E. (1977b). Phylogenetic structure of the prokaryotic domain: the primary kingdoms. *Proc. Nat. Acad. Sci., U.S.A.* **74**, 5088-5090.

Wu, J.C. and Manuelidis, L. (1980). Sequence definition and organization of a human repeated DNA. *J. Mol. Biol.* **142**, 363-386.

Young, M.W. (1979). Middle-repetitive DNA: a fluid component of the *Drosophila* genome. *Proc. Nat. Acad. Sci., U.S.A.* **76**, 6274-6278.

Young, M.W. and Schwartz, H.E. (1981). Nomadic gene families in *Drosophila*. *Cold Spring Harbor Symp. Quant. Biol.* **45**, 629-640.

Transposable DNA Sequences in Eukaryotes

DAVID J. FINNEGAN, BARBARA H. WILL, ALEXEI A. BAYEV*,
ANNE M. BOWCOCK and LESLEY BROWN

Department of Molecular Biology, University of Edinburgh

INTRODUCTION

One half of the moderately repetitive DNA in the genome of
Drosophila melanogaster (5 to 10% of the whole genome) is
comprised of about 30 families of transposable elements
(Rubin *et al.*, 1981). The best studied families (known as
copia, 412, 297, mdg1 and mdg3 after the first member of each
to be studied) have a number of properties in common,
although there is no detectable homology between them (Finnegan *et al.*, 1978; Strobel *et al.*, 1979; Ilyin *et al.*, 1980).
These sequences are sometimes referred to generically as
"*copia*-like" elements. The members of each family are well-
conserved and are located at 20 to 40 sites distributed
throughout the genome. The number and locations of elements
in each family vary between strains of *D. melanogaster* and
between embryonic and tissue culture cell DNA, where there
may be up to 150 copies per haploid genome (Tchurikov *et al.*,
1978; Potter *et al.*, 1979). This suggests that, even if
these elements cannot integrate at random, then they must at
least be able to do so at many sites.

At both ends of an element there are long direct repeats,
the length and sequence of which are specific for each family
of elements. At the extreme ends of each element are short
(about 10 base-pairs) inverted repeat sequences which, except
for the mdg3 family, are found at both ends of the long
direct repeats. Immediately before and after each element is
a short direct repeat with a characteristic length (4 and 5
base-pairs have been found so far). The lengths of these
short repeats are the same for all members of a particular
family but their sequences are not (Dunsmuir *et al.*, 1980;
Levis *et al.*, 1980; Bayev *et al.*, 1980; B.M. Will, A.A. Bayev

---

*Present address: Institute for Molecular Biology, Academy of
Science of USSR, Moscow

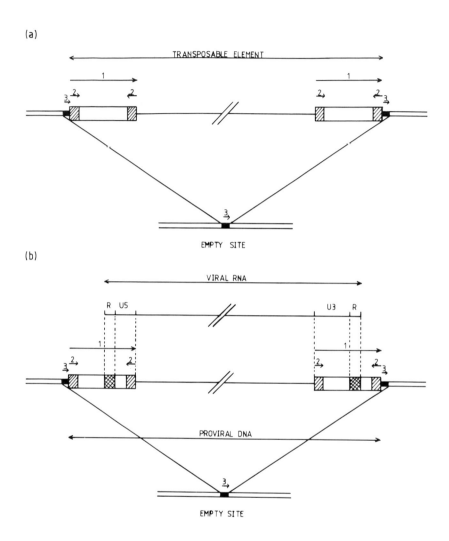

FIG. 1 (a) *Diagram of a typical eukaryotic transposable element, indicating the positions and orientations of the various
repeat sequences.   The relationship between a full and empty
site is shown.   The repeats are 1, long direct repeat; 2,
short inverted repeat; 3, repeat of a host sequence present
once only at an empty site.* **(b)** *Diagram showing the relationships between sequences in retroviral RNA and proviral DNA and
between an integrated provirus and a site into which it has
inserted.   U5, R and U3 are described in the text.   Repeat
sequences are described in* **(a)** *above.*

TABLE I

Characteristics of the Various Repeat Sequences associated with Eukaryotic Transposable Elements and Proviruses

Element	Length of direct repeat	Terminal inverted repeat	Length of flanking repeat	Reference or source
copia	276	13/17	5	Levis *et al.* (1980)
				Dunsmuir *et al.* (1980)
297	412	3	4	Young and Rubin (unpublished results)
				Spradling and Rubin, (1981)
412	481	8/10	4	Will, Bayev and Finnegan (unpublished results)
mdg3	269	15/18	5	Bayev *et al.* (1980)
Ty1	334	2	5	Farabaugh and Fink, (1980)
				Gafner and Philippsen, (1980)
SNV	569	3	5	Shimotohno *et al.* (1980)
MMTV	1327	6	–	Donehower *et al.* (1981)
MSV	588	11	4	Dhar *et al.* (1980)
ev1	273	5/7	6	Hishinuma *et al.* (1981)

The lengths of repeat sequences are given in base-pairs.

and D.J. Finnegan, unpublished results).

In the case of the *copia*, 297 and 412 families, comparisons have been made between corresponding sites, one containing and one lacking a transposable element, in the genomes of different strains of *D. melanogaster* (Dunsmuir *et al.*, 1980; E. Young and G.M. Rubin, unpublished results; B.M. Will and D.J. Finnegan, unpublished results). These indicate that the bases of the short repeats occur only once at the site into which an element inserts. Figure 1(a) shows a typical *copia*-like element and Table I summarises the properties of the best-characterised elements from *D. melanogaster*.

TRANSPOSABLE ELEMENTS IN OTHER ORGANISMS

Transposable DNA elements are not confined to the genome of *D. melanogaster*. In the yeast *Saccharomyces cerevisiae*, for example, there is a family of transposable elements (Ty1 elements) with properties very similar to those in *D. melanogaster* (Table I), and in *Escherichia coli*, transposable elements have been known for some time (Calos and Miller, 1980). These are of two broad classes, insertion sequences (IS elements) and transposons. IS elements are generally about 1000 bases long and resemble the long direct repeat sequences of *copia*-like elements, in that they have inverted repeat sequences at their ends. Transposons are several thousand bases long and are distinguished by carrying one or more antibiotic resistance genes. They are of two types, the first being like IS elements in just having short inverted repeats at their ends. Those of the second type are similar to transposable elements in *D. melanogaster*, and have long terminal repeats, although in this case they are most commonly in inverted orientation. These long repeats also have short inverted repeats at their ends, and in several cases can transpose independently as IS elements.

Several authors have noted the similarity between the sequence organisation of transposable elements in *D. melanogaster* and of the proviruses of vertebrate retroviruses (Levis *et al.*, 1980; Bayev *et al.*, 1980). These two classes of elements (1) are of similar length, (2) are bounded by direct repeats a few hundred base-pairs long, which are themselves terminated by short inverted repeats, and (3) are flanked by a small number of bases, which occur once at the site of insertion in the absence of the element. The similarity extends to the first and last two base-pairs of each element. All proviruses that have been analysed start with the sequence T-G and end with C-A, and the same is true of *copia*, 412 and mdg3 (but not 297) elements.

The terminal direct repeats of proviruses, known as long

terminal repeats (LTRs), are formed during the events that take place after a retrovirus has infected a host cell and contain sequences from both ends of the viral RNA. At the left-hand end of an LTR is a unique sequence from the 3' end of the viral RNA. Adjacent to this is the sequence repeated at both ends of the viral RNA and a unique sequence from its 5' end. These are shown in Figure 1(b), where they are designated U3, R and U5, respectively, and the properties of some vertebrate proviruses are given in Table I. Viral RNA molecules are first reverse transcribed into linear double-stranded DNAs with LTRs at both ends. These are converted to circular molecules with one copy of the LTR or two copies in tandem. It is not known which of these DNAs integrates into host chromosomes but some evidence favours a circular form (Shoemaker *et al.*, 1980). A detailed discussion of this is given by Gilboa *et al.* (1979). Synthesis of progeny viral RNA is presumed to start at the beginning of R in the left-hand LTR of the provirus and to terminate at the end of R in the right-hand LTR. Vertebrates do not have to be infected with a retrovirus to acquire sequences of this type. The genomes of most, if not all, vertebrates contain endogenous proviruses (Teich and Weiss, 1977), which can make up a substantial proportion of their DNA, and Todaro *et al.* (1980) estimate that about 0.1% of the genomes of rodents and primates may be virus-related sequences.

The similarity between proviruses and transposable elements extends beyond their sequence organisation, since Varmus and his colleagues (Cohen and Varmus, 1979; Hughes *et al.*, 1979) have shown that the chromosomal locations of endogenous proviruses differ between individuals within populations of chickens and between inbred strains of mice. Whether these similarities indicate that, as Temin (1980) contends, proviruses arise from transposable elements or whether transposable elements are degenerate proviruses, we cannot tell.

THE MECHANISM OF TRANSPOSITION

Little is known of the mechanism of transposition in eukaryotes. A circular molecule consisting of a transposable element with only one direct repeat could be inserted *via* a single crossover event into a chromosomal site containing a second repeat. The product would be a filled site that is terminally redundant. Deletion of an element from a filled site could occur by the reverse reaction, giving a circular DNA and an empty site (Finnegan *et al.*, 1978). The many potential sites of integration demonstrated for 412, *copia* and 297 elements (Strobel *et al.*, 1979) lead one to predict that if this is the mechanism of transposition, then there should be many empty sites, each containing a copy of the direct repeat. Levis *et al*. (1980) have shown by analysis of digests of total *D. melanogaster* DNA that this is not the case for *copia* elements, and that there are few if any free copies of the direct repeat in the *D. melanogaster* genome. We have done similar experiments and have found the same to be true for 412 direct repeats (L. Brown and D.J. Finnegan, unpublished results). This is consistent with the base sequence comparisons that have been made between full and empty sites. In no case has an empty site been found to contain a copy of a direct repeat. Reciprocal recombination between direct repeats therefore cannot play a significant role in insertion of these elements.

Free copies of the direct repeat (known as the δ sequence) of Tyl do occur in the genome of *S. cerevisiae* (Cameron *et al*., 1979), yet even here they do not appear to be targets for insertion. Farabaugh and Fink (1980) have sequenced a site of insertion in the His4 locus and have shown that it does not contain a δ sequence. The free δ sequences that do exist could indicate that they can transpose independently of the rest of Tyl. Alternatively, they could be left behind by reciprocal recombination events between the ends of Tyl elements but unrelated to transposition. Accumulation of free repeats by similar events may be precluded in *D. melanogaster*.

One can easily imagine how proviruses could change their locations by going through a complete viral life-cycle (transcription of viral RNA, formation of extracellular virions, infection, reverse transcription, integration) or a partial life-cycle going directly from viral RNA to proviral DNA and bypassing virions. The topological similarities between proviruses and transposable elements might indicate that they also move *via* a complete or partial viral life-cycle. If this were the case, then in *D. melanogaster* cells one might expect to find RNA complementary to entire transposable elements and circular DNA molecules like those found after retroviral in-infection.

Flavell and Ish-Horowicz (1981) have detected circular DNAs containing *copia*-like sequences in *D. melanogaster* tissue culture cells. On the basis of restriction site mapping and DNA hybridisation experiments, these appear to be analogous to circular proviral DNAs having all the DNA from the body of a *copia* element plus one terminal repeat or two terminal repeats in tandem.

These circular molecules need not, as Flavell and Ish-Horowicz point out, be derived from reverse transcription of *copia* RNA. They could be produced by recombination events involving either *copia* direct repeats or the five base-pairs of non-*copia* DNA repeated immediately before and after each element. This may well be their origin and need not preclude their being intermediates in transposition.

The mechanism by which extrachromosomal proviral DNA integrates into the genome of the host cell is not known. The same mechanism nevertheless may well act directly on proviruses or transposable elements, allowing them to move from one position in the genome to another. This might resemble transposition in *E. coli* and could account for some of the structural similarities between prokaryotic and eukaryotic transposable elements.

Different transposition mechanisms may be used by different transposable elements and any one element may have more than one possibility open to it. The only structural requirement for transposition might be short terminal inverted repeats to serve as recognition sequences (although it is difficult to imagine how this could be true of repeats as short as the 2 base-pairs of Ty1 and the 3 base-pairs of spleen necrosis virus). The simplest illustration of this would be the IS elements of prokaryotes and transposons that just have short terminal inverted repeats (Calos and Miller, 1980). The terminal inverted repeats of the remaining transposons are part of longer repeats that lie in direct or inverted orientation. The terminal inverted repeats of the best-studied eukaryotic transposable elements are also provided by longer repeats that are always in direct orientation (transposable elements with long inverted terminal repeats have been reported by Potter *et al.*, 1980). If eukaryotic transposable elements are related to retroviruses, then this direct orientation would result from the nature of the retroviral life-cycle. There is no such constraint on the long terminal repeats of transposons and, in this case, there might even be a selective advantage in an inverted orientation, since this would preclude excision of the intervening DNA by recombination. If short terminal inverted repeats are the only structural requirement for transposition, then any sequence flanked by copies of the same transposable element may itself be transposable. This is true of segments of *E. coli* DNA lying between copies of the terminal repeats of the transposon Tn*10* (Foster *et al.*, 1981).

BIOLOGICAL EFFECTS OF TRANSPOSITION

Each transposition is a mutation that may or may not have a detectable effect. Transposable elements should inactivate any gene into which they insert, and many spontaneous mutations previously regarded as point mutations could be insertions. This certainly occurs in *Escherichia coli* (Calos and Miller, 1980). Gehring and Paro (1980) noticed that labelled DNA containing *copia*-specific sequences would hybridise *in situ* to the position of the white eye locus (*w*) on chromosomes carrying the white apricot mutation but not if they carried the wild-type allele. On the strength of this, they suggested that the white apricot mutation could be due to insertion of a *copia* element into the white eye gene. This has recently been confirmed genetically and at the DNA level by Bingham, Judd and Rubin (see Spradling and Rubin, 1981). Several mutations of the bithorax locus are also insertions, although the DNA sequences involved are not known to be related to any family of transposable elements characterised so far (Bender, Lewis and Hogness, quoted by Spradling and Rubin, 1981).

Deletions, inversions and translocations could result from recombination between different copies of the same transposable element on different chromosomes or at non-homologous regions of the same chromosome. An unstable mutation *his4-912* in *S. cerevisiae* is due to insertion of a Ty1 element in the region of the His4 locus. Chromosomes carrying this mutation have a high incidence of gross chromosomal rearrangements with one of their endpoints within the Ty1 element. Similarly, the unstable mutations w^c and w^{DZ1} of the white eye gene in *D. melanogaster* are due to insertions (Bingham, Collins, Levis and Rubin, quoted by Spradling and Rubin, 1981).

All *D. melanogaster* transposable elements studied in any detail are transcribed in tissue culture cells, and at least some stages in the life-cycle of the organism (Finnegan *et al*., 1978; Tchurikov *et al*., 1978; Flavell *et al*., 1980). This transcriptional activity of transposable elements could affect the expression of nearby sequences. A newly arrived element might switch off an active gene or switch on a sequence not normally expressed. An example of the former effect has been detected in yeast, in which the *his4-912* mutation is due to insertion of a Ty1 element just upstream of the putative *his4* promoter (Farabaugh and Fink, 1980). Expression of the His4 locus is restored in chromosomes from which the body of the Ty1 elements has excised (presumably by recombination), leaving behind a copy of the direct repeat δ. Farabaugh and Fink (1980) suggest that this effect is due to transcription initiated within the Ty1 element interfering with that from the *his4* promoter.

The opposite effect (a chromosomal sequence being switched on by a DNA element inserting next to it) has been found recently in chickens. Payne *et al*. (1981) and Neel *et al*. (1981) have studied avian leukosis virus (ALV) proviruses in the DNA of tumours formed in ALV-infected chickens. In cells from different tumours, ALV proviral DNA was inserted next to a specific host sequence (a potential oncogene), which was then transcribed, apparently from a proviral promoter. Events of this sort may not require a recent viral infection, and Cairns

(1981) has argued persuasively that many cancers are the result of genetic transpositions.

If genes can be switched on and off by transposable elements in their vicinity, one might ask whether they are involved in the patterns of gene expression normally seen during the life of an individual. This is unlikely in *D. melanogaster*, since the number and locations of *copia*, 412 and 297 transposable elements are different in chromosomes of the same tissue (the larval salivary gland) of individuals from different laboratory strains (Potter *et al.*, 1979). Similarly, transposition cannot generally be an integral part of differentiation, since the distribution of 412 elements seems to be the same in DNA from embyros, larval brains and adults of the same strain of *D. melanogaster* (A.M. Bowcock and D.J. Finnegan, unpublished results).

The evidence now available thus suggests that transposable DNA elements are widespread and that their activity can be responsible for major genetic events. There is no reason, however, to suppose that these effects are anything but fortuitous consequences of their ability to transpose.

ACKNOWLEDGEMENTS

This work was supported by project grants from the Medical Research Council. One author (B.M.W.) is the holder of a Medical Research Council postgraduate studentship and another (A.M.B.) was a British Petroleum research student.

REFERENCES

Bayev, A.A., Krayev, A.S., Lyubomirskaya., Ilyin, Y.V., Skryabin, K.G. and Georgiev, G. (1980). The transposable element mdg3 in *Drosophila melanogaster* is flanked with perfect direct repeats and mismatched inverted repeats. *Nucl. Acids Res.* **8**, 3263-3273.

Cairns, J. (1981). The origin of human cancers. *Nature (London)* **289**, 353-357.

Calos, M.P. and Miller, J.H. (1980). Transposable elements. *Cell* **20**, 579-595.

Cameron, J.R., Loh, E.Y. and Davis, R.W. (1979). Evidence for transposition of dispersed repetitive DNA families in yeast. *Cell* **16**, 739-751.

Cohen, J.C. and Varmus, H.E. (1979). Endogenous mammary tumour virus DNA varies among wild mice and segregates during inbreeding. *Nature (London)* **278**, 418-423.

Dhar, R., McClements, W.L., Enquist, L.W. and Vande Woude, G.F. (1980). Nucleotide sequences of integrated Moloney sarcoma provirus long terminal repeats and their host and viral junctions. *Proc. Nat. Acad. Sci., U.S.A.* **77**, 3937–3941.

Donehower, L.A., Huang, A.L. and Hager, G.L. (1981). Regulatory and coding potential of the mouse mammary tumour virus long terminal redundancy. *J. Virol.* 37, 226–238.

Dunsmuir, P., Brorien, W.J., Simon, M.A. and Rubin, G.M. (1980). Insertion of the *Drosophila* transposable element *copia* generates a 5 base pair duplication. *Cell* 21, 576–579.

Farabaugh, P.J. and Fink, G.R. (1980). Insertion of the eukaryotic transposable element Ty1 creates a 5 base-pair duplication. *Nature (London)* **286**, 352–356.

Finnegan, D.J., Rubin, G.M., Young, M.W. and Hogness, D.S. (1978). Repeated gene families in *Drosophila melanogaster*. *Cold Spring Harbor Symp. Quant. Biol.* **42**, 1053–1063.

Flavell, A.J. and Ish-Horowicz, D. (1981). Extrachromosomal circular copies of the eukaryotic transposable element *copia* in cultured *Drosophila* cells. *Nature (London)* **292**, 591–594.

Flavell, A.J., Ruby, S.W., Toole, J.J., Roberts, B.E. and Rubin, G.M. (1980). Translation and developmental regulation of RNA encoded by the eukaryotic transposable element *copia*. *Proc. Nat. Acad. Sci., U.S.A.* **77**, 7107–7111.

Foster, T.J., Davis, M.A., Roberts, D.E., Takeshita, K. and Kleckner, N. (1981). Genetic organisation of transposon Tn*10*. *Cell* 23, 201–213.

Gafner, J. and Philippsen, P. (1980). The yeast transposon Ty1 generates duplications of target DNA on insertion. *Nature (London)* **286**, 414–418.

Gehring, W.J. and Paro, R. (1980). Isolation of a hybrid plasmid with homologous sequences to a transposing element of *Drosophila melanogaster*. *Cell* **19**, 897–904.

Gilboa, E., Mitra, S.W., Goff, S. and Baltimore, D. (1979). A detailed model of reverse transcription and tests of crucial aspects. *Cell* 18, 93–100.

Hishinuma, F., DeBona, P.J., Astrin, S. and Skalka, A.M. (1981). Nucleotide sequence of accepter site and termini of integrated avian endogenous provirus ev1: integration creates a 6bp repeat of host DNA. *Cell* 23, 155–164.

Hughes, S.H., Payvor, F., Spector, D., Schimke, R.T., Robinson, H.L., Payne, G.S., Bishop, J.M. and Varmus, H.E. (1979). Heterogeneity of genetic loci in chickens: analysis of endogenous viral and nonviral genes by cleavage of DNA with restriction endonucleases. *Cell* 18, 347–359.

Ilyin, Y.V., Chmeliauskaite, V.G. and Georgiev, G. (1980). Double stranded sequences in RNA of *Drosophila melanogaster*: relation to mobile disperse genes. *Nucl. Acids Res.* 8, 3439–3457.

Levis, R., Dunsmuir, P. and Rubin, G.M. (1980). Terminal repeats of the *Drosophila* transposable element *copia*: nucleotide sequence and genomic organisation. *Cell* 21, 581–588.

Neel, B.G., Hayward, W.S., Robinson, H.L., Fang, J. and Astrin, S.M. (1981). Avian leukosis virus-induced tumors have common proviral integration sites and synthesise discrete new RNAs: oncogenesis by promoter insertion. *Cell* 23, 323–334.

Payne, G.S., Courtneidge, S.A., Crittenden, L.B., Fadly, A.M., Bishop, J.M. and Varmus, H.E. (1981). Analysis of avian leukosis virus DNA and RNA in bursal tumors: viral gene expression is not required for maintenance of the tumor state. *Cell* 23, 311–322.

Potter, S.S., Brorien, W.J., Dunsmuir, P. and Rubin, G.M. (1979). Transposition of elements of the 412, *copia* and 297 dispersed repeated gene families in *Drosophila*. *Cell* 17, 415–527.

Potter, S., Truett, M., Phillips, M. and Maher, A. (1980). Eucaryotic transposable elements with inverted terminal repeats. *Cell* 20, 639–647.

Rubin, G.M., Brorien, W.J., Dunsmuir, P., Flavell, A.J., Levis, R., Strobel, J.J., Toole, J.J. and Young, E. (1981). "*copia*-like" transposable elements in the *Drosophila* genome. *Cold Spring Harbor Symp. Quant. Biol.* **45**, 619–628.

Shimotohno, K., Mizutani, S. and Temin, H.M. (1980). Sequence of retrovirus provirus resembles that of bacterial transposable elements. *Nature (London)* 285, 550–554.

Shoemaker, C., Goff, S., Gilboa, E., Pashkind, M., Mitra, S.W. and Baltimore, D. (1980). Structure of a cloned circular moloney murine leukemia virus DNA molecule containing an inverted segment: implications for retrovirus integration. *Proc. Nat. Acad. Sci., U.S.A.* 77, 3932–3936.

Spradling, A.C. and Rubin, G.M. (1981). *Drosophila* genome organisation: conserved and dynamic aspects. *Annu. Rev. Genet.* 15, in the press.

Strobel, E., Dunsmuir, P. and Rubin, G.M. (1979). Polymorphisms in the chromosomal locations of elements of the 412, *copia* and 297 dispersed repeated gene families in *Drosophila*. *Cell* 17, 429–439.

Tchurikov, N.A., Ilyin, Y.V., Ananiev, E.V. and Georgiev, G.P. (1978). The properties of gene Dm225 a representative of dispersed repetitive genes in *Drosophila melanogaster*. *Nucl. Acids Res.* **6**, 2169–2187.

Teich, N.M. and Weiss, R.A. (1977). Beware the lurking virogene. *In* "Recombinant Molecules: Impact on Science and Society" (Beers, R.F., and Bassett, E.G., eds), pp. 471–483, Raven Press, New York.

Temin, H.M. (1980). Origin of retroviruses from cellular
   moveable genetic elements. *Cell* 21, 599-600.
Todaro, G.J., Callahan, R., Rapp, V.R. and DeLarco, J.E.
   (1980). Genetic transmission of retroviral genes and
   cellular oncogenesis. *Proc. Roy. Soc. ser. B* 210, 367-385.

Sequencing and Manipulating Highly Repeated DNA

GEORGE L. GABOR MIKLOS

Research School of Biological Sciences, Canberra

We have addressed our data to the three broad issues suggested
by the organizers of this meeting.

1. What are the changes at the nucleotide sequence level
   that have affected the organization of a species genome as well as the genomes of closely related species?
2. How have changes in organization and position of
   sequences affected the biology of the organism?
3. What might be the relationship (if any) between
   nucleotide sequence change and spectiation?

NUCLEOTIDE SEQUENCE ANALYSES

The consensus sequences of satellite DNAs or the most common
restriction fragments of some highly repeated DNAs have now
been determined for a number of invertebrates and vertebrates
(Southern, 1970; Salser *et al.*, 1976; Pech *et al.*, 1979a,b;
Cooke and Hindley, 1979; Mullins and Blumenfeld, 1979; Brutlag, 1980; Christie and Skinner, 1980a,b; Donehower *et al.*,
1980; Poschl and Streeck, 1980; Rubin *et al.*, 1980; Wu and
Manuelidis, 1980; Miklos and Gill, 1981a,b; Thayer *et al.*
1981). Sequence complexities vary from 2 to at least 2350
base-pairs and, in spite of claims to the contrary, the
sequences exhibit extensive variation both within and between
species. It is the *within* species variation that is of para-
mount importance, since it is only an accurate estimate of
this that can lead to meaningful discussions of between species variation.

We present two examples of within species variation. The
first is from cloned restriction fragments of the Hawaiian fly
Drosophila gymnobasis; the second is from cloned fragments of
the 1.706 satellite of the calf.

Figure 1 illustrates the *within* species variation from ten
clones of a laboratory stock of *D. gymnobasis* (Miklos and Gill,

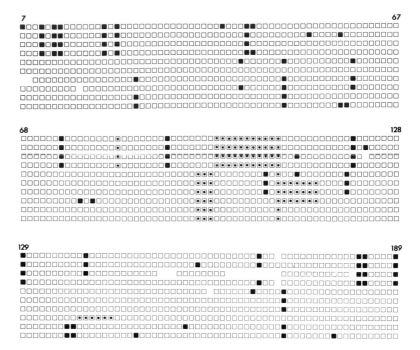

FIG. 1 *Visual display of the sequence heterogeneity between 10 clones of* D. gymnobasis. *Alterations from the consensus sequence are as follows; filled squares, base changes; dot in square, deletions. The nucleotide sequences themselves are discussed by Miklos and Gill (1981a,b).*

1981a). The 189 base-pair prototype sequence shows extensive variation, with base changes (in black) and deletion/addition events (circles) being prominent. In conventionally accepted molecular terms, the total variation is approximately 11% (number of altered bases/total number of bases). The base changes and deletional events each contribute about equally to this variation. Since the DNA used for cloning was isolated from a specific region of an acrylamide gel, the total variation found here is a minimum. This is so, since any similar sequences present as oligomers or dispersed elsewhere in the genome in non-tandem packs would not have been cloned and any deviations from a monomer sequence can formally be considered as components of variation. Since the DNA was also from a large number of individuals (~ 500), we are unable to partition the variation into within and between individuals. It should be noted, however, that the minimum level of variation found here is far in excess of the figure of 1 to 2% commonly derived from the insensitive melting experiments using

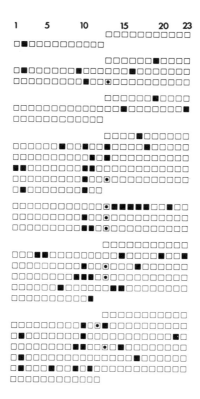

FIG. 2 *Sequence heterogeneity in a portion of the calf 1.706 satellite. Base changes are denoted as filled squares, and deletions as dots in squares. Data are from Pech* et al. *(1979b).*

satellites of short repeat length and RNase-treated hybrid structures.

The types of variation are also important. The sizes of the deletion events are from one to 11 base-pairs, not in multiples of three as occur for variants in protein coding sequences. There are at least five stop codons in each reading frame and it is unlikely that sensible translation products would emerge if this DNA were ever to be transcribed.

Figure 2 shows the *within* species variation exhibited by eight clones from the 1.706 satellite of the calf (Pech *et al.*, 1979b). The deviations from the 23 base-pair consensus sequence are shown in black. It is again quite clear that there is extensive variation of the order of 12%, most of it being due to base changes, rather than the deletion/addition events, at positions 12 and 13. Since the DNA has been pooled from an unknown number of individuals, it is not possible to

partition the total variation. However, unlike the case of
the Hawaiian *Drosophila*, where only monomer units were cloned,
the calf clones consist of a number of adjacent 23 base-pair
sequences, which must be contiguous on a single chromosome.
Thus, as can be gauged from the last clone of Fig. 2, there
is already significant variation between adjacent repeats on
a single chromosome.

Further examination of the variation revealed in Fig. 2
shows that of the 23 base-pair sequences, only three of the
24 are the same. Thus, unless this was a highly non-random
sample, there are 21 different variants. There are, furthermore, roughly 5×10^6 copies of the "23 base-pair like"
sequences in the haploid genome, so that the overall number of
variants must be quite staggering. The number of "23 base-
pair like" sequences in the calf genome is thus enormous,
since not only is the 1.706 satellite made up of subsegments
that have an underlying 22 to 23 base-pair periodicity (Pech
et al., 1979b) but the 1.720 satellite has a 46 base-pair
repeating unit, which is really a 23 base-pair dimer (Poschl
and Streeck, 1980). The total amount of variation must also
exceed 12%, since the related 23 base-pair sequences not
present at a density near 1.706 g/cm^3 would not have been
cloned.

These examples from Hawaiian *Drosophila* and the calf show
that cloning and sequencing techniques have unearthed far more
variation *within* a species than has previously been found
using less sensitive techniques. The reality of the situation, however, has been overly coloured by the case of the
1.688 satellite of *D. melanogaster*, where a lack of variation
has been emphasised. However, it should be recalled that the
variation of 4% in this particular case refers to 15 adjacent
359 base-pair repeats (Hsieh and Brutlag, 1979a). Carlson and
Brutlag (1977) originally showed that, in addition to the
monomers, dimers and short oligomers revealed by restriction
analysis, there were long oligomers, DNA fragments which migrated between the oligomeric series, and undigested material.
Such a situation is hardly indicative of homogeneity.

The total variation in the "359 base-pair like" sequences
is extensive. One of its variants is 254 base-pairs in length
and differs from it by one massive 98 base-pair deletion (or
addition), six smaller deletions and numerous base changes
(Fig. 3). In total, these two sequences share only about 60%
homology (Carlson and Brutlag, 1979), reinforcing the issue
that the variation is far greater than 4%.

One further feature of DNA sequence structure is often
alluded to as being of potential significance to the biology
of the organism in terms of protein/DNA interactions. This
is a region of potential dyadic symmetry that may be

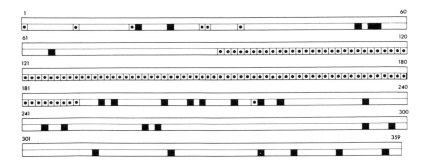

FIG. 3 *Comparison of the average sequences of the 359 (above) and 254 base-pair (below) variants of the 1.688 satellite of* D. melanogaster. *Base changes are denoted as filled squares and deletion/addition events as dots in squares. Data are from Carlson and Brutlag (1979).*

recognized by a DNA binding protein. The extent of the *within* species variation in such potential structures may be examined in the case of the 359 and 254 base-pair variants that have just been discussed. Figure 4 illustrates a potential dyadic symmetry in the 359 base-pair repeat and the corresponding nucleotide run in the 254 base-pair repeat. In the 359 base-pair repeat there are eight mismatches in 37 positions, whereas in the corresponding run in the 254 base-pair repeat there are 14 mismatches. Another symmetry element, this time in the 254 base-pair sequence, has 15 mismatches in 45 positions. The corresponding sequence in the 359 base-pair repeat has 21 mismatches. Thus there is quite some variation even *within* a laboratory stock in the fidelity of base-paired potential secondary structures within a family of "359 base-pair" sequences.

Wu and Manuelidis (1980) have previously drawn attention to this feature in both *within* as well as *between* species comparisons of primates. Their computer searches revealed that dyadic sequences were not maintained with great fidelity. We have reached a similar conclusion from between species comparisons of Hawaiian *Drosophila* (Miklos and Gill, 1981a).

The cloning and sequencing results have drawn us to the following conclusions on DNA sequence structure *within* a species.

    1. Sequences are remarkably variable, the sources of this variation being base changes, additions/deletions and very probably rearrangements.

    2. Changes are non-randomly localised within a sequence.

```
                                          A·T        C      T
                                          A·T          A·T
                                          A·T        C      T
                                          C·G          C·G
                                          G·C          G·C
                                          G·C          G·C
                                       C           T
                                       T·A          T·A
        T·A              A          T      C     T      C
        T·A            T·A         T·A          T·A
        T·A            T·A         T·A          T·A
        T·A            T·A       T      C     T      C
    A           A               A·T          A      A
    A           A             C      T     A·T
        A·T            A·T       A·T          A      A
        C·G            C·G       T·A          T·A
        G·C            G·C       T·A          T·A
              G              G    A      A   A      A
        T·A              A    A      A   A      A
        T·A            T·A    A      A   A      A
        T·A            T·A       A·T        G      T
        T·A            T·A    G      A   G      A
        A·T            A·T       T·A          T·A
    C           N               A·T          A·T
        A·T            A·T       A·T          A·T
        C·G        C      A       C·G        A      G
    C      A             A       A·T          A·T
    G      T      G·C                 C              C
        G·C        G      G       A·T          A·T
    T      C    C      C       C·G          C·G
        T·A            T·A       T·A          T·A
        T·A            T·A       T·A          T·A
        T·A            T·A    G      T   G      T
        T·A            T·A       T·A          T·A
        T·A            T·A    G      T   G      T
    A           A               T      G   T      G
        A·T        C      T       C·G          C·G
        A·T            A·T             A              A
        T·A        T      G             A              A
        T·A            T·A       A·T          A·T
        A·T            A·T    A      G   A      G
        A·T            A·T       A·T          A·T
        A·T            A·T       G·C        C      C
        G·C        A      C       T·A          T·A
        G·C            G·C       A·T          A·T

        (a)            (b)           (c)            (d)
```

FIG. 4   For legend see opposite.

3. Potential secondary structures exhibit fair variation *within* a species.

*As unpalatable as it may seem, sequence analyses
have not provided us with any sensible biolgical connection.*

Repeated sequences appear to undergo recurrent cycles of
amplification and deletion by mechanisms for which few hard
data exist. Evidence from yeast indicates that unequal
exchange or intrachromosomal conversion may have profound
influences on the homogeneity of tandem packs (Petes, 1980;
Szostak and Wu, 1980; Klein and Petes, 1981). Whatever the
mechanisms, sequences may be altered non-randomly through time
and they may well be without functional attributes, even if,
as in the case of the 1.688 satellite, a portion of it does
bind a protein (Hsieh and Brutlag, 1979b). It is not without
significance that no reasonable hypothesis has yet been put
forward in a biological context relating the protein binding
properties of the 1.688 DNA. It is also not without significance that 1.688 type sequences are undetectable in some
closely related *Drosophila* genomes by sensitive blotting
techniques (Strachan *et al.*, 1982).

If we erect the null hypothesis that sequence *per se* is neutral (by analogy to Kimura's (1979) theory for allelic variation), then we would be hard-pressed to fail it on the available data. Note that we are discussing the potential attributes of sequence *per se*, not the amount of that sequence or
its variants. The presence, and in some cases the amount, of
localised repetitive DNA is certainly known to have large
effects in the germ line (John and Miklos, 1979; Miklos and
John, 1979; Yamamoto, 1979; John, 1981).

THE SOMATIC EFFECTS OF CHROMOSOME MANIPULATIONS
USING HIGHLY REPETITIVE DNA

The data of a number of authors on cell cycle times and nucleotypic effects led us to ask about somatic effects of satellite
DNA. Barlow (1973) originally demonstrated a significant

FIG. 4 *Potential dyadic symmetries in (a) the 359 base-pair
repeat and (b) its corresponding nucleotide run in the 254
base-pair repeat. Potential dyadic symmetries in (c) the 254
base-pair repeat and (d) its corresponding nucleotide run in
the 359 base-pair repeat. Data are from Brutlag (1980). Only
base-pairing is indicated; the sequence hyphens have been
omitted for clarity.*

effect of extra X chromosome material on cell cycle time in
human cell lines. Furthermore, increasing attention is being
paid to various aspects of Bennett's (1971) "nucleotype" hypothesis (e.g. see Nagl, 1978; Olmo and Morescalchi, 1978;
Cavalier-Smith, 1978,1980a,b; Hutchinson *et al.*, 1979). Quantitative evaluation of somatic effects also bears directly on
some aspects of the selfish DNA argument (Orgel and Crick,
1980; Doolittle and Sapienza, 1980).

In an attempt to determine the relevance and generality of
all this evidence, we took the following approach. Do size-
able phenotypic effects occur when the *amount* of heterochromatin in the *Drosophila* genome is varied in situations where the
genetic background is strictly controlled? Many of the data
available from the literature deal with comparisons *between*
species and these data, in general, all suffer from the same
criticism; it is not possible to tie the observed effects to
heterochromatin *per se*, since other components of the genome
have also altered since speciation. It is only when deletions
and additions of heterochromatin are analysed in rigidly controlled genetic backgrounds that meaningful conclusions concerning the putative somatic inertness of this class of DNA
can be made.

We have therefore sought for perturbations during two different stages of *Drosophila* development when chromosomes consisting overwhelmingly of satellite sequences are added or
subtracted from the genome. First, is the developmental timing of embryogenesis upset? From fertilisation onwards,
the *Drosophila* embryo must undergo a number of complex processes culminating in a larva. If satellite DNA *per se* were
to have gross somatic effects, they ought to be readily demonstrable by upsets in such a delicate developmental package.
Second, what distortions occur during larval development of
the optic imaginal disc that ultimately determine the number
of facets in the compound eye?

1. DNA Alterations and Embryonic Developmental Timing

A visual display of the proportion of heterochromatin in a
normal male genome with a haploid genome size of 165 million
base-pairs (mbp) is shown in Fig. 5. The sizes of the chromosome fragments are as follows: 2 mbp (*mini 1187*), 5 mbp (*mini
164*), 6 mbp (*mini 118YM*), 10 mbp (*minis 1492* and *1514*), 20 mbp
(*mini 3*) and 40 mbp for the Y chromosome.

The normal diploid male genome contains approximately 214
mbp of euchromatin and 116 mbp of heterochromatin. By adding
singly the chromosomes depicted in Fig. 5, we constructed a
series of males of the constitution (XY)(XY + *mini 1187*).....
(XYY). Similarly, we manufactured a female series consisting

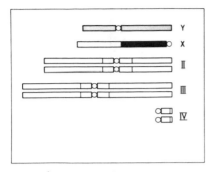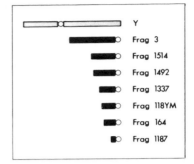

FIG. 5 *Diagrammatic representation of a diploid* D. melanogaster *genome together with the* mini *chromosomes used in this study.*

of (XX), (XX + *mini 1187*)...(XXY). Appropriate crosses also yielded X0 males without a Y chromosome. Thus in males we had variation in diploid genome size from 290 mbp (214 *eu* + 76 *het*) to 370 mbp (214 eu + 156 *het*). The heterochromatin content of these genomes thus varies from 76 to 156 mbp.

Following such manipulations, we enquired whether the time of development in the egg was significantly altered in genotypes with more or less heterochromatin. We stress that the genotypes (control and experimental) arise from the same cross. In method A, XY males were mated to XX *mini* females and produced XY, XY *mini*, XX and XX *mini* progeny, all of which developed under identical conditions. In method B, we produced the same genotypes but passed the *mini*s through the male, in case of potential maternal effects. The XY *mini* males when mated to XX females yield six classes of progeny from the same

TABLE I

Differences in Hatching Times between XY mini *and XY Males, and between XX* mini *and XX Females when* mini *Chromosomes are Transmitted* via *the Female*

	DIFFERENCES IN MEDIAN EMERGENCE TIME (IN MIN)	
CHROMOSOME	XY *mini* − XY	XX *mini* − XX
mini 1187	−1	−4
mini 1492	+2	+1
mini 1514	+1	+10
mini 3	+11	+11
Y	+30	+15

TABLE II

Lifetable Characteristics of Genotypes in which mini *Chromosomes are Transmitted via the Female*

Cross containing extra chromosome	Eggs laid	Larvae hatched	Adults emerged	Adult phenotypes			
				XY	XY *mini*	XX	XX *mini*
mini 1187	899	830	760	0.26	0.22	0.27	0.25
mini 1492	1166	774	676	0.27	0.23	0.24	0.26
mini 1514	856	668	537	0.24	0.21	0.26	0.29
mini 3	1229	959	871	0.25	0.24	0.25	0.26
Y	1405	1264	1105	0.23	0.25	0.27	0.25

TABLE III
Death During Different Developmental Stages

CROSS	EGG–LARVA	MAXIMUM % INVIABILITY	LARVA–ADULT	% INVIABILITY
mini 1187	1.00 - 0.92	8	0.92 - 0.85	7
mini 1492	1.00 - 0.66	34	0.66 - 0.58	8
mini 1514	1.00 - 0.78	22	0.78 - 0.63	15
mini 3	1.00 - 0.78	22	0.78 - 0.71	7
Y	1.00 - 0.90	10	0.90 - 0.79	11

TABLE IV
Differences in Hatching Times between Genotypes where Extra Chromosomes are Transmitted via *the Male*

CHROMOSOME	MEDIAN EMERGENCE TIME (IN MIN)		TOTAL ADULT PROGENY
	XY *mini* - XY	XX *mini* - XX	
mini 1187	0	7	281
mini 164	0	0	1011
mini 118 YM	2	4	1478
mini 1337	18	9	1620
mini 1492	3	-5	977
mini 1514	6	-1	638
mini 3	3	5	335

*Unpublished data of M. Yamamoto

cross: XY, XY *mini* and X *mini* males, XX, XX *mini* and XXY females.

Tables I to IV as well as Fig. 6 reveal the extent of the perturbations caused by the extra chromosomes relative to their controls. The average time to complete early embryogenesis and to hatch from the egg is about 1250 minutes (21 hours). The largest difference between control and experimental is only 30 minutes longer for XYY males *versus* their XY siblings (Table I and Fig. 6). Thus the differences in developmental timing are extremely small, being less than 2.5% of the average developmental time.

Table II reveals the losses during the various stages of the the life cycle. An examination of the inequalities of adult phenotypes shows that no one genotype suffers much loss. The percentage hatching from the egg is not a reliable indicator of embryonic mortality, since egg hatch percentages vary

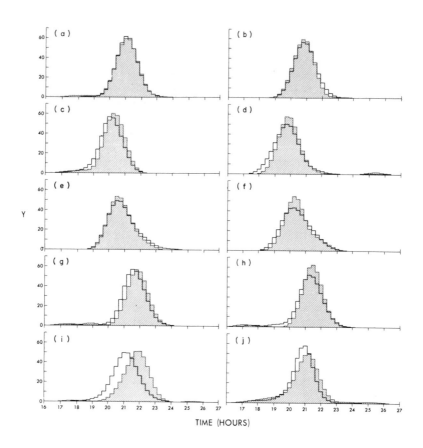

FIG. 6 *Hatching times from the egg of different genotypes;*
XY (open) versus *XY* mini *(shaded) or XX (unshaded)* versus *XX*
mini *(hatched). (a) and (b)* mini 1187; *(c) and (d)* mini 1492;
(e) and (f) mini 1514; *(g) and (h)* mini 3; *(i) and (j)* Y. *The*
data points have been smoothed and analysed by computer in
order to remove high frequency noise from the data. The y
axis is a composite function that sets both curves to a total
area of 100 units (see Miklos et al., *1980).*

drastically for the same females sampled on successive occasions on the same day. A significant contribution to this figure may well come from unfertilised eggs. The most reliable
indicator of larval and adult somatic effects is the loss in
the larva to adult phase. The overall percentage losses of
7, 8, 15, 7 and 11 are unrelated to increasing heterochromatic

content of the genome.

Although the chromosomes we have added are made up largely of satellite sequences, most have a few genes that serve as visible markers. Some carry the 18 S and 28 S ribosomal genes as well as an unknown number of gene loci sequestered within the heterochromatin. The sum total of these sequences produces negligible effects on the timing of early development and has little (if any) somatic effects during the larva to adult transition.

When similar genotypes are constructed by transmitting the *mini* chromosomes through males, analogous results are found. There is a non-significant effect of either X or Y heterochromatin (Table IV). It is noteworthy that the *mini* chromosome that has the largest effect (*mini 1337*) is a chromosome that carries more euchromatic material than any other in this series. Furthermore, whilst it has been relatively easy to construct fertile individuals with two *mini* chromosomes, such has not been possible with *mini 1337* (G.L.G. Miklos and M. Yamamoto, unpublished results). It is quite likely that the increased euchromatic content of this particular chromosome is the cause of the perturbation. If this is indeed the case, then it may well be that *all* the small effects that we have attempted to measure are due to the genic content of these chromosomes and the satellite content is without significant input.

Owing to the vagaries of the *Drosophila* meiotic system, it is possible to produce males that are X *mini* in genotype. When their developmental times are compared to their XY siblings from the same cross, it is found that X *mini* males develop slightly faster than XY controls. The differences are again of a trivial nature, being 14, 22 and 14 minutes for *mini*s *1492*, *1514* and *3*. This represents a difference of about 1% relative to the control.

Lastly, we have produced XO males by two different methods; first, the use of *attached* XY/O males mated to normal XX females; and second, by utilising an X chromosome ($bb^{\ell 158}$) which, by non-disjunction, gives rise to X, Y, *nullo* and XY sperm with high frequency. In both these cases, the differences in developmental time are again very small, less than 1% of the overall developmental time.

It seems an unavoidable consequence of these data that if there are any somatic effects due to increases or decreases of heterochromatin in rigidly controlled genetic backgrounds, they are miniscule.

2. *Facet Number in the Compound Eye.*

One of the problems associated with the embryo of *Drosophila*

is that mitotic activity ceases in most parts by the 10th to
11th hour. Thus our probability of affecting the rate of cell
division is confined to the first half of embryogenesis. Consequently, we sought to examine a system in which division continued, and for this we chose the development of the optic
imaginal disc in the larva (Baker, 1978; Morata and Lawrence,
1979). The cells continue to multiply as simple undifferen-
tiated cells throughout the early stages of optic imaginal
disc growth, later giving rise to the components of the compound eye. The effective period of sensitivity during which
facet number can be altered is a relatively short period of
about 15 hours in mid larval development (Luce, 1935; Luce *et
al.*, 1951; de Marinis, 1952). A sound groundwork has been
laid concerning the *Drosophila* eye, particularly by using combinations of various alleles at the *Bar* locus. Most authors
have argued that it is cell division in the developing optic
bud that is the major process concerned with facet production
during the effective period.

We attempted to perturb this system by again carefully constructing genotypes that differed from each other by the presence or absence of an extra chromosome rich in satellite
sequences. We reduced the facet number from that of the normal 800 or so, to an easily quantifiable number using a *Bar*
allele carried on a Y chromosome. We measured the facet number of different genotypes under the scanning electron microscope.

The results are shown in Fig. 7 and Table V. Except for
mini 3, the effects on facet number are very small. There is
no correlation between the amount of heterochromatin added and
the magnitude of the perturbation. The Y chromosome is twice
the size of *mini 3*, yet the Y has an effect comparable to *mini
1187*, the smallest chromosome. It is noteworthy that the
largest effect is caused by *mini 3*, and in this particular
series it has the largest euchromatic content. *mini 1187* and
the Y carry the genes y^+ and ac^+ as markers. *mini*s *1492* and
1514 carry y^+, ac^+, sc^+ and svr^+. *mini 3*, however, not only
carries y^+, ac^+ and sc^+ but at least ten complementation
groups from the euchromatic base of the X, including genes
such as wap^+, $uncl^+$ and $su(f)^+$.

There is no evidence from these data that the satellite DNA
per se on these chromosomes has any significant effect. The
small differences may all be explicable by the total genic
content of these chromosomes; that is, the visible markers as
well as any other genic or transcribed sequences sequestered
within the heterochromain itself.

In summary, it is plain that when the genetic background
is kept well-controlled, and the amount of heterochromatin
is varied by a factor of 2, no somatic effects of consequence

TABLE V

Facet Number in Genotypes of Differing Heterochromatin Content

Genotype	Experimental series 1		Experimental series 2		Experimental series 3	
	Facet number	Number of eyes counted	Facet number	Number of eyes counted	Facet number	Number of eyes counted
XY	102	42	73	42	71	60
XY *mini 1187*	94	44	78	56	61	61
XY	105	42	82	45	–	–
XY *mini 1492*	103	41	77	51	–	–
XY	107	44	43	40	–	–
XY *mini 1514*	93	49	46	53	–	–
XY	115	36	88	53	84	54
XY *mini 3*	89	45	74	51	68	47
XY	27°C 42	77	25°C 49	88	18°C 124	99
XYY	44	74	50	69	122	101
XY			25°C 47	72		
XYY			49	92		

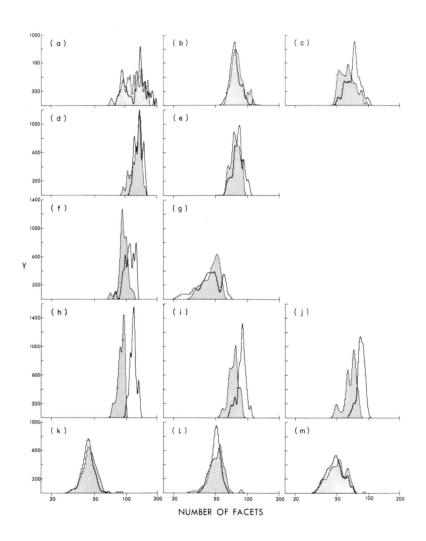

FIG. 7 *Facet numbers in different genotypes; XY versus XY mini. (a) to (c)* mini 1187; *(d) and (e)* mini 1492; *(f) and (g)* mini 1514; *(h) to (j)* mini 3; *(k) to (m) Y. The data points are treated as for Fig. 6.*

are detected that can be attributed to satellite sequences *per se*. An important point is that the limits of heterochromatin variation *within* a species do not exceed the limits we have used in our laboratory situations.

When truly huge additions are made to a *Drosophila* genome, such as the construction of XYYY males and XXYY females, then a number of somatic effects become apparent (Cooper, 1956). The uniformity of the eye coloration may be disturbed to yield a mottled effect, and the legs may become thickened and deformed. However, these effects may have nothing whatsoever to do with satellite DNA *per se*, but could be due to the presence of genic sequences within the heterochromatin. The genome could well be overdosing on ribosomal genes, for example, of which it now has four tandem arrays instead of the normal two.

We are unaware of data from other organisms in which the experimental protocols are similar to our own, and where predominantly simple sequence DNA is added to a normal diploid genome. Analogous experiments have been attempted in *between* species situations and conclusions similar to our own have been reached. Hutchinson *et al.* (1979) analysed different species of *Lolium* that differed radically in their amounts of nuclear DNA. An analysis of morphological characteristics in the parent, backcross and F_2 families revealed few somatic effects of consequence.

VARIATION, GERM LINE EFFECTS AND SPECIATION

1. Polymorphisms and the Recombination System

A major problem with investigations into the molecular or biological properties of DNA sequences is that attention is focused almost exclusively on *between* species comparisons. The most serious objection to the interpretations based on such comparisons, however, is that whatever results are found, it is not possible to assign the causes of an effect with certainty to any particular component of the genome.

Analysis of *within* species variation, or better still *within* population variation offers better prospects. In the case of the human Y chromosome, a 3.4 kilo bp *Hae*III fragment is polymorphic between individuals (McKay *et al.*, 1978). Gosden *et al.* (1981) have described a cloned sequence from human satellite III and its distribution in polymorphic variants of chromosomes 1, 9 and 16. There is extensive between individual variation. In the Orthoperan *Atractomorpha similis*, chromosomes can be polymorphic for large blocks of repeated sequences (Miklos and Nankivell, 1976). In another species, *Cryptobothrus*, C-banding and length measurements reveal extensive differences between some homologues (John and King, 1977). A similar situation occurs in mice (Forejt, 1973) and in humans (Kurnit, 1979).

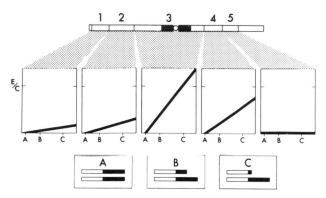

FIG. 8 *Recombination levels (experimental/control) in 5 regions of chromosome III when the X chromosomes are heterozygous
for heterochromatic deletions of varying sizes.   Data are from
Yamamoto (1979).*

This variation cannot be essential for basic metabolism,
differentiation or development, since polymorphic individuals
within a population are segregating for the presence or total
absence of these DNA sequences on particular chromosomes
(John, 1981).

When "artificial" polymorphisms are created in laboratory
situations, repeated sequences can have profound effects in
the germ line. For example, when a series of heterochromatic
deletions is made for the X chromosome in *Drosophila melanogaster*, not only are the recombinational characteristics of
that chromosome altered (Yamamoto and Miklos, 1978), but so
also is recombination on chromosome III (Fig. 8) (Yamamoto,
1979). Thus, when the genetic background is again carefully
controlled, the data reveal germ line effects on the recombination system. Similarly, when a heterochromatic portion
of the Y chromosome is added to the tip of the X chromosome,
the euchromatic region next to this predominantly simple
sequence DNA becomes effectively *rec⁻* (Miklos and John, 1979).

These data are worth reiterating, since they show us that
the germ line is the place in which highly repeated DNA can
be put into a sensible biological context (see also Bostock,
1980). We do not intend to belabour these *Drosophila* data
unnecessarily; they are clean and supported by extensive comparable data from natural populations. We do stress that
we are unable to *prove* that in our case the effects of modify-
ing the recombination system in this manner are of adaptive or
evolutionary significance. However, it would seem an anathema
to us in view of the extensive nature of the data, if none of
the effects we have described would contribute to variation

within a population.

There are certainly excellent data from plant populations that show demonstrable interrelations between chiasmata, variability and response to selection (Rees and Dale, 1974). For example, the higher the chiasma frequency, the lower is the phenotypic and genetic variance for polygenic characters such as flowering time. Furthermore, the response to selection for such characters is less effective. Thus, whilst these data do not prove that the effects of the recombination system are adaptive, they are certainly very suggestive. We should also point out that the presently more topical ways of introducing variation, such as by means of nomadic elements, suffer from precisely the same constraints. We are unaware of any data from natural populations that prove that these elements are major contributors to adaptive or evolutionary change. For example, there is no information on how their redeployment affects the variability of phenotypes within populations or whether their movement influences the response to selection of quantitative characters. This is the crux of the problem for virtually any repeated DNA sequence one wishes to nominate, there are no data proving its adaptive or evolutionary significance.

2. Speciation

We now turn to the significance of repeated sequence change to speciation and again we shall commence with data.

In the case of the Hawaiian *Drosophila*, we compared the cloned sequences of three closely related species, *D. gymnobasis*, *D. silvarentis* and *D. grimshawi* (Miklos and Gill, 1981a). These species are homosequential and the morphological characteristics of the first two are exceedingly similar. All three species have approximately 40% of their genomes as the same buoyant density satellite and as variants of a very similar 189 base-pair prototype sequence. The cloned sequences of *D. gymnobasis* and *D. silvarentis* are almost identical. It would seem a sensible proposition to us that these particular sequences were not causal factors in speciation, since their amount and sequence is hardly altered. Even if we make the slightly more distant comparison to *D. grimshawi*, this conclusion is still valid. One would need to argue that the sequence changes involved from the 189 base-pair consensus sequence to the 177, 179 and 185 base-pair variants were instrumental in reproductive isolation. This is hardly a plausible, let alone a viable suggestion.

The situation in the Hawaiian *Drosophila* is that there are well over 800 species that have radiated on a small number of islands in a time span of less than 6×10^6 years. There

have been very few gross chromosome rearrangements such as
Robertsonian fusions or fissions. A large number of species
groups are homosequential, with differences between groups
being confined at the cytological level to euchromatic inversions and loss or gain of heterochromatic segments. In the
group in which we have carried out our sequencing analyses, it
seems that differences of a genic nature were undoubtedly the
major factors in any speciation events.

In the *Drosophila virilis* phylogeny, some idea can also be
gleaned of the role of highly repetitive sequences in speciation. The data of Oshima on the fertility of interspecific
hybrids in this group has been summarised by us (John and
Miklos, 1979). There the viability and fertility of hybrids
does not depend on similarities or differences in the satellite content of the parental species.

There are no data that to our knowledge singularly point
to an involvement of satellite sequences in speciation; in
fact, the Hawaiian data point to the opposite. It seems to
us that the evidence on the differences distinguishing present
day species leads to the notion of a plurality of causes for
speciation events. *Proving* which of these possibilities was
the prime mover in any speciation event is the major stumbling
block (see Miklos *et al.*, 1980). This point has been made
explicitly by Dover (1980).

> *"Naturally, we cannot reconstruct the myriad of events
> that have gone into shaping the contemporary pattern of
> relationships of a species group; nor can we realistic-
> ally test....our hypotheses as to the prime determina-
> tive forces that have been involved."*

3. *Selfish DNA*

An attractive escape from the vexed problems in this field is
apparently provided by the concept of selfish DNA. In the
short time that has elapsed since the theory was propounded
(Orgel and Crick, 1980; Doolittle and Sapienza, 1980) it has
already become bogged down in the semantics of definitions
and clarifications, such as "parasitic DNA", "junk DNA",
"dead DNA", etc. (Orgel *et al.*, 1980; Dover and Doolittle,
1980). It has become necessary to distinguish between tandemly repeated and dispersed sequences, the former evolving
by mechanisms that might be more "ignorant" than "selfish".

To determine which sequences within a genome are *parasitic*
is quite difficult, since the theory has no clear predictions.
Instead, Orgel and Crick (1980) have drawn attention to three
general areas of research that may be useful in producing
relevant data. These are given below.

1. *"....it is important to know where DNA sequences occur which have little obvious function."*
2. *"....if the increase of selfish DNA and its movement around the chromosome are not rare events in evolution, it may be feasible to study, in laboratory experiments, the actual molecular mechanisms involved in these processes."*
3. *"....a careful study of all the nonspecific effects of extra DNA would give us a better idea of how it affected different aspects of cellular behaviour."*

The feature that distinguishes selfish or parasitic DNA from other more conventional theories is that it stresses *competition between replicators of differential efficiency*. It thus appears more suited to transposable elements than to other portions of the genome. At first sight the arguments of Doolittle and Sapienza (1980) seem compelling, especially since these authors have emphasised the potential selfishness of transposable elements. However, it should be stressed that there is no evidence that transposable elements conform to the basic requirement of the theory; that is, competition on the basis of replicative ability. Similarly, there is no evidence that highly repeated DNAs compete on the basis of differential replication with other portions of the genome. In fact, one can take a quite different viewpoint. What if there is no competition between different replicators? What if the mechanisms of replication and recombination themselves are the real culprits and the DNA *per se* is a passive polymer? What if these mechanisms are continually amplifying and deleting all portions of the genome, genic as well as non-genic? In this case most events will be deleterious, (since they delete or amplify gene sequences), but non-transcribed arrays ought to have a higher probability of survival. The extent of the genetic variation that impinges on the replicative and recombinative mechanisms will determine the extent to which repeated arrays are formed.

One beneficial effect of the theory is that it defuses the issue of mandatory assignment of functions to every piece of DNA. However, it replaces this with an emphasis on *cellular* phenomena (Orgel and Crick, 1980) as well as on the potential usefulness of DNA sequence information in helping to decide between the relative amounts of "parasitic DNA", "ignorant DNA", "dead DNA", etc. in a genome (Orgel *et al.*, 1980). However, as I have tried to point out here, the salient lesson that has emerged from the sequencing analyses of highly repeated DNA is that this avenue is an inappropriate one to pursue if the aim is to decide between the alternatives just discussed.

Furthermore, the theory only allows for significant interpretations at the *cellular* level, yet it is an assumption that
important answers will be found here. Consider for example
the meiotic recombination system in eukaryotes, which is
invariable discussed in *evolutionary* terms (Felsenstein and
Yokoyama, 1976; Rees and Dale, 1974; Maynard Smith, 1977;
John, 1981). There is a good reason for an *evolutionary*
rather than a cellular perspective in such discussion. The
events involved in shuffling DNA sequences in the germ line of
an organism do not intrude on its phenotype. Their significance (or lack thereof) is tested only in *succeeding generations*.

Consider the data which show that there are recombinational
effects exerted by highly repeated DNA sequences. The consequences of making a region effectively rec^- do not intrude on
the phenotype of the organism, yet the DNA sequence arrays in
the next generation *may* be altered. If the effects of recombination in a particular region are of evolutionary significance, then DNA that modifies the position of recombination
may play as much part in evolucionary strategy as the individual effects of particular genes. Alternatively, if recombination in a given region leads to no change or to change that
is not selected, then the satellite DNA that influences the
position of this recombinational event will be evolutionarily
unimportant.

The extent to which *selfishness*, *neutrality* and *selection*
apply to the question about which parts of any genome can be
designated as "parasitic", "dead" or "useful" will depend on
how well they cope with the data. The following questions
will provide useful data against which predictions of the
various hypotheses can be tested.

1. Are there different replicators and is there competition between them?
2. What is the predicted variation at the DNA sequence
   level for genic and non-transcribed sequences?
3. Is there any adaptive significance to the effects of
   the *amounts* of satellite sequences in altering recombination patterns?
4. Do the correlations between total genome size, cell
   volume and cell growth rate have generality and, if
   so, do they have any adaptive significance (Cavalier-
   Smith, 1978, 1980a,b)?
5. How well does each theory cope with the data that
   certain organisms delete or underreplicate massive
   parts of their genomes in somatic cells, but do not
   do so in the germ line?

CONCLUDING REMARKS

At the beginning of this article I addressed myself to two fundamental issues. First, what insights have been provided by DNA sequencing studies into important biological problems? And second, what phenotypic effects occur when the amount of highly repeated DNA is manipulated experimentally?

I have come to the following conclusions.

(1) DNA sequence analyses have unearthed extensive variation *within* a species but the sequences *per se* have not provided any solutions to biological problems. Furthermore, the available data cannot distinguish between *selective* and *neutral* modes of satellite DNA evolution. The selfish DNA theory is largely irrelevant to highly repeated DNA, its major handicap being its lack of strict predictive ability. If highly repeated sequences were to conform to the basic requirement of the theory, namely to be the result of *competition between replicators of differential efficiency*, then an extraordinarily diverse array of sequences of different complexities would seem to be capable of such abilities. The lack of any common sequence structure between satellites from the same species or from diverse species groups makes such a case unlikely.

(2) Gross distortions of the satellite DNA content of a genome have negligible somatic effects. In the germ line, however, the recombination system can be perturbed, leading to the appearance of potentially novel recombinants. Data bearing on whether these new types are of adaptive or evolutionary significance are unavailable. The existence of natural polymorphisms severely constrains most hypotheses on function, particularly those dealing with cellular roles. Certainly, highly repeated DNA cannot be essential for processes that are highly canalised, such as basic metabolism, differentiation or development. Its effects in the germ line, however, where the important parameter is the *amount* of condensed chromatin, seem to provide a natural means of altering the potential genetic variation by other than strict genetic means.

(3) Lastly, the data from the three species of Hawaiian *Drosophila* indicate that the satellite sequences themselves were unlikely to have been the major causative factors in the various speciation events.

MATERIALS AND METHODS

1. Chromosomes

The chromosome fragments consist of varying portions of the

heterochromatin of X chromosomes. They are *Dp 1187*, *Dp 164*, *Dp 118 YM*, *Dp 1337*, *Dp 1492*, *Dp 1514* and *Dp 3* heretofore termed *mini 1187*, etc. The 2 Y chromosomes used were the B^SY and the y^+Y. The heterochromatically deficient X used to generate XO males was $bb^{\ell 158}$. The *attached* XY used for a similar purpose was an X X^L Y^S. All chromosomes are adequately described by Lindsley and Grell (1968) or Yamamoto and Miklos (1978).

2. *Flies*

The genotypes in any given experiment were generated from one cross; for example, XY males when mated to XX *mini* females yield XY, XY *mini*, XX and XX *mini* progeny.

3. *Egg Laying Conditions*

Eggs were collected over 1h on Petri dishes containing banana food with the addition of a colouring agent to facilitate egg visualisation. They were kept at 25°C and larvae collected every 15 min till all had emerged. The larvae were kept on banana food till adulthood and scored for phenotypic characteristics, which allowed unambiguous assignation of their genotypes.

4. *Scanning electron microscopy*

Flies of the appropriate genotypes were collected and prepared for scanning electron microscopy. The heads were removed, split, and the number of facets in each eye was counted. The eye was photographed and the number of facets checked on the photograph. The logistics of such experiments are not feasible with normal eyes, since each contains about 800 facets. Consequently, we introduced the eye mutant *Bar* into our stocks (by using it on a Y chromosome). This reduces the facet number to a pleasing (to humans) 100 or so per eye. The 4 X chromosome fragments were each sequentially added to XY control males to yield the desired increases in X heterochromatin. The y^+Y chromosome furnished the means for Y heterochromatin addition.

ACKNOWLEDGEMENTS

I am very grateful to the following people for their help in this research. Margaret Kovacs for her conscientious and diligent efforts in carrying out all the electron microscopic work; Julie Higginbotham for her painstaking labours on the emergence time experiments; Garry Brown and Sandy Smith for their unrestrained artistic advice and illustrations under

pressure, and to Maureen Whittaker for her excellent photo-
graphic reproductions.

Finally, I express deep appreciation to my colleague,
Bernard John, for his guidance and incisive criticisms.

REFERENCES

Baker, W.K. (1978). A clonal analysis reveals early developmental restrictions in the *Drosophila* head. *Devel. Biol.*
62, 447–463.

Barlow, P. (1973). The influence of inactive chromosomes on
human development. *Humangenetik* 17, 105–136.

Bennett, M.D. (1971). The duration of meiosis. *Proc. Roy.
Soc. ser. B* 178, 259–275.

Bostock, C.J. (1980). A function for satellite DNA? *Trends
Biochem. Sci.* 5, 117–119.

Brutlag, D.L. (1980). Molecular arrangement and evolution of
heterochromatic DNA. *Annu. Rev. Genet.* 14, 121–144.

Carlson, M. and Brutlag, D. (1977). Cloning and characterisation of a complex satellite DNA from *Drosophila melanogaster*. *Cell* 11, 371–381.

Carlson, M. and Brutlag, D. (1979). Different regions of a
complex satellite DNA vary in size and sequence of the
repeating unit. *J. Mol. Biol.* 135, 483–500.

Cavalier-Smith, T. (1978). Nuclear volume control by nucleo-
skeletal DNA, selection for cell volume and cell growth
rate, and the solution of the DNA C-value paradox. *J. Cell
Sci.* 34, 247–278.

Cavalier-Smith, T. (1980a). r- and k-tactics in the evolution
of protist developmental systems: cell and genome size,
phenotype diversifying selection, and cell cycle patterns.
BioSystems 12, 43–59.

Cavalier-Smith, T. (1980b). How Selfish is DNA? *Nature (London)* 285, 617–618.

Christie, N.T. and Skinner, D.M. (1980a). Evidence for non-
random alterations in a fraction of the highly repetitive
DNA of a eukaryote. *Nucl. Acids Res.* 8, 279–298.

Christie, N.T. and Skinner, D.M. (1980b). Selective amplification of variants of a complex repeating unit in DNA of a
crustacean. *Proc. Nat. Acad. Sci., U.S.A.* 77, 2786–2790.

Cooke, H.J. and Hindley, J. (1979). Cloning of human satellite III DNA: different components are on different chromosomes. *Nucl. Acids Res.* 6, 3177–3197.

Cooper, K.W. (1956). Phenotypic effects of Y chromosome
hyperploidy in *Drosophila melanogaster*, and their relation
to variegation. *Genetics* 41, 242–264.

de Marinis, F. (1952). Action of the Bar series in relation
to temperature studied by means of Minute-N mosaic

technique. *Genetics* **37**, 75–89.

Donehower, L., Furlong, C., Gillespie, D. and Kurnit, D. (1980). DNA sequence of baboon highly repeated DNA: evidence for evolution by non-random unequal crossovers. *Proc. Nat. Acad. Sci., U.S.A.* **77**, 2129–2133.

Doolittle, W.F. and Sapienza, C. (1980). Selfish genes, the phenotype paradigm and genome evolution. *Nature (London)* **284**, 601–603.

Dover, G.A. (1980). Problems in the use of DNA for the study of species relationships and the evolutionary significance of genomic differences. *In* "Chemosystematics: principles and practice (Bisby, F.A., Vaughn, J.G. and Wright, C.A., eds), pp. 241–268, Academic Press, London, New York and San Francisco.

Dover, G. and Doolittle, W.F. (1980). Modes of genome evolution. *Nature (London)* **288**, 646–647.

Felsenstein, J. and Yokoyama, S. (1976). The evolutionary advantage of recombination. II Individual selection for recombination. *Genetics* **83**, 845–859.

Forejt, J. (1973). Centromeric heterochromatin polymorphism in the house mouse. *Chromosoma (Berlin)* **43**, 187–201.

Gosden, J.R., Lawrie, S.S. and Cooke, H.J. (1981). A cloned repeated DNA sequence in human chromosome heteromorphisms. *Cytogenet. Cell. Genet.* In the press.

Hsieh, T.-S. and Brutlag, D. (1979a). Sequence and sequence variation within the 1.688 g/cm^3 satellite DNA of *Drosophila melanogaster*. *J. Mol. Biol.* **135**, 465–481.

Hsieh, T.-S. and Brutlag, D. (1979b). A protein that preferentially binds *Drosophila* satellite DNA. *Proc. Nat. Acad. Sci., U.S.A.* **76**, 726–730.

Hutchinson, J., Rees, H. and Seal, A.G. (1979). An assay of the activity of supplementary DNA in *Lolium*. *Heredity* **43**, 411–421.

John, B. (1981). Heterochromatin variation in natural populations. *In* "Proceedings of the 7th international chromosome conference", Oxford. In the press.

John, B. and King, M. (1977). Heterochromatin variation in *Cryptobothrus Chrysophorus*. II Patterns of C-banding. *Chromosoma (Berlin)* **65**, 59–79.

John, B. and Miklos, G.L.G. (1979). Functional aspects of satellite DNA and heterochromatin. *In* "International review of cytology (Bourne, G.H. and Danielli, J.F., eds), pp. 1–114, Academic Press, New York, San Francsico and London.

Kimura, M. (1979). The neutral theory of molecular evolution. *Sci. Amer.* **241**, 94–104.

Klein, H.L. and Petes, T.D. (1981). Intrachromosomal gene conversion in yeast. *Nature (London)* **289**, 144–148.

Kurnit, D.M. (1979). Satellite DNA and heterochromatin variants: the case for unequal mitotic crossing over. *Human Genet.* 47, 169–186.

Lindsley, D.L. and Grell, E.H. (1968). Genetic variations of *Drosophila melanogaster*. *Carnegie Inst. Washington,* Pub. no. 627.

Luce, W.M. (1935). Temperature studies on Bar-Infrabar. *J. Expt. Zool.* 71, 125–147.

Luce, W.M., Quastler, H. and Chase, H.B. (1951). Reduction in facet number in Bar-eyed *Drosophila* by X-rays. *Genetics* 36, 488–499.

Maynard Smith, J. (1977). Why the genome does not congeal. *Nature (London)* 268, 693–696.

McKay, R.D.G., Bobrow, M. and Cooke, H.J. (1978). The identification of a repeated DNA sequence involved in the karyotype polymorphism of the human Y chromosome. *Cytogenet. Cell Genet.* 21, 19–32.

Miklos, G.L.G. and Gill, A.C. (1981a). The DNA sequences of cloned complex satellite DNAs from Hawaiian *Drosophila* and their bearing on satellite DNA sequence conservation. *Chromosoma* 82, 409–427.

Miklos, G.L.G. and Gill, A.C. (1981b). Nucleotide sequences of highly repeated DNAs: compilation and comments. *Genet. Res.* In the press.

Miklos, G.L.G. and John, B. (1979). Heterochromatin and satellite DNA in man: properties and prospects. *Amer. J. Human Genet.* 31, 264–280.

Miklos, G.L.G. and Nankivell, R.N. (1976). Telomeric satellite DNA functions in regulating recombination. *Chromosoma (Berlin)* 56, 143–167.

Miklos, G.L.G., Willcocks, D.A. and Baverstock, P.R. (1980). Restriction endonuclease and molecular analyses of three rat genomes with special reference to chromosome rearrangement and speciation problems. *Chromosoma (Berlin)* 76, 339–363.

Morata, G. and Lawrence, P.A. (1979). Development of the eye-antenna imaginal disc of *Drosophila*. *Devel. Biol.* 70, 355–371.

Mullins, J.I. and Blumenfeld, M. (1979). Satellite 1c: a possible link between the satellite DNAs of *D. virilis* and *D. melanogaster*. *Cell* 17, 615–621.

Nagl. W. (1978). "Endopolyploidy and polyteny in differentiation and evolution". North-Holland, Amsterdam.

Olmo, E. and Morescalchi, A. (1978). Genome and cell size in frogs: a comparison with salamanders. *Experientia* 34, 44–46.

Orgel, L.E. and Crick, F.H.C. (1980). Selfish DNA: the ultimate parasite. *Nature (London)* 284, 604–607.

Orgel, L.E., Crick, F.H.C. and Sapienza, C. (1980). Selfish
    DNA. *Nature (London)* **288**, 645-646.
Pech, M., Igo-Kemenes, T. and Zachau, H.G. (1979a). Nucleo-
    tide sequence of a highly repetitive component of rat DNA.
    Nucl. Acids Res. **7**, 417-432.
Pech, M., Streeck, R.E. and Zachau, H.G. (1979b). Patchwork
    structure of a bovine satellite DNA. *Cell* **18**, 883-893.
Petes, T.D. (1980). Unequal meiotic recombination within tan-
    dem arrays of yeast ribosomal DNA genes. *Cell* **19**, 765-774.
Poschl, E. and Streeck, R.E. (1980). Prototype sequence of
    bovine 1.720 satellite DNA. *J. Mol. Biol.* **143**, 147-153.
Rees, H. and Dale, P.J. (1974). Chiasmata and variability in
    Lolium and *Festuca* populations. *Chromosoma (Berlin)* **47**,
    335-351.
Rubin, C.M., Deininger, P.L., Houck, C.M. and Schmid, C.W.
    (1980). A dimer satellite sequence in bonnet monkey DNA
    consists of distinct monomer subunits. *J. Mol. Biol.* **136**,
    151-167.
Salser, W., Bowen, S., Browne, D., El Adli, F., Federoff, N.,
    Fry, K., Heindell, H., Paddock, G., Poon, R., Wallace, B.
    and Whitcome, P. (1976). Investigation of the organization
    of mammalian chromosomes at the DNA sequence level. *Fed.
    Proc. Fed. Amer. Soc. Exp. Biol.* **35**, 23-35.
Southern, E.M. (1970). Base sequence and evolution of guinea-
    pig α satellite DNA. *Nature (London)* **227**, 794-798.
Strachan, T., Coen, E.S., Webb, D.A. and Dover, G.A. (1982).
    Modes and rates of change of abundant DNA families in
    Drosophila. *Cell* (submitted).
Szostak, J.W. and Wu, R. (1980). Unequal crossing over in the
    ribosomal DNA of *Saccharomyces cerevisiae*. *Nature (London)*
    284, 426-430.
Thayer, R.E., Singer, M.F. and McCutchan, T.F. (1981).
    Sequence relationships between single repeat units of
    highly reiterated African green monkey DNA. *Nucl. Acids
    Res.* **9**, 169-181.
Wu, J.C. and Manuelidis, L. (1980). Sequence definition and
    organisation of a human repeated DNA. *J. Mol. Biol.* **142**,
    363-386.
Yamamoto, M. (1979). Interchromosomal effects of heterochro-
    matic deletions on recombination in *Drosophila melanogas-
    ter*. *Genetics* **93**, 437-448.
Yamamoto, M. and Miklos, G.L.G. (1978). Genetic studies on
    heterochromatin in *Drosophila melanogaster* and their impli-
    cations for the functions of satellite DNA. *Chromosoma
    (Berlin)* **66**, 71-98.

Changes to Genomic DNA in Methotrexate-resistant Cells

CHRISTOPHER J. BOSTOCK and CHRISTOPHER TYLER-SMITH

Medical Research Council, Mammalian Genome Unit,
King's Buildings, West Mains Road, Edinburgh, EH9 3JT, Scotland

INTRODUCTION

1. Selection for Methotrexate Resistance is an Evolutionary Process

It may seem strange that a paper on the acquisition of resistance to methotrexate by mammalian cells can find itself amongst a collection on the evolution of eukaryotic genomes. Methotrexate (MTX) is a widely used anti-tumour drug, whose principal mode of action is to inhibit the activity of cellular dihydrofolate reductase (DHFR), thereby preventing the cell from endogenously synthesising thymidylate, purines or glycine. The inhibition of DHFR is achieved by the stoichiometric binding of MTX, the DHFR-MTX complex being more stable than the binding of DHFR with its natural substrate. Resistance to MTX has been ascribed to one or more of three different mechanisms. Since MTX is taken up by cells *via* an active transport mechanism, alterations to this transport system can reduce the intracellular concentration of the drug (see e.g. Sirotnak *et al.*, 1968). A second mechanism involves structural mutations in the gene for DHFR such that the affinity of DHFR for MTX is reduced (see e.g. Flintoff *et al.*, 1976; Jackson and Niethammer, 1977). A third mechanism involves "over-production" of DHFR such that there is an intracellular excess of enzyme molecules over MTX molecules. The more that this latter mechanism is studied *in vitro*, the more we can identify features that are directly relevant to the questions to which this meeting is addressing itself; often large-scale alterations in the genome structure that result in distinct changes to the phenotype.

2. Methotrexate Resistance Involves Amplification of Dihydrofolate Reductase Genes

The last of the three mechanisms, that cells can develop resistance to MTX by overproduction of DHFR, has been known for over 20 years (Fischer, 1961; Hakala et al., 1961). It is the more recent studies on a number of model in vitro systems that have shown that the overproduction results from elevated levels of DHFR messenger RNA (Chang and Littlefield, 1976; Kellems et al., 1976), which in turn reflect an increased number of DHFR genes within drug-resistant cells (Alt et al., 1978; Nunberg et al., 1978; Dolnick et al., 1979; Kaufman et al., 1979; Melera et al., 1980). Thus, in cells that overproduce DHFR in response to MTX selection, there has been an amplification of the genes that code for that enzyme. This genomic change has a marked effect on the ability to survive under the selective conditions; they might be said to have evolved.

3. Large Genomic Segments may be Involved in Gene Amplification

Little is known about the mechanism of amplification, although two points have been established to date. First, the structure of the amplified copies of DHFR genes appear to be the same as the structure of normal DHFR genes in the haploid genome (Nunberg et al., 1980). Second, where it has been estimated, the unit of DNA that becomes amplified is at least an order of magnitude larger than the DHFR gene. Estimates of the size of the DHFR gene show it to be 32 kilobase-pairs (Abelson, 1980) or 42 kilobase-pairs (Nunberg et al., 1980) in size. By contrast, the average size of the amplified unit in MTX-resistant L5178Y mouse lymphoma cells is estimated to be 800 kilobase-pairs (Dolnick et al., 1979), whereas in MTX-resistant PG19 mouse melanoma cells the average size is as large as 3000 kilobase-pairs (Bostock and Clark, 1980). Thus the amplification of the gene in response to MTX-selection results in the duplication of many DNA sequences in addition to those directly involved in the structure of the DHFR gene.

4. Cytogenetic Changes Accompany Methotrexate Resistance

The large size of the amplified unit explains why additional abnormal chromosome forms can be observed cytologically in many resistant cells. These aberrant forms range from long regions of homogeneously staining material (hsr) attached to, or integrated into, normal chromosomes, through to the multiple non-centromeric fragments called "double minutes" (dm: Kaufman *et al.*, 1979; Bostock *et al.*, 1979; Biedler *et al.*, 1980). The link between these chromosome structures and extra copies of DHFR genes has been strengthened by the localisation of DHFR coding sequences, using hybridisation, to the hsr of chromosomes (Nunberg *et al.*, 1978; Dolnick *et al.*, 1979) and demonstration of enrichment of DHFR coding sequences in DNA purified from dm (Kaufman *et al.*, 1979).

In this paper we describe some experiments on a series of MTX-resistant mouse cells in which extensive rearrangement of DNA sequences has taken place, indicating that gene amplification can be a complex process. As far as is known, MTX does not interact with DNA, nor is it incorporated into DNA, which probably explains the lack of any documented carcinogenic effects of the drug (Bagshawe, 1977; Bailin *et al.*, 1975). We might suppose that rather than inducing directly the molecular processes that result in gene amplification, MTX is selecting for the rare amplification events that occur naturally at many genetic loci at a low frequency. This view is supported by the observation that amplification of genes for different target proteins follows upon selection for resistance to *N*-(phosphonacetyl)-L-asparate (Wahl *et al.*, 1979), Vincristine (Meyers and Biedler, 1981), or cadmium (Walters *et al.*, 1981; Beach and Palmiter, 1981); quite different target proteins and quite different selective agents.

SOME EXPERIMENTAL RESULTS

1. *Resistant Cells Contain More Dihydrofolate Reductase and More Dihydrofolate Reductase Genes than Wild-Type Cells*

We have examined the properties of four independently selected MTX-resistant clones of mouse EL4 lymphoma cells (kindly given to us by Dr Thomas Alderson), and two independently selected lines of mouse PG19 melanoma cells. As Table I shows, each of these grows in concentrations of methotrexate at least four orders of magnitude greater than that which would inhibit sensitive cells; each has a large increase in the activity of DHFR, and each has a large increase in the number of DHFR genes per haploid genome equivalent.

2. *DHFR Gene Amplification Can Result in Rearrangements Near the Ends of Genes.*

Examination of the genomic arrangement of DHFR coding sequences in these cells by hybridisation of a Southern blot of *Eco*RI-digested DNA with ^{32}P nick-translated DHFR complementary DNA, shows a number of features (Fig. 1). First, considering

TABLE I

*Characteristics of MTX-resistant Cells**

CELL LINE	FINAL RESISTANCE LEVEL: MTX CONCN[†] (M)	SPECIFIC ACTIVITY OF DHFR RELATIVE TO WILD-TYPE	NUMBER OF DHFR GENES PER 3×10^9 BASE-PAIRS[#]
EL4/3	2.2×10^{-3}	5800	750 – 1120
EL4/8	1.1×10^{-3}	1980	690 – 1330
EL4/11	1.1×10^{-3}	886	600 – 825
EL4/12	1.1×10^{-3}	1276	650 – 1360
PG19T3:S1	1×10^{-4}	1000	~ 100
PG19T3:S2	1×10^{-3}	–	~ 1000

[*]From Bostock *et al.* (1979), Tyler-Smith and Alderson (1981), and unpublished data.
[†]Wild-type cells are effectively killed by 10^{-8}M-MTX.
[#]Wild-type cells have approx. 1 DHFR gene/3×10^9 DNA base-pairs.

that equal amounts of DNA were loaded in the gel tracks, comparison of the times of exposure of parent DNA hybrids with resistant cell DNA hybrids gives a good indication of the degree of amplification. Second, all four genomic fragments in *Eco*RI digests of parent non-amplified DNA that hybridise DHFR cDNA are also found amplified in all resistant cells. Of the six resistant cells listed in Table I, EL4/12, PG19T3:S1 and PG19T3:S2 contained only these bands. The remaining three cell lines, EL4/3, EL4/8 and EL4/11, each had additional bands (other than those generated by non-specific endonucleases). The sizes of the new fragments in each of these cells is characteristic for each cell. Using isolated subportions of the cDNA probe, these new bands have been shown to represent genomic rearrangements close to or within the 3' end of the gene in a subportion of the amplified genes in EL4/3, EL4/8 and EL4/11 cells and additionally, a rearrangement near the 5' end of some other amplified copies of the gene in EL4/3 cells (Tyler-Smith and Alderson, 1981). Thus in three out of the six MTX-resistant cell lines, rearrangements close to one end (EL4/8 and EL4/11) or to both ends of the gene had occurred at some stage during the amplification process. This is the first piece of evidence that suggests that gene amplification in these cells involves DNA rearrangement and, therefore, recombination.

3. DNA Sequences other than DHFR Genes are Amplified

It is a striking exercise to simply view ethidium bromide-
stained, high-resolution gels of restriction endonuclease-
digested DNAs of the highly MTX-resistant cell lines. Figure
1 shows one such example. The most immediately apparent feature is that, whereas the parental DNA shows up as the usual
continuous "smear" of fragments, the DNAs from all the resistant lines show complex patterns of fragments, representing
sequences that are repeated many times per haploid genome,
superimposed upon the wild-type smear. Much of the repetition
must derive from the fact that these sequences have been
amplified along with the DHFR genes as part of the large unit,
but some must be additionally repeated within the unit. We
will refer to these sequences collectively as amplified DNA
sequences. In one of the PG19 sublines, pre-existing repeated
sequences have been amplified during selection, since the percentage cellular content of mouse satellite DNA increases with
increased resistance, and satellite DNA sequences can be
localised to hsr of MTX-resistant cells (Bostock and Clark,
1980). In these cells, examination of the structure of satellite DNA-containing blocks in hsr suggests that chromosomal
rearrangements must have taken place during the formation of
the amplified unit.

The second, and most significant point about the ethidium
bromide-stained gels of Fig. 1 is the fact that the pattern of
bands observed is different in each cell line that has been
independently selected from wild-type. This means that in
each of the five lines shown in Fig. 1, the DHFR genes were
amplified as part of a unit that had a different structure.
(We will show later that the complex ethidium bromide-stained
pattern derives from a single structure common to all amplified units within a cell line, and does not reflect varying
heterogeneities in the cell population.)

The existence of distinct sets of amplified DNA sequences
in each of the five cell lines shows that amplification could
not have proceeded by the simple notion that a pre-existing
segment of genomic DNA, which contains the DHFR gene, was
excised and subsequently differentially amplified *without further rearrangement*. This follows from the fact that each of
the lines has DNA sequences that are uniquely amplified only
in those cells. Even allowing for the possibility that the
DHFR gene is flanked by different sequences in each of the
two homologues that contain it, and that the DHFR gene on
either or both these homologues may be amplified, the maximum
number of distinct patterns that can be generated, without any
subsequent patterns being composed entirely of sequences present in one or more of the other cell lines, is four. Thus

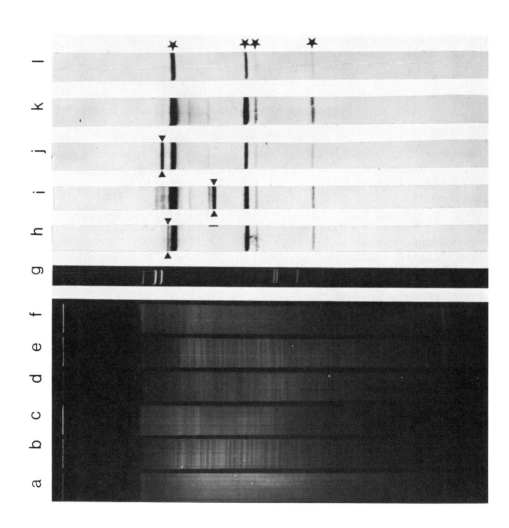

FIG. 1 *Amplified DNA and DHFR gene sequences in MTX-resistant cells. Ethidium bromide-stained 0.75% agarose gels (tracks (a) to (g)) and Southern transfers (tracks (h) to (l)) probed with ^{32}P nick-translated pDHFR11 DNA. The DNAs were digested with EcoRI restriction endonuclease and approx. 10µg of digested DNAs were loaded into each well of a 17 cm × 3 mm × 36 cm gel and run at a voltage gradient of 2 V/cm. After staining with ethidium bromide and photographing, the DNA fragments were denatured and transferred to nitrocellulose by the method of Southern (1975). Fragments containing DHFR coding sequences were detected by hybridisation to ^{32}P nick-translated pDHFR11, a cloned DHFR cDNA kindly given to us by Dr R.T. Schimke. (a) EL4 wild-type; (b) and (h) EL4/3; (c) and (i) EL4/8; (d) and (j) EL4/11; (e) and (k) EL4/12; (f) and (l) PG19T3:S2; (g) λDNA digested with both EcoRI and HindIII. The sizes of the λDNA fragments used as size markers were 25.2, 21.8, 5.24, 5.05, 4.21, 3.41, 1.98, 1.90, 1.57, 1.32, 0.93, and 0.84 kilobases. The asterisks denote the sizes of fragments containing DHFR coding sequences found in normal diploid cells and wild-type cell lines of mouse; their sizes are approximately 15.0, 6.3, 5.8 and 3.4 kilobase-pairs. Also shown with arrows are a band at 15 kilobase-pairs in EL4/3, one at 8.6 kilobase-pairs in EL4/8 and another at 18.5 kilobase-pairs in EL4/11 DNA. These are the bands that represent rearrangements at the ends of the genes.*

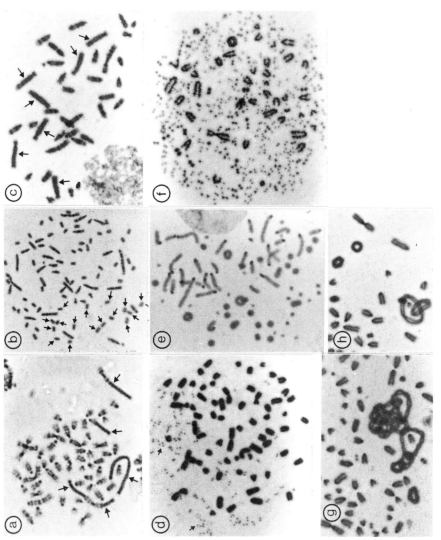

FIG. 2 For legend see opposite.

once again we see that amplification in these cells must have involved recombination events resulting in the DHFR gene being joined to sequences other than those to which it is normally joined. Whether these sequences are derived from pre-existing parts of the genome, or are generated by ligation events, or a *de novo* process during amplification, we cannot say.

4. Gene Amplification is Accompanied by the Acquisition of Extra Abnormal Chromosome Forms

Figure 2 shows a selection of the kinds of chromosome forms found in metaphase spreads of the MTX-resistant cells under study here. The wild-type EL4 cell has a modal chromosome number of 40 and, although there are marker chromosomes, the karyotype is superficially not too different from diploid. All the MTX-resistant subclones of EL4 have basically the same complement of normal chromosomes but, in addition, they have a varied collection of abnormal structures. EL4/8 and EL4/12 have variable numbers of double minutes, which vary in size up to forms that appear as rings. About one-third of the cells of these sub-lines contain some form of ring chromosome, some of which can become quite bizarre. EL4/3 contains variable numbers of acentric fragments about the size of small chromosomes. Most EL4/11 cells contain, on average, seven large chromosomes, which are stained entirely homogeneously but which contain centromeres. The most significant chromosomal change in PG19T3:S1 is the presence of four or five large marker chromosomes that contain hsr. It is these hsr that contain satellite DNA, giving the chromosomes their C-banding property. The other MTX-resistant PG19T3 cell, PG19T3:S2, contains a very large but variable number of dm, and about 10% of the cells contain one (rarely more) ring chromosome.

On the basis of our studies on the chromosomes of these MTX-resistant cells, we suggest that the usual path of chromosomal change is as follows. Double minutes containing DHFR genes are formed first, perhaps as relatively small circular

---

FIG. 2 *Metaphase chromosome spreads of MTX-resistant cell lines. (a) PG19T3:S1; (b) EL4/3; (c) EL4/11; (d) PG19T3:S2; (e) EL4/8; (f) EL4/12; (g) and (h) EL4/8. (a) hsr of PG19T3:S1 integrated into normal-type chromosome material. (b) The acentric chromosome fragments, some of which are arrowed. (c) The entire homogeneously staining chromosomes of EL4/11. (d), (e) and (f) Various examples of dm. In (e) the range in size up to circular forms is particularly apparent. (g) and (h) Examples of twisted and interlocked circles commonly found in EL4/8 and EL4/12 cells.*

chromosomal structures. Since they have no centromere, they
will segregate unequally at mitosis, giving the opportunity
for one daughter cell to possess more DHFR genes than its sister, thereby giving it a selective advantage. If a double
minute is a circular structure, or can become one, it would
enable gene amplification to proceed by a second mechanism;
duplication and enlargement of the ring (by sister chromatid
exchange or by replication errors). Although we have been
unable to observe directly a circular structure in dm, we
believe the continuous variation in the size of dm up to
"spherical" structures and then to chromosomes that clearly
do appear as rings is suggestive of this possibility (Fig.
2(e)). The ring chromosomes then take on the various inter-
locking forms predicted for ring chromosomes (see e.g. Douthart, 1972; Sandler, 1965) but they do appear to be unstable,
because fragmented rings are occasionally observed. We suggest, therefore, that the short acentric forms seen in EL4/3,
or the longer forms seen in EL4/11, were originally derived
from the breakdown of a ring chromosome. In at least one
case we have followed, as closely as is possible, the conversion of a ring into linear forms that replace dm and become
stabilised in the population (Bostock and Tyler-Smith, 1981;
and section 7, below).

Cells that contain dm, rings or acentric fragments lose
DHFR activity and resistance rapidly during growth in the
absence of MTX. This loss is accompanied by loss of the
abnormal chromosome forms; for example, after 14 days growth
in the absence of MTX, EL4/8 cells lose over 90% of their dm.
In contrast, EL4/11 cells retain the abnormal chromosomes
(they have centromeres), DHFR activity and their ability to
grow in MTX. Although the EL4/8 cells lose 90% of their dm
over 14 days growth in the absence of MTX, they lose only
about half their copies of DHFR genes and other amplified
sequences (Tyler-Smith and Bostock, 1981). This suggests
that many amplified units are stabilised by linkage in some
way to normal chromosomes, perhaps being integrated singly or
in small blocks into normal-type chromosomes, and thus go un-
undetected cytogenetically. This is supported by density gradient separation of double minutes from normal chromosomes of
EL4/8 cells; amplified DHFR genes and DNA sequences are found
associated with both chromosomal components (Tyler-Smith and
Bostock, 1981).

5. *Amplified DNA Sequence Patterns Reflect a Common Structure of all the Amplified Units within a Cell or Cell Population*

The complexity of the ethidium bromide staining patterns in
gels of restriction endonuclease-digested DNAs isolated from

highly MTX-resistant cells could, at one extreme, reflect all
the varied sequences that comprise each and every amplified
unit in the cell population. At another extreme, the complexity could derive from a vast sequence heterogeneity between
different amplified units within the same cell or within different cells in the same population. The differenes in the
patterns observed between cell types would, in the first possibility, represent differences in the DNA sequence composition of the amplified unit in different cell lines. In the
second possibility, the difference in patterns would reflect
the particular heterogeneous collection of sequences at the
time of harvesting the cells for isolation of DNA. These possibilities can be distinguished on the basis of cloning single
cells and single or sub-populations of double minutes or chromosomes containing hsr.

6. All Cells in a Population Contain the Same Complex Pattern

PG19 cells were selected for high levels of resistance to MTX
by threefold increases in the MTX concentration following a
period of growth at each newly established MTX level. Each
increase in the MTX concentration results initially in a large
majority of the cells dying. Thus at each selection only a
limited sub-population of cells survives to grow and to popu-
late the culture at the new MTX level. If the complexity of
the ethidium bromide banding pattern represents heterogeneity
between cells, the changing cell population at each selection
should be manifested in changing gel banding patterns of DNAs
isolated at different resistance levels. That such is not the
case is shown in Fig. 3; the pattern characteristic of
PG19T3:S2 is qualitatively the same in cells growing at 10^{-4}M,
3×10^{-4}M and 10^{-3}M-MTX. All that changes is that the pattern
becomes more distinct with an increase in MTX concentration,
reflecting an increased abundance of the DNA sequences present
in the bands due to their amplification.

DNA isolated from three single cell-derived clones of
PG19T3:S2 cells growing at 10^{-3}M-MTX show qualitatively and
quantitatively identical gel banding patterns (see Fig. 4).
It is clear, therefore, that the complex banding pattern is a
property common to each cell in the population. Heterogeneity
could still derive from a collection of different amplified
units common to all cells in the population. We are left,
therefore, to distinguish between a complex sequence composition common to all amplified units within all cells or a complex sequence composition derived from different sequences in
different amplified units or, at least, limited sub-
populations of amplified units.

Fig. 3 *Ethidium bromide-stained 0.75% agarose gel of EcoRI-digested DNAs from the 3 final selection levels of PG19T3:S2 line. (a) λDNA size markers as in Fig. 1; (b) wild-type PG19T3 cells; (c) PG19T3:S2 growing at $10^{-4}M$-MTX; (d) PH19T3:S2 growing at $3 \times 10^{-4}M$-MTX; (e) PG19T3:S2 growing at $10^{-3}M$-MTX.* (From Tyler-Smith and Bostock, 1981.)

7. *The Complex Ethidium Bromide Gel Staining Pattern is a Property of a Single Double Minute*

We have taken two approaches to examining the DNA sequence structure of individual amplified units. The first is based on the hypothesis that stable forms of chromosomes carrying hsr may be derived by fragmentation from large ring chromosomes, which in turn may develop by multiple duplications from single dm. During the course of studies on the stability

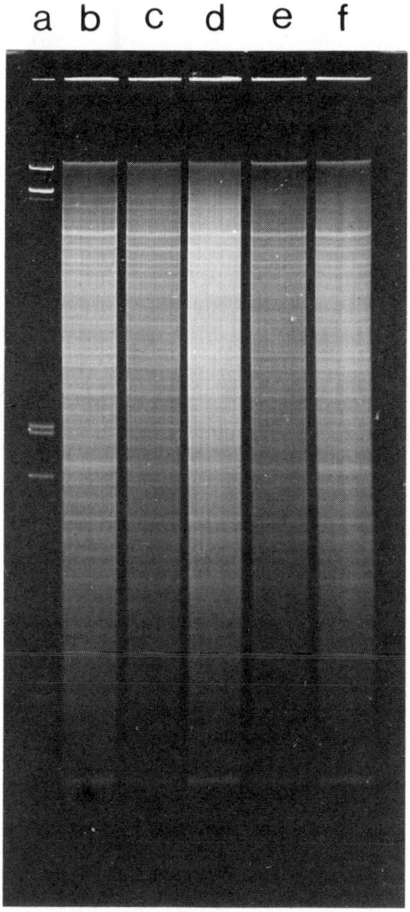

FIG. 4 *Ethidium bromide-stained 0.75% agarose gel of EcoRI-digested DNAs from individual single-cell derived clones of PG19T3:S2 growing at 10^{-3}M-MTX. (a) λDNA size markers as in Fig. 1; (b) the uncloned PG19T3:S2; (d), (e) and (f), three separate clones of PG19T3:S2; (c) the uncloned PG19T3: S2 after a period of growth necessary to clone and grow up those shown in tracks (d), (e) and (f).* (From Tyler-Smith and Bostock, 1981.)

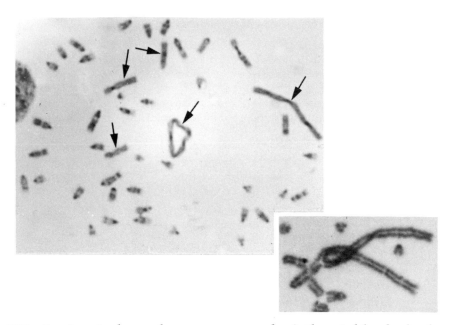

FIG. 5 *A metaphase chromosome spread of the stable derived sub-line of EL4/8 cells stained by the trypsin-Giemsa method. Within a single chromosome spread there is a large ring chromosome, a linear form of this and 3 short fragments derived from further breakdown. Notice that this cell does not contain any dm. In the inset is an example of how fragmented chromosomes may become associated with others and thus acquire a centromere for stability.*

of EL4/8 cells (a dm sub-line), we noticed by chance that the the dm-containing cells were being replaced by cells that had lost all dm; instead, these cells contained two large acro-centric chromosomes and one large sub-metacentric chromosome, all with hsr. A small proportion of the cells in the popula-tion had, in addition, a long homogeneously staining linear chromosome which, on the basis of chromosome banding proper-ties, could have been the precursor of the three shorter ones, and a large circular form of the linear chromosome (see Fig. 5; and Bostock and Tyler-Smith, 1981). It is pos-sible, therefore, that all the amplified units in the newly derived stable form of EL4/8 had been derived by duplication and subsequent fragmentation from a single dm; at the very least, it should contain a sub-population of all amplified units in the original cell. Tracks (a) and (b) of Fig. 6 show a comparison of the ethidium bromide staining patterns of *Eco*RI-digested DNAs from the original EL4/8 cells and the

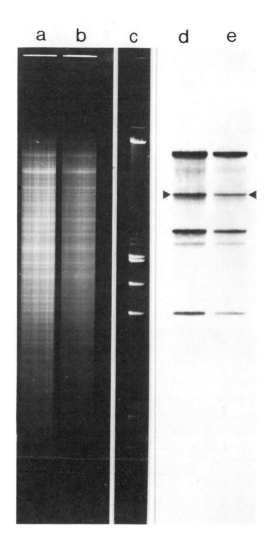

FIG. 6 *A comparison of the amplified DNA sequences of the stable derived line of EL4/8 and its parent EL4/8 cell line. (a) and (d) EcoRI-digested EL4/8 DNA; (b) and (e) EcoRI-digested EL4/8 derived line; (c) λDNA size markers as in Fig. 1. (a) and (b) Ethidium bromide-stained 0.75% gel tracks; (d) and (e) Southern transfers of these as described in Fig. 1. The arrows point to the extra fragment representing a rearrangement to gene sequences that is characteristic of EL4/8 cells.*

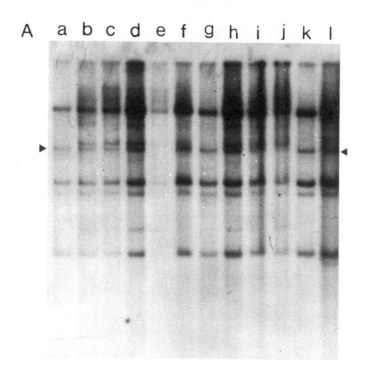

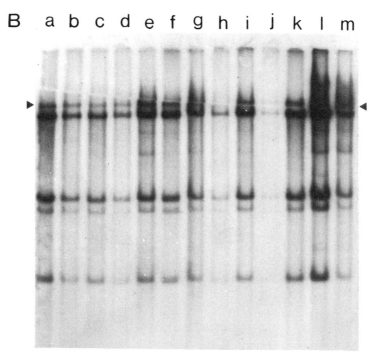

FIG. 7 For legend see opposite.

stable derivative, and tracks (d) and (e) of Fig. 6 compare
the structure of the DHFR genes. At both levels of analysis,
the two DNAs are essentially identical, suggesting that the
complex sequence composition of cells is a true reflection
of the composition of each and every amplified unit.

The second approach to cloning of amplified units has
been to use chromosomes isolated and purified from MTX-
resistant EL4/8 and EL4/11 cells to transform mouse L cells
for MTX-resistance. In this experiment we can follow the
transforming DNA because both EL4/8 and EL4/11 have characteristic rearrangements close to the DHFR gene that can be
identified with the cDNA probe. Individual colonies of
transformants resistant to 3×10^{-8}M-MTX were picked and
grown up for the preparation of DNA, which was then analysed
by Southern transfers and hybridisation. As Figure 7A and B
show, all transformants contain a few extra copies of DHFR
genes, over that number present in the parent L cell. More
importantly, at least 90% of transformants contain the
arrangements of the gene sequences that are characteristic
of the transforming agent rather than the L-cell. Thus each
transformed L cell grows at 3×10^{-8}M-MTX by virtue of it
having taken up a small number of exogenous DHFR genes.
They must therefore contain only a small number of amplified
units of EL4/8 or EL4/11 DNA. What happens if these cells
are now selected for high levels of MTX-resistance?
Interestingly, the exogenously supplied DHFR genes become
amplified, and the amplified units remain intact. As Fig. 8
shows, both the arrangement of gene sequences and the patterns of ethidium bromide-stained bands are the same in the
highly MTX-resistant transformed L cells as they are in the
transforming EL4 cells. L cells transformed with EL4/8

FIG. 7 *Southern transfers that have been hybridised with*
^{32}P nick-translated pDHFR11 of EcoRI-digested DNAs isolated
from cloned isolates of TK⁻mouse L cells that had been
transformed with isolated chromosomes from (A) EL4/8 and
(B) EL4/11 cells. The cells were transformed by the calcium
phosphate precipitate method and selected for enhanced MTX
resistance at 3×10^{-8}M-MTX. At this concentration of MTX,
less than 1 per 10^6 control TK⁻mouse L cells survive. Trans-
formation occurred at a frequency of about 10^{-5}. The DNA
fragments representing rearrangements of the DHFR genes
characteristic of EL4/8 and EL4/11 cells are arrowed, cf.
Fig. 1. Notice that 10 out of 12 clones representing TK⁻L
cells transformed with EL4/8 chromosomes (A) and 11 out of
13 which were transformed with EL4/11 chromosomes (B) clearly
contain the "marker" DHFR gene fragment.

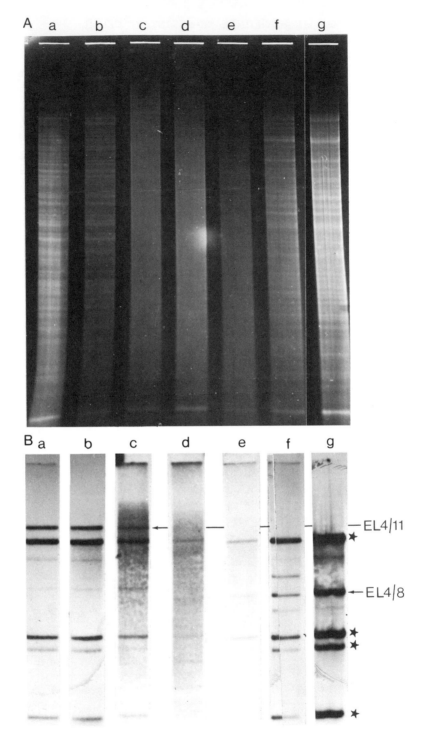

FIG. 8 For legend see opposite.

chromosomes have the pattern characteristic of EL4/8 DNA
and contain dm; whereas L cells transformed with EL4/11
chromosomes have the pattern characteristic of EL4/11 DNA
and contain homogeneously staining chromosome fragments.
Since we know that only a few EL4 genes were present in any
of the original transformants, this result strongly suggests
that all amplified units in the EL4 cells have the same
large and complex structure of DNA sequences.

DISCUSSION

1. *Characteristics of Dihydrofolate Reductase Gene Amplification*

The cellular response to strong stepwise selection for
resistance to MTX is an increase in DHFR activity, the basis
for which is the repeated duplication (amplification) of the
gene for DHFR. Duplications have played an important role
in the evolution of genes, either in the production of multigene families (e.g. structural RNA genes, histone genes,
etc.), or in the creation of related single copy genes (e.g.
the haemoglobin gene system). The relative ease with which

FIG. 8 *Comparison of amplified DNA sequences in MTX-resistant transformed TK⁻ mouse L cells and MTX-resistant EL4
cells. A, Ethidium bromide-stained 0.8% agarose gel of
EcoRI-digested DNAs from (a) EL4/11 cells growing at
1.1×10^{-3}M-MTX; (b) clone 5 of TK⁻mouse L cells transformed
with EL4/11 chromosomes growing at 10^{-4}M-MTX; (c) clone 5 of
TK⁻mouse L cells transformed with EL4/11 chromosomes growing
at 3×10^{-8}M-MTX; (d) non-transformed TK⁻mouse L cells growing in absence of MTX; (e) clone 6 of TK⁻mouse L cells transformed with EL4/8 chromosomes growing in 3×10^{-8}M-MTX;
(f) clone 6 of TK⁻mouse L cells transformed with EL4/8
chromosomes growing at 3×10^{-4}M-MTX; (g) EL4/8 cells growing
at 1.1×10^{-3}M-MTX. Notice that the pattern of bands in (a)
is the same as in (b) and that the pattern in (f) is the same
as in (g).
B, Southern blot of the gel shown in A hybridised to ^{32}P
nick-translated pDHFR11 DNA. The autoradiographs in (a),
(b), (f) and (g) were exposed for 0.025 the interval for
tracks (c), (d) and (e). The position of the "marker"
fragments for EL4/8 and EL4/11 DNAs are indicated by arrows.
The normal fragments containing DHFR coding sequences are
indicated by the asterisks. Minor additional bands are
visible in some tracks. These probably result from endogenous nuclease action at hypersensitive sites.*

duplications of the DHFR gene can be selected therefore provides a system in which we can begin to characterise the molecular mechanisms of gene duplication in somatic cells growing in culture. What can we say about DHFR gene duplication in cells that have been selected for resistance to MTX?

(a) The amplified unit is much larger than the dihydrofolate reductase gene. The first point is that in all cases where it can be estimated, the unit of amplification is very large. This is true for Chinese hamster ovary cells (Nunberg *et al.*, 1978), mouse L5178Y cells (Dolnick *et al.*, 1979), mouse PG19 cells (Bostock *et al.*, 1979) and mouse EL4 cells (Bostock and Tyler-Smith, 1981). In the case of the EL4 cells described here, the size of the unit can be estimated as follows. The extra abnormal chromosome material constitutes about 15% of the total nuclear DNA, i.e. about 10^9 base-pairs of DNA. Since there are about 2000 DHFR genes per cell, this means that, on average, about 500 kilobase-pairs of DNA is duplicated along with every copy of the DHFR gene. In EL4 and PG19 cells, we have been unable to identify any protein other than DHFR that is synthesised in resistant cells in amounts greater than in sensitive cells (Bostock *et al.*, 1979; Tyler-Smith and Alderson, 1981). This suggests that the 500 kilobase-pairs of non-DHFR DNA that is amplified does not contain gene sequences that are expressed as proteins at a level greater than wild-type.

(b) There are two phases to gene amplification. The second point is that DHFR gene amplification must occur in two quite distinct phases; formation of a unit, and its subsequent multiplication. We make this distinction because the structure of amplified DNA is different in each cell line that has been separately selected from the start, whereas sub-clones of already partially resistant or fully resistant lines have the same structure of amplified DNA.

(c) Amplification must involve recombination. The third point is that the formation of the amplified unit must involve recombinational processes. In some cell lines, a portion of the gene sequences has rearrangements near one or other or both ends of the gene and the composition of non-DHFR gene sequences is different in the amplified units of each cell line. Although the end result, duplication of DHFR genes, is the same in each MTX-resistant cell line, the precise recombinational events, which result in a stable duplicated unit containing a DHFR gene, are different each time they occur. The unit must contain more than one copy

of the DHFR coding sequences, since in EL4/3, EL4/8 and EL4/11 there are both "normal" and "rearranged" genes, and the relative ratio of these remain essentially the same even after transformation of L cells with EL4 chromosomes.

(d) The amplified unit is stable. The fourth point is that, once formed, the amplifiable unit appears to be very stable. Its basic unitary structure does not change during growth and further selection in the cell line in which it originated, nor does it change during transformation, growth and selection for amplification in a different host cell.

Once formed, the amplifiable unit can increase in number in the cell, either by unequal segregation at mitosis or by tandem duplication of a ring structure. Double minutes, ring chromosomes and acentric linear fragments allow the rapid increase in gene number, but ensure that resistance is unstable and is lost when the selection pressure is removed. The resistant phenotype can be stabilised and retained in the absence of selection by the integration of amplified units into normal-type chromosomes, either in tandem arrays to form hsr, or randomly and singly, or in blocks that are too small to be visualised in the light microscope. The resistant phenotype can also be stabilised by the large, homogeneously staining fragments acquiring a centromere. Integration, like the initial duplication event, must involve recombinational processes but, interestingly, these recombination events do not result in any detectable alterations to the structure of the unit. The pattern of amplified DNA sequences remains the same after stabilisation, and in those copies that remain integrated into normal-type chromosomes of EL4/8 after all the dm have been lost (Tyler-Smith and Bostock, 1981).

2. *Significance of Dihydrofolate Gene Amplification to other Evolutionary Phenomena*

We should consider to what extent the kinds of processes involved in amplification of the DHFR gene are a good model for the formation of gene families during the course of evolution. The later steps in DHFR gene amplification, where there is simply multiplication of the amplified unit, occur in a highly artificial situation where there is very strong stepwise selection for overproduction of a single protein. While there may be analogous situations in nature, e.g. selection for insecticide resistance (Devonshire and Sawicki, 1979), such situations are probably rare. In contrast, the early steps in DHFR gene amplification, where there is a low level of amplification and extensive DNA rearrangement leading

eventually to the formation of an amplifiable unit, may
correspond more closely to naturally occurring situations
such as the origin of dispersed multigene families or the
generation of isolated members of tandemly repeated gene
families. In this context, our studies of DHFR gene amplification suggest a number of speculations of interest.

First, the duplication events are staggeringly frequent.
At the initial selection level for MTX resistance, resistant
colonies arise at a frequency of about 1 in 10^6 cells. While
some may be resistant because of the alternative mechanisms
outlined in the Introduction, most of them have amplified
copies of the DHFR gene. If we suppose that duplications of
portions of the genome occur at random, and MTX is merely
selecting for the cells that have by chance duplicated the
DHFR gene, we can make a rough estimate of the frequency with
which a duplication event occurs within a cell. The DHFR
gene is about 30 kilobase-pairs and therefore represents
only some 10^{-5} part of the haploid mouse genome (about 3×10^6
kilobase-pairs). Thus, if a cell picks at random a 30
kilobase-pair DNA segment for duplication, there is a 1 in
10^5 chance that it will be the DHFR gene. As noted above,
the frequency with which duplications of the DHFR gene are
detected is 1 in 10^6, which is only an order of magnitude
less than its proportion in the entire genome. This means
that about 1 in 10 of the cells in the population must have
a duplication of some part of the genome, or that the DHFR
gene is particularly frequently duplicated. While duplication events may occur less frequently in whole organisms, and
may be selected against more rigorously, they may still be
frequent compared to other kinds of mutation.

Secondly, the size of the DNA fragment involved in the rearrangement and amplification is variable but is always small
compared with a mammalian chromosome. In the PG19T3:S1
cells, the stretch of DNA containing the DHFR gene retains
the wild-type structure for $<10^5$ base-pairs, while in the
EL4/3 cells, sequences close to the 5' and 3' ends of the
gene are rearranged: thus it is possible that in the latter
cells the amplified stretch of DNA containing the DHFR gene
is less than 40 kilobase-pairs long. From the estimates of
the total size of the amplified unit given earlier (500 to
3000 kilobase-pairs) we can therefore conclude that some
tens of DNA fragments are assembled to form the amplified
unit.

Thirdly, the consequence of such an assembly process is
that the gene-sized pieces of DNA are scrambled: they end up
in a different environment of neighbouring DNA sequences and,
after integration, in a different region of the chromosome.
The result of this could be expression of the duplicated

gene at a different rate or under a different control system. Such alterations in the expression of cellular proto-oncogenes seem to be important among the genetic changes undergone by tumour cells, where the chromosomal manifestations of amplification events, double minutes and hsrs, have also been observed frequently. How important such events have been in the evolution of eukaryotic genomes remains to be seen.

If such scrambling has been the means by which members of gene families have been dispersed around the genome, we can predict that the other sequences in the million base-pairs in the vicinity of the duplicated gene will be a diverse set originating from many parts of the genome. In contrast, most other models for gene duplication would predict a simpler pattern of neighbouring sequences: they would derive either from the region of the duplicated gene or the integration site.

ACKNOWLEDGEMENTS

We wish to thank Dr Thomas Alderson, who generously gave us the MTX-resistant EL4 cells; and Elma Clark, who has given many hours of expert technical help. We also wish to thank Lederle Laboratories for giving us much of the Methotrexate used in these experiments.

REFERENCES

Abelson, J. (1980). A revolution in biology. *Science* **209**, 1319-1321.

Alt, F.W., Kellems, R.E., Bertino, J.R. and Schimke, R.T. (1978). Selective multiplication of dihydrofolate reductase genes in methotrexate-resistant variants of cultured murine cells. *J. Biol. Chem.* **253**, 1357-1370.

Bagshawe, K.D. (1977). Lessons from choriocarcinoma. *Proc. Roy. Soc. Med.* **70**, 303-306.

Bailin, P.L., Tindall, J.P., Roenigk, H.H. and Hogan, M.D. (1975). Is methotrexate therapy for Psoriasis carcinogenic? *J. Amer. Med. Assoc.* **4**, 359-362.

Beach, L.R. and Palmiter, R.D. (1981). Amplification of the metallothionine I gene in cadmium-resistant mouse cells. *J. Supramol. Struct. Cell. Biochem.* (Suppl.) **5**, 437.

Biedler, J.L., Melera, P.W. and Spengler, B.A. (1980). Specifically altered metaphase chromosomes in antifolate-resistant Chinese hamster cells that overproduce dihydro-folate reductase. *Cancer Genet. Cytogenet.* **2**, 47-60.

Bostock, C.J. and Clark, E.M. (1980). Satellite DNA in large marker chromosomes of methotrexate-resistant mouse cells.

Cell **19**, 709–715.

Bostock, C.J. and Tyler-Smith, C. (1981). Gene amplification in methotrexate-resistant mouse cells. II Rearrangement and amplification of non-dihydrofolate reductase gene sequences accompany chromosomal changes. *J. Mol. Biol.* In the press.

Bostock, C.J., Clark, E.M., Harding, N.G.L., Mounts, P.L., Tyler-Smith, C., van Heyningen, V. and Walker, P.M.B. (1979). The development of resistance to methotrexate in a mouse melanoma cell line. I Characterisation of the dihydrofolate reductases and chromosomes in sensitive and resistant cells. *Chromosoma* **74**, 153–177.

Chang, S.E. and Littlefield, J.W. (1976). Elevated dihydrofolate reductase messenger RNA levels in methotrexate-resistant BHK cells. *Cell* **7**, 391–396.

Devonshire, A.L. and Sawicki, R.M. (1979). Insecticide-resistant *Myzus persicae* as an example of evolution by gene duplication. *Nature (London)* **280**, 140–141.

Dolnick, B.J., Berenson, R.J., Bertino, J.R., Kaufman, R.J., Nunberg, J.H. and Schimke, R.T. (1979). Correlation of dihydrofolate reductase elevation with gene amplification in a homogeneously staining chromosomal region in L5178Y cells. *J. Cell Biol.* **83**, 394–402.

Douthart, R.J. (1972). Topographical constraints during replication. II Complex forms generated by multiple replication forks or track switching. *J. Theoret. Biol.* **35**, 337–358.

Fischer, G.A. (1961). Increased levels of folic acid reductase as a mechanism of resistance to amethopterin in leukemic cells. *Biochem. Pharmacol.* **7**, 75–77.

Flintoff, W.F., Davidson, W.V. and Siminovitch, L. (1976). Isolation and partial characterisation of three methotrexate-resistant phenotypes from Chinese hamster ovary cells. *Somat. Cell Genet.* **2**, 245–261.

Hakala, M.T., Zakrewski, S.F. and Nichol, C.A. (1961). Relation of folic acid reductase to amethopterin resistance in cultured mammalian cells. *J. Cell Biol.* **236**. 952–958.

Jackson, R.C. and Niethammer, D. (1977). Acquired methotrexate resistance in lymphoblasts resulting from altered kinetic properties of dihydrofolate reductase. *Euro. J. Cancer* **13**, 567–575.

Kaufman, R.J., Brown, P.C. and Schimke, R.T. (1979). Amplified dihydrofolate reductase genes in unstable methotrexate-resistant cells are associated with double minute chromosomes. *Proc. Nat. Acad. Sci., U.S.A.* **76**, 5669–5673.

Kellems, R.E., Alt, F.W. and Schimke, R.T. (1976). Regulation of folate reductase synthesis in sensitive and methotrexate-resistant sarcoma S-180 cells. *In vitro* translation and characterisation of folate reductase mRNA. *J. Biol. Chem.* **251**, 6987-6993.

Melera, P.W., Lewis, J.A. Biedler, J.L. and Hession, C. (1980). Antifolate-resistant Chinese hamster cells. Evidence for dihydrofolate reductase gene amplification among independently derived sublines overproducing different dihydrofolate reductases. *J. Biol. Chem.* **255**, 7024-7028.

Meyers, M.B. and Biedler, J.L. (1981). Increased synthesis of a low molecular weight protein in vincristine-resistant cells. *Biochem. Biophys. Res. Commun.* In the press.

Nunberg, J.H., Kaufman, R.J., Schimke, R.T., Urlaub, G. and Chasin, L.A. (1978). Amplified dihydrofolate reductase genes are localised to a homogeneously staining region of a single chromosome in a methotrexate-resistant Chinese hamster ovary cell line. *Proc. Nat. Acad. Sci., U.S.A.* **75**, 5553-5556.

Nunberg, J.H., Kaufman, R.J., Chang, A.C.Y., Cohen, S.N. and Schimke, R.T. (1980). Structure and genomic organisation of the mouse dihydrofolate reductase gene. *Cell* **19**, 355-364.

Sandler, L. (1965). The meiotic mechanics of ring chromosomes in female *Drosophila melanogaster*. *Nat. Cancer Inst. Monog.* **18**, 243-272.

Sirotnak, F.M., Kurita, S. and Hutchison, D.J. (1968). On the nature of a transport alteration determining resistance to amethopterin in the L1210 leukemia. *Cancer Res.* **28**, 75-80.

Southern, E.M. (1975). Detection of specific sequences among DNA fragments separated by gel electrophoresis. *J. Mol. Biol.* **98**, 503-517.

Tyler-Smith, C. and Alderson, T. (1981). Gene amplification in methotrexate-resistant mouse cells. I DNA rearrangement accompanies dihydrofolate reductase gene amplification in a T-cell lymphoma. *J. Mol. Biol.* In the press.

Tyler-Smith, C. and Bostock, C.J. (1981). Gene amplification in methotrexate-resistant mouse cells. III Interrelationships between chromosome changes and DNA sequence amplification or loss. *J. Mol. Biol.* In the press.

Wahl, G.M., Padgett, R.A. and Stark, G.R. (1979). Gene amplification causes overproduction of the first three enzymes of UMP synthesis in N-(Phosphoacetyl)-L-aspartate-resistant hamster cells. *J. Biol. Chem.* **254**, 8679-8689.

Walters, R.A., Enger, M.D., Hildebrand, C.E. and Griffith, J.K. (1981). Genes coding for metal induced synthesis of RNA are differentially amplified and regulated in mammalian cells. *J. Supramol. Struct. Cell. Biochem.* (Suppl.) **5**, 439.

Conserved and Divergent Sequences of Bovine Satellite DNAs

GÉRARD PHILIPPE ROIZÈS and MICHEL PAGÈS

Equipe de Recherche de Biophysique 140 CNRS, Université des Sciences et Techniques du Languedoc, Montpelier, France

INTRODUCTION

The highly repeated elements of eukaryotic genomes have been extensively studied in the past and both restriction and sequence analysis has enabled us to describe their structural organisation in great detail; particularly those elements that are tandemly arranged.

It is not yet certain if these various DNA sequences have any function at all in the cell, or whether they are selfish or parasitic DNA (Orgel *et al.*, 1980), spreading within genomes and populations without making any contribution to the phenotype.

However, arguments developed against such an hypothesis imply that the precise DNA sequence in itself is of no importance, and that the eventual phenotypic effects would arise from the mere presence or position of the sequence within the genome as postulated by Dover (1980) who introduced, instead, the concept of ignorant DNA.

It has been demonstrated that some DNA sequences in higher organisms, particularly in *Drosophila* (Potter *et al.*, 1979), are mobile. In a similar fashion, all kinds of arrangements, rearrangements, amplifications or reductions and replacement of DNA sequences are postulated to occur in the genomes of many higher organisms (Flavell, 1981; Dover *et al.*, 1981).

It is apparent that the eukaryotic genome can be considered as a fluid entity through which there is a constant flux of DNA sequences. Despite this, we always find that for any set of DNA sequences there is relative homogeneity within each family and within each species. It is generally thought that the mechanism by which this arises involves a continuous process of amplification and elimination of sequences by

unequal exchange in a stochastic manner. This has been described as coincidental, horizontal or concerted evolution
(Hood *et al*., 1975). Computer simulations have demonstrated
that accumulation of random intrachromosomal (sister chromatid) crossovers would lead to such a situation (Smith, 1974,
1976).

In this paper, we shall question the completely random
nature of such processes and the apparently non-specific character of the highly repetitive elements of eukaryotic genomes.

HIGHLY REPETITIVE ELEMENTS IN THE CALF GENOME

Although structural studies on the repetitive elements of the
calf genome cannot be compared with other closely related species, as has been done for primates or rodents, the calf genome in itself offers a variety of highly repeated sequences
which, interestingly enough, allow us to ask questions about
their evolution and their eventual biological significance.

Eight satellite DNAs, representing almost 25% of the genomic DNA, have been detected in the bovine genome (Macaya *et
al*., 1978). Five of them have been extensively studied by
restriction or sequencing analysis (Roizes, 1976; Roizes *et
al*., 1980; Pech *et al*., 1979; Poschl and Streeck, 1980; Skowronski and Plucienniczak, 1981; Streeck, 1981). Three,
although they reveal quite different buoyant densities and
structural organisations in their repetition units, appear to
have evolved from a common ancestor. The data published concerning these satellite DNAs show all sorts of sequence
arrangements and amplifications that have been postulated to
occur within other genomes.

The original common ancestor to these satellite DNAs is a
23 base-pair repeat which, by successive postulated steps of
duplications and amplifications, gave rise to the three satellite DNAs referred to in Fig. 1.

This short original sequence at the stage of 46 base pairs
has been amplified to give rise to a satellite DNA component
separable in cesium chloride gradients at a buoyant density
of 1.720 g/cm^3.

These duplications and amplifications led to the stage
where the 2.35 kb,* thus arranged or rearranged has been
amplified to 50 000 copies, presumably by accumulation of
unequal crossovers.

Therefore, as in many other species, the calf genome is
also capable of arranging, or rearranging, DNA sequences and

---

*Abbreviation used: kb, 10^3 bases or base-pairs where approp-
  riate.

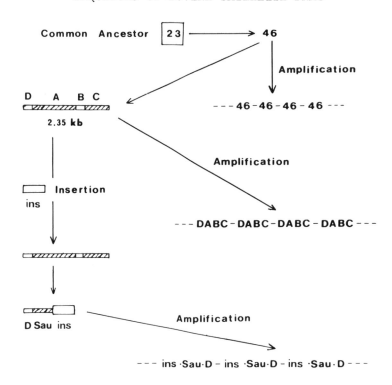

FIG. 1 *The evolutionary relationships of the 3 calf satellite DNAs shown have been compiled from Püschl and Streeck (1980) for 46.46.46 (100 000 copies of a 46 base-pair long unit ; buoyant density : 1.720 g/cm³) ; from Pech et al. (1979) for DABC . DABC . DABC (50 000 copies of a 2.35 kb unit; buoyant density : 1.706 g/cm³) ; and from Streeck (1981) for the third (30 000 copies of 1413 base-pair unit ; buoyant density : 1.711 g/cm³). Their formation has been postulated by these authors to be the result of duplications, deletions and amplifications of a dodecanucleotide sequence that gave rise to the 23 base-pair unit common to all 3 satellite DNAs. A and C on the one band, and B and D, on the other band, are called* Sau *and* Pvu *segments, respectively. ins is an insertion element 611 base-pairs long.*

of amplifying them to various extents. Insertion elements may have been introduced as well within DNA sequences already amplified and been subjected to a new set of amplifications. This will be discussed at the end of the paper.

However, some arrays of tandemly repeated sequences do not show short-range periodicities at all, either because the divergence was such that no homology at all is detectable any-

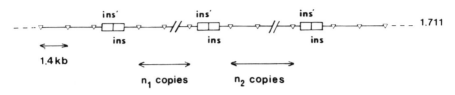

FIG. 2 *Two long-range periodicity satellite DNAs can be sep-arated on high-resolution cesium chloride gradients at 1.715 and 1.711 g/cm³ buoyant densities. The structure of the 1.711 component has been deduced from Streeck (1981) and Skowronski and Plucienniczak (1981). ins and ins' are 2 unrelated inser-tion elements. (∇) EcoRI.*

more or, most probably, because more complex DNA sequences present in the genome have been taken for amplification.

This has actually occurred in the calf genome (Fig. 2), since the bovine genome has amplified a sequence of approxi-mately 1.4 kb, to about 100 000 copies in a tandem fashion on all the autosomes of this organism (Kurnit *et al*., 1973).

A MECHANISM FOR MAINTAINING SEQUENCE HOMOGENEITY IS ACTING

It is noteworthy that the 23 base-pair repeats, which are com-mon to the four subfragments of the 2.35 kb unit, have diverged considerably from each other, showing that, unless a mechanism for maintaining homogeneity between the repeats pre-vails, these repeats would diverge rapidly.

The same is true with the mouse satellite DNA, which has recently been sequenced (Hörz and Altenburger, 1981). The eight subunits that primitively constituted the basic repeti-tion unit show considerable divergence between themselves while, when amplified to 10^6 copies or so, the 234 base-pair repetition units, taken as a whole, show much less divergence.

Again, the same is true with rat satellite DNA, which has developed a 370 base-pair repeat from four originally identi-cal 92 or 93 base-pair subunits, which have diverged consider-ably from each other (Pech *et al*., 1979).

In man also, two subunits of 169 and 171 base-pairs show 27% base variations while, in contrast, base variations are of the order only of 1% when dimers are compared with each other (Wu and Manuelidis, 1980).

STRUCTURAL ANALYSIS OF CALF SATELLITE DNA I

Calf satellite DNA I is made of 100 000 copies of a 1.4 kb
long repeat and has a density of 1.715 g/cm³ in cesium chloride.   It was already obvious from restriction analysis and
complementary RNA fingerprints that this satellite DNA is not
internally repetitious (Roizès, 1976).   This has since been
confirmed by direct sequencing of a copy of this DNA after
cloning (in both Dr Bernardi's and Dr Plucienniczak's laboratories (unpublished results)).

From the restriction analysis of this satellite DNA, a
number of structural aspects are evident and are summarised
in Fig. 3.

Obviously mutation rates, including deletions and additions,
are highly variable along the DNA sequence of the repeat.

It is particularly striking that, while the G-A-T-C- site
(*Mbo*I) is present in almost all the copies, the T-G-A-T-C-A

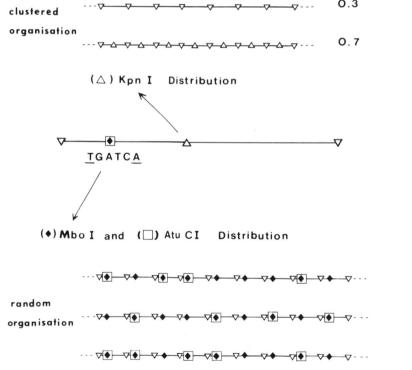

FIG. 3 *On this 1.4 kb unit of calf satellite DNA I, only a
few sites are shown, which are either present in almost all
the repeats, EcoRI (▽) and MboI (♦), or less frequently present, KpnI (△) and AtuCI (□).   For more details, see Roizès
et al. (1980).*

site (*Atu*CI) is absent in about half of them and randomly distributed when present.

This, obviously, comes from the high alteration rate of one
of the two external bases of the restriction site as compared
to the four internal bases, which do not vary at all: this is
confirmed in the two sequenced cloned repeats mentioned above.

A second observation is that some sites, like *Kpn*I, are
clustered in a non-random fashion when present or altered in
the repeats.

We can add some light to the structural organisation of
this satellite DNA by comparing a series of clones of this
DNA, as shown in Fig. 4.

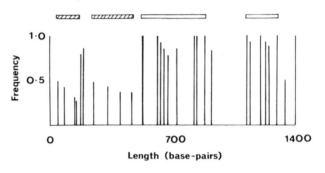

FIG. 4 *A number of restriction sites have been localised on
a series of cloned 1.4 kb repeats of calf satellite DNA I in
pBR322 (EcoRI, PstI, MboII, HpaII, HhaI, XmaI, KpnI, SfaNI,
SacI). The frequencies (0 to 1.0) of the sites at each position are indicated by vertical bars. Highly variable and
highly conserved regions are indicated by hatched and open
boxes, respectively.*

It can easily be seen that some regions of the repeats have
accumulated more changes (base changes, deletions and additions) than others. This is easy to notice on such a long-
range periodicity repeat but hard to distinguish in short
repeats.

This was already clearly visible, however, in the 359 base-
pair repeat of one satellite DNA of *Drosophila melanogaster*
(Hsieh and Brutlag, 1979a), where the authors concluded that
the variations from one repeat to the next are not random,
and that certain positions in the repeat can be highly variable while most of the satellite repeat is conserved.

Again, the points of variations in the rat satellite DNA
component are not randomly distributed along the 370 base-
pair repeat (Pech *et al.*, 1979).

A MECHANISM FOR MAINTAINING HOMOGENEITY IN SOME REGIONS OF THE REPEATS: SEQUENCE CONVERSION

The existence of regions in the DNA sequence of some satellite DNAs that are more stable than others poses a problem that is hard to answer by simply saying that strong selection pressures are being exerted at the nucleotide level, although this cannot be excluded *a priori*. It is rather more satisfactory to think that some type of molecular mechanism exists whereby these amplified sequences have avoided evolutionary divergence in some of the regions of their repetition unit.

Such a mechanism has been postulated in order to explain the similarity of DNA sequences between the fetal β-like non-allelic human globin G_γ and A_γ genes in one individual, where the genes on the same chromatid share exactly identical sequences in their coding parts and in non-coding regions of the whole gene, particularly in their second intervening sequence; whereas the two allelic A_γ genes have diverged more. Slightom *et al.* (1980) have applied to this situation the model developed by Meselson and Radding (1975) and by Radding (1978), which implies mismatch repair and replication of the converted sequence after strand transfer. This mechanism leads to a conversion of a gene, which explains perfectly well this differential divergence in the DNA regions of the duplicated genes.

Similarly, Klein and Petes (1981) have shown a mechanism in yeast by which repeating genes on a chromosome can be made homogeneous. They suggest that this mechanism, intrachromosomal gene conversion, may have a general relevance to the maintenance of sequence homogeneity within families of repeated eukaryotic genes.

Scherer and Davis (1980) have produced evidence of non-homologous interchromosomal gene conversion, again in yeast, by which deleterious mutations could be corrected and beneficial mutations could spread rapidly among the family.

We propose that such mechanisms, in addition to unequal crossing over, would apply to some tandemly repeated sequences, such as those of calf satellite DNA I, at some stages of their evolution (Fig. 5).

Sequence non-reciprocal conversion could occur during mitosis in germ line cells, between misaligned sister chromatids and consequently between non-homologous repeats, one converting its sequence to the other, or both doing so, as shown in Fig. 5 (A).

It is known that intrachromosomal sister chromatid exchanges within constitutive heterochromatin are much more frequent during mitosis than are those due to homologous recombination during meiosis within the same heterochromatin

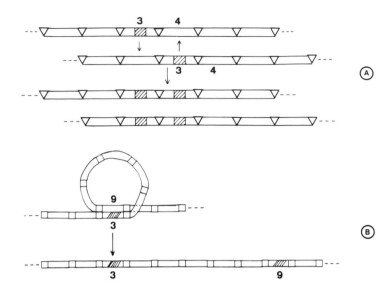

FIG. 5 *Sequence conversion between misaligned sister chromatids and, therefore, between non-homologous tandem repeats (a) during mitosis. The possibility that the exchanges occur within single DNA molecules folded back on themselves (b) is not excluded. This could lead to the so-called type B patterns in satellite DNAs.*

where satellite DNAs are concentrated. Although rare, inter-chromosomal exchanges do also occur and could explain the spread of the converted sequences through the different chromosomes of a karyotype (Kurnit, 1979). Unequal exchanges could then involve large arrays of tandem repeats that subsequently spread through the whole genome, the conserved regions becoming flanked by the variable regions.

The spread of repeats through a population would have to be rapid as is found for the non-transcribed repeat of the ribosomal DNA cluster of *D. melanogaster*. This is the last stage of so-called concerted evolution (Dover and Coen, 1981).

Another structural feature of satellite DNAs could be explained by mechanisms like gene conversion (Fig. 5(B)). Type B patterns are obtained, particularly in mouse satellite DNA, when restriction sites are present in only a fraction of the repeats, in contrast to type A patterns, which reveal restriction sites in all the repeats (Hörz and Zachau, 1977).

Experiments have shown that these patterns do not overlap and that the B type sites are located at the same place in all the repeats when they are present (Hörz and Zachau, 1977; Brown and Dover, 1980a).

The amounts of satellite DNA vary between *Mus musculus* and *Mus spretus* species, and so do the abundances of type B patterns between chromosomes and species (Brown and Dover, 1980b). This shows that single chromosomes or whole genomes in the *Mus* genus have addressed themselves, during recent amplifications, to patterns that were already present within the genome, and have not randomly generated new ones.

Gene conversion of certain repeats, because they may provide advantage to the cell, as we will discuss now, could allow the survival in different chromosomes and within each chromosome of the variants that show type B patterns, while unequal crossing over alone would lead to a randomization of these variants or to large differences in the type B patterns in the *Mus* species studied.

POSSIBLE BIOLOGICAL SIGNIFICANCE OF THE CONSERVED REGIONS

We would now like to take up the question of the eventual significance of regions more conserved than others or preferentially amplified within non-coding repeats organised in tandem arrays.

Most approaches consider the families of repeated sequences, particularly those which are tandemly arranged, as the result of a range of random processes that are capable of generating all sorts of imaginable pattern organisation and, moreover, which are in constant flux.

Despite this, consideration of a number of data implies that these sorts of sequences may be involved with processes like chromosome mechanics, the frequency and localisation of recombination, or the expression of nearby genes. All these sorts of phenotypic effects would arise from the mere presence or position of the sequence within the genome, and Dover (1980) therefore introduced the concept of ignorant DNA, meaning that the sequence itself has no specific effect at all.

However, DNA in higher organisms has to fulfil its functions, whatever they are, through a nucleoprotein complex, which has to be arranged into nucleosomes in a highly specific manner. These are susceptible to conformational changes when genes are activated, as shown by the preferential sensitivity to DNAase I of chromatin when it is actively transcribed.

Secondly, the DNA of higher organisms is susceptible to modifications which, in the vertebrates and in plants, consist essentially of the methylation of some of the cytosine residues in a manner that is not yet well-described and hence not yet understood, but which seems to follow some specific rules (Razin and Riggs, 1980). In particular, the methylation affects the cytosine residues that are in the 5' position of a guanine.

It is largely proved now that genes are hypomethylated when active as compared to their degree of methylation in tissues where they are not expressed or not active (McGhee and Ginder, 1979; Mandel and Chambon, 1979).

It should be noticed that this is not a rule without exceptions, since there are contradictory results and, moreover, base modifications in *Drosophila* are not of the type described above (News and Views, 1981).

We are prepared to accept such results because we know that coding sequences have to be turned on and off and because we do not know how this is effected. Therefore, we would like to draw correlations, for instance, between nucleosome conformation, DNA methylation and gene expression. But we are not expecting any correlation of that sort between satellite DNA, taken as a totally ignorant molecule, and its methylation for instance or its chromatin organisation.

Satellite DNAs, however, are also arranged into nucleosomes and susceptible to modifications like the methylation of cytosine into 5-methyl cytosine.

In our laboratory, we have followed the methylation of the CpG in calf satellite DNA I by restriction analysis with enzymes that have C-G dinucleotides in their recognition sequences and are unable to cut these sites when they are methylated.

Unpublished results can be summarised briefly as follows. Firstly, it is clear that methylation is arranged in a more or less even, but not random, distribution. Secondly, some of the sites are highly variable from one tissue to the other. These two aspects are illustrated by Fig. 6.

The arrangement of the 5-methyl cytosine residues when present within the restriction sites of *Hha*I (G-C-G-C) found in most repeats of calf satellite DNA I, at three positions in the 1.4 kb unit, is particularly interesting. Given the methylation of some of the sites along the DNA molecules, a complete digest by *Hha*I looks like a partial digest with normal enzymes and, thus, multimers of the basic fragments are obtained. A distribution of these 5-methyl cytosine residues within the *Hha*I sites in a random fashion would lead to fragments of 1.4 kb long, ± 39 base-pairs; 39 being the distance between these two sites. This does not occur in the four tissues examined. Fragments of 1.4 kb - 39 are actually obtained in only two instances, brain and sperm.

It can be concluded, therefore, that the arrangement of some 5-methyl cytosine residues in this satellite DNA is indeed highly specific and variable with tissues.

A correlation can be tentatively made with the arrangement of nucleosomes along the DNA fiber, since it has been suggested that the nucleosomes are in phase in the chromatin corresponding to this satellite DNA (Musich *et al*., 1977).

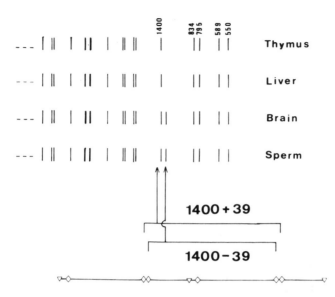

FIG. 6 *The distribution of 5-methyl cytosine within* HhaI *sites was examined by complete digestion with this enzyme of total calf DNA from different tissues of one individual except for the sperm, which was obtained from another animal. After transfer to a nitrocellulose filter, hybridization was performed with nick-translated cloned calf satellite DNA I. HhaI sites (◊) were located on a series of cloned repeats of calf satellite DNA I in pBR322, by comparison with EcoRI sites (∇). The basic fragment lengths are indicated (in base-pairs). Larger fragments are multimers obtained because of the methylation of certain sites.*

Although not unique, such a phase has been established in rat chromatin, in which the nucleosomes were found predominantly in three defined positions on the 370 base-pair satellite I monomer unit (Igo-Kemenes *et al.*, 1980).

The consequences of such a non-random arrangement of nucleosomes in heterochromatin, where the satellite DNAs of calf or rat are concentrated, is not known. This is comparable, however, with the non-random arrangement of nucleosomes along the underlying DNA as shown for the repeated genes of 5 S RNA in *D. melanogaster* and *Xenopus laevis*, or of histones in *D. melanogaster* (Louis *et al.*, 1980; Gottesfeld and Blommer, 1980; Samal *et al.*, 1981).

The authors point out, as a probable reason for such a strict positioning of nucleosomes, that the DNA sequence itself is probably of essential importance.

Therefore, we think that the calf satellite DNA I cannot

be entirely ignorant, in that it has developed primitive DNA
sequence variants during its evolution; or DNA regions may
have acquired some biological significance through the methylation of sites that position nucleosomes on the DNA.

It is noteworthy that these sites, susceptible to methyl
groups, are precisely in a region of the repeat that does not
seem to diverge much (Fig. 4).

We do not wish to imply that all repeated sequences carry
such signals, which would have necessarily arisen during evolution by base substitutions and then their fixation. This
may not be the case for a number of satellite DNAs, where it
has been proved, as in the seven sibling species of D. melanogaster that their satellite DNA sequences are completely unrelated (Barnes et al., 1978).

What we mean is, that satellite DNAs being present for
whatever reasons, opinion is still largely open on this subject, they have the possibility of adapting themselves to
their environment, sometimes acquiring signals that may be of
help to the genome to fulfil certain functions, particularly
some germ line functions. These functions may be favored by
fixation of relatively specific sequences within completely
ignorant DNA.

Other examples may be relevant to this idea. The first is
taken from the D. melanogaster satellite DNA of density 1.688
g/cm^3 in cesium chloride. It has developed a repeat of 359
base-pairs which, in a particular DNA region, binds specifically to a protein extracted from the embryo of this organism
and, again, this cannot be taken as non-specific DNA (Hsieh
and Brutlag, 1979b).

The second comes from the calf itself in which, as Streeck
(1981) described, a satellite component contains an insertion
element of 611 base-pairs carrying a perfect T-A-T-A box,
inverted repeats of eight base-pairs at the extreme ends, a
region of more than 150 base-pairs open for reading in all
frames, and ATG codons at the beginning.

As Streeck (1981) points out, the insertion has occurred in
a 23 base-pair repeat unit that is highly variable as compared
to the prototype 23 base-pair sequence.

This raises the possibility that the insertion occurred in
a region of the already amplified 23 base-pair unit, where a
DNA sequence suited to accept an insertion had arisen by base
substitutions and/or other minor changes.

This satellite DNA may have accepted such an insertion
element either for functional reasons (it has been reported
that transcription of satellite DNAs is actually occuring in
some tissues (Varley et al., 1980)) or for accidental reasons.
It is noteworthy that insertion elements that normally occur
within some of the 28 S rDNA genes of D. melanogaster have

been located in the centromeric heterochromatin, where satellite DNAs are located (Peacock *et al.*, 1981), and moreover,
are flanked by the same 28 S gene segments that flank the
insertion element in the 28 S gene itself (Roiha *et al.*,
1981). This shows that these insertion elements within heterochromatic DNA originate from the rDNA 28 S gene and not the
contrary (Dover and Coen, 1981).

Therefore, some satellite DNAs may have acquired insertion
elements for the reasons developed above, and could fill the
role of a reservoir from which such elements could be supplied
when necessary, or of a waste disposal for the undesirable
elements.

Thirdly, it is interesting to point out Brutlag's (1980)
finding that satellite DNAs of very different origins, like
those in *Drosophila*, African green monkey, calf, man and rat,
have homologies that are statistically significant.

Similarly, these satellite DNAs have developed regions of
dyad symmetry within their repeat units and this may also be
of biological significance (Brutlag, 1980).

We would like to mention another interesting point that
may have some importance for generating DNA sequences within
ignorant molecules, and which may also have biological significance. Purely alternating dG.dC polynucleotides (Wang *et
al.*, 1979) or, likely, purely alternating Pu-Py sequences
(Arnott *et al.*, 1980), even flanked by any other natural DNA,
are susceptible *in vitro* at high salt levels to a change in
their conformation from the B structure, which is a right-
handed helix, to a Z structure, which is a left-handed helix.
In vivo, some ligands or proteins could have the same effect
as salt. Moreover, it has been shown that m^5 dC in purely
alternating dC.dG polynucleotides is able to stabilize such a
Z structure at a much lower salt concentration than is
required *in vitro* by the unmethylated residue (Behe and Felsenfeld, 1981).

The effects of such transitions are large on the overall
topography of the genome, since the Z structure is capable of
unwinding supercoiled DNA to a considerable extent (Klysik *et
al.*, 1981).

Since a tiny portion of a chromosome in such a conformation
has large effects, non-specific DNAs such as satellite DNAs
are more likely to accept such sequences than would the coding
sequences or their neighbours.

As it has been suggested that satellite DNAs could be
involved, at a distance, in both the localisation and frequency of recombination during meiosis, it may be thought that
such transitions could be of importance in such events (Yamamoto and Miklos, 1978).

We do not wish to make an exhaustive list of these sorts

of data, which are scattered throughout the literature, or to add other possible signals that can be thought of as being of interest to this discussion. We only want to mention, finally, that J. Maio is suggesting that another determinant of the evolution and organisation of these highly repeated sequences, at least in mammals, could be a site-specific mammalian endo-nuclease present in all mammals examined by him, which cleaves mammalian repetitive DNAs at sequences containing interrupted, hyphenated or imperfect palindromes.

CONCLUSION

Satellite DNAs, and maybe some interspersed DNA sequences as well, surely constitute long molecules of non-specific sequences, or ignorant DNA. They might be subjected to all sorts of amplifications, reductions and rearrangements. However, we think that, in spite of these apparently totally random processes, these ignorant molecules may become punctuated by sequences that have acquired some specificity and which are sometimes fixed because they help some biological functions at the molecular level.

In conclusion, even if the variations that are observed in a number of satellite DNAs do not all arise through the same mechanism, they may have been produced by base substitutions, and sometimes short deletions or additions, and their fixation.

These newly arisen sequences may carry favorable phenotypic effects, at least in some instances, and may be spread by mechanisms including sequence conversions as well as unequal crossovers.

Since DNA has been subject to a number of new qualifications for the past months we now say:

Ignorant DNA, with time and patience, can learn from its environment and may become more or less wise. It then has a pronounced tendency, as in humans, to try to convert its neighbourhood, thus becoming proselytic DNA.

REFERENCES

Arnott, S., Chandrasekaran, R., Birdsall, D.L., Leslie, A.G.W. and Ratliff, R.L. (1980). Left-handed DNA helices. *Nature* (London) 283, 743-745.

Barnes, S.R., Webb, D.A. and Dover, G.A. (1978). The distribution of satellite and main-band DNA components in the *melanogaster* species subgroup of *Drosophila*. *Chromosoma* 67, 341-363.

Behe, M. and Felsenfeld, G. (1981). Effects of methylation on a synthetic polynucleotide: The B-Z transition in poly(dG-m^{5}dC).poly(dG-m^{5}dC). *Proc. Nat. Acad. Sci., U.S.A.*

78, 1619–1623.

Brown, S.D.M. and Dover, G. (1980a). The specific organization of satellite DNA sequences on the X-chromosome of *Mus
musculus*: partial independence of chromosome evolution.
Nuc. Acids Res. **8**, 781–792.

Brown, S.D.M. and Dover, G. (1980b). Conservation of segmental variants of satellite DNA of *Mus musculus* in a related
species: *Mus spretus. Nature (London)* **285**, 47–49.

Brutlag, D.L. (1980). Molecular arrangement and evolution of
heterochromatic DNA. *Annu. Rev. Genet.* **14**, 121–144.

Dover, G. (1980). Ignorant DNA? *Nature (London)* **285**, 618–620.

Dover, G. and Coen, E. (1981). Springcleaning ribosomal DNA:
a model for multigene evolution? *Nature (London)* **290**, 731–
732.

Dover, G.A., Strachan, T. and Brown, S.D.M. (1981). The evolution of genomes in closely related species. *Evolution
Today (Proc. II Int. Congr. Syst. and Evol. Biol.)*, 337–349.

Flavell, R.B. (1981). Molecular changes in chromosomal DNA
organisation and origins of phenotypic variation. *Chromosomes Today* **7**, 42–54.

Gottesfeld, J.M. and Bloomer, L.S. (1980). Non-random alignment of nucleosomes on 5 S RNA genes of *X. laevis. Cell* **21**,
751–760.

Hood, L., Campbell, J.H. and Elgin, S.C.R. (1975). The organization, expression, and evolution of antibody genes and
other multigene families. *Annu. Rev. Genet.* **9**, 305–353.

Hörz, W. and Altenburger, W. (1981). Nucleotide sequence of
mouse satellite DNA. *Nucl. Acids Res.* **9**, 683–696.

Hörz, W. and Zachau, H.G. (1977). Characterisation of distinct segments in mouse satellite DNA by restriction nuc-
leases. *Eur. J. Biochem.* **73**, 383–392.

Hsieh, T.S. and Brutlag, D.L. (1979b). A protein that pre-
variation within the 1.688 g/cm^3 satellite DNA of *Drosophila melanogaster. J. Mol. Biol.* **135**, 465–481.

Hsieh, T.-S. and Brutlad, D.L. (1979b). A protein that preferentially binds *Drosophila* DNA. *Proc. Nat. Acad. Sci.
U.S.A.* **76**, 726–730.

Igo-Kemenes, T., Omori, A. and Zachau, H.G. (1980). Non-
random arrangement of nucleosomes in satellite I containing
chromatin of rat liver. *Nucl. Acids Res.* **8**, 5377–5390.

Klein, H.L. and Petes, T.D. (1981). Interchromosomal gene
conversion in yeast. *Nature (London)* **289**, 144–148.

Klysik, J., Stirdivant, S.M., Larson, J.E., Hart, P.A. and
Wells, R.D. (1981). Left-handed DNA in restriction fragments and a recombinant plasmid. *Nature (London)* **290**, 672–
677.

Kurnit, D.M. (1979). Satellite DNA and heterochromatin variants: the case for unequal mitotic crossing over. *Hum.*

Rev. Genet. **47**, 169–186.

Kurnit, D.M., Shafit, B.R. and Maio, J.J. (1973). Multiple satellite DNAs in the calf and their relation to the sex chromosomes. *J. Mol. Biol.* **81**, 273–284.

Louis, C., Schedl, P., Samal, B. and Worcel, A. (1980). Chromatin structure of the 5 S RNA genes of *D. melanogaster. Cell* **22**, 387–392.

Macaya, G., Cortadas, J. and Bernardi, G. (1978). An analysis of the bovine genome by density gradient centrifugation. *Eur. J. Biochem.* **84**, 179–188.

Mandel, J.L. and Chambon, P. (1979). DNA methylation: organ specific variations in the methylation pattern within and around ovalbumin and other chicken genes. *Nucl. Acids Res.* **7**, 2081–2103.

McGhee, J.D. and Ginder, G.D. (1979). Specific DNA methylation sites in the vicinity of the chicken β-globin genes. *Nature (London)* **280**, 419–420.

Meselson, M.S. and Radding C.M. (1975). A general model for genetic recombination. *Proc. Nat. Acad. Sci., U.S.A.* **72**, 358–361.

Musich, P.R., Maio, J.J. and Brown, F.L. (1977). Subunit structure of chromatin and the organization of eukaryotic highly repetitive DNA: indications of a phase relation between restriction sties and chromatic subunits in African green monkey and calf nuclei. *J. Mol. Biol.* **117**, 657–677.

News and Views. (1981). DNA methylation and control of gene expression. *Nature (London)* **290**, 363–364.

Orgel, L.E., Crick, F.H.C. and Sapienza, C. (1980). Selfish DNA. *Nature (London)* **288**, 645–646.

Peacock, W.J., Appels, R., Endow, S. and Glover, D. (1981). *Genet. Res. Camb.* **37**, 209.

Pech, M., Igo-Kemenes, T. and Zachau, H.G. (1979). Nucleotide sequence of a highly repetitive component of rat DNA. *Nucl. Acids Res.* **7**, 417–432.

Pöschl, E. and Streeck, R.E. (1980). Prototype sequence of bovine 1.720 satellite DNA. *J. Mol. Biol.* **143**, 147–153.

Potter, S.S., Brorein, W.J. Jr, Dunsmuir, P. and Rubin, G.M. (1979). Transposition elements of the 412, copia and 297 dispersed repeated gene families in *Drosophila. Cell* **17**, 415–427.

Radding, C.M. (1978). Genetic recombination: strand transfer and mismatch repair. *Annu. Rev. Biochem.* **47**, 847–880.

Razin, A. and Riggs, A.D. (1980). DNA methylation and gene function. *Science* **210**, 604–610.

Roiha, H., Miller, J.R., Woods, L.C. and Glover, D.H. (1981). Arrangements and rearrangements of sequences flanking the two types of rDNA insertion in *D.melanogaster. Nature (London)* **290**, 749–753.

Roizes, G. (1976). A possible structure for calf satellite DNA I. *Nucl. Acids Res.* **3**, 2677-2696.

Roizes, G. Pages, M. and Lecou, C. (1980). The organisation of the long range periodicity calf satellite DNA I variants as revealed by restriction enzyme analysis. *Nucl. Acids. Res.* **8**, 3779-3792.

Samal, B., Worcel, A., Louis, C. and Schedl, P. (1981). Chromatin structure of the histone genes of *D. melanogaster*. *Cell* **23**, 401-409.

Scherer, S. and Davis, R.W. (1980). Recombination of dispersed repeated DNA sequences in yeast. *Science* **209**, 1380-1384.

Skowrónski, J. and Plucienniczak, A. (1981). The family of sequences with regional homology to satellite 1.715 repeated DNA exists in the calf genome. *Nucl. Acids Res.* In the press.

Slightom, J.L., Blechl, A.E. and Smithies, O. (1980). Human fetal $^{G}\gamma$ - and $^{A}\gamma$ - globin genes: complete nucleotide sequences suggest that DNA can be exchanged between these duplicated genes. *Cell* **21**, 627-638.

Smith, G.P. (1974). Unequal crossover and the evolution of multigene families. *Cold Spring Harbour Symp. Quant. Biol.* **38**, 507-513.

Smith, G.P. (1976). Evolution of repeated DNA sequences by unequal crossover. *Science* **191**, 528-535.

Streeck, R.E. (1981). Inserted sequences in bovine satellite DNAs. *Science* in the press.

Varley, J.M., MacGregor, H.C. and Erba, H.P. (1980). Satellite DNA is transcribed on lampbrush chromosomes. *Nature (London)* **283**, 686-688.

Wang, A.H.J., Quigley, G.J., Kolpak, F.J., Crawford, J.L., Van Boom, J.H., Van Der Marel, G. and Rich, A. (1979). Molecular structure of left-handed double-helical DNA fragment at atomic resolution. *Nature (London)* **282**, 680-686.

Wu, C.J. and Manuelidis, L. (1980). Sequence definition and organization of a human repeated DNA. *J. Mol. Biol.* **142**, 363-386.

Yamamoto, M. and Miklos, G.L.G. (1978). Genetic studies on heterochromatin in *Drosophila melanogaster* and their implications for the functions of satellite DNA. *Chromosoma* **66**, 71-98.

Evolution of Primate DNA Organization

DAVID GILLESPIE[1], LARRY DONEHOWER[2] and
DAVID STRAYER[1]

[1] *Barry Ashbee Laboratories, Orlowitz Cancer Institute,
Hahnemann Medical College, Philadelphia, U.S.A.*

[2] *Laboratory of Tumor Virus Genetics,
National Cancer Institute, Bethesda, U.S.A.*

INTRODUCTION

We propose in this article that portions of animal genomes
are reorganized or "reset" periodically during evolution.
Genome reorganization means an alteration of the long-range
organization of DNA sequences and may be distinct in principle and mechanisms from chromosomal rearrangements that
have occurred during evolution. Genome resetting in this
model is mediated by newly amplified repeated DNA. Though
genome resetting may have a role in genetic programming, e.g.
during development, the evolutionary consequence of successful genome reorganization is speciation. The manifestations
of genome resetting are new orientations between repeated and
infrequent DNA elements at a microscopic level and possibly
new karyotypic features at a macroscopic level. It is argued
that these reorganizations serve to establish new genetic
programs in development and differentiation, which define the
morphological and physiological characteristics of the new
species. Implicit in this model is the idea that contemporary evolution is driven by major genetic rearrangements,
rather than by the slow accumulation of point mutations.

FACTS AND INTERPRETATION

Like other evolutionary arguments, the genome resetting hypothesis derives from correlative data. There are four pieces
of evidence suggesting that major genomic reorganizations
have occurred recently in primate evolution. The evidence is
concerned solely with repeated DNA evolution. The correlation

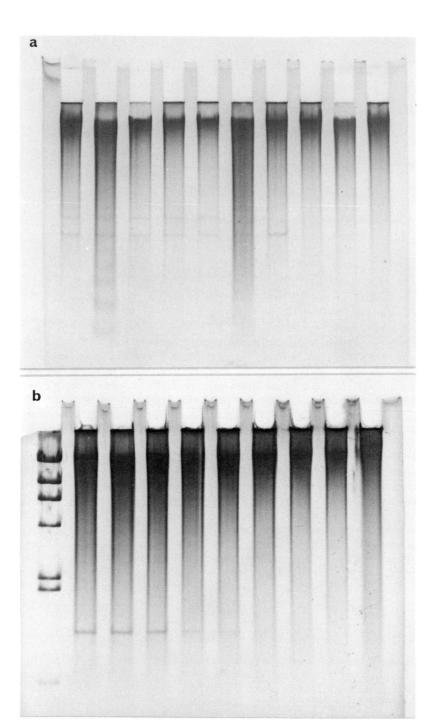

FIG. 1 For legend see opposite

in timing between certain events in repeated DNA evolution and speciation as determined by the fossil record has prompted the idea that repeated DNA evolution may be causally linked to speciation.

The four aspects of repeated DNA evolution that bear on the genome resetting theory are as follows.

1. Human "Alu family" Members are More Closely Related to One Another than to Mouse Alu Family Analogues

The predominant interspersed repeated DNA element in humans is an approximately 300 base-pair DNA sequence discovered by Schmid and co-workers (see Rubin *et al.*, 1980). It is called the Alu family because a large fraction of the 300 000 related sequences possess a cleavage site for the restriction endonuclease *Alu*I. Alu family members occur roughly once every 10 000 bases, on the average, in human DNA. Rodents carry an interspersed repeated DNA that is genetically related to human Alu family sequences (Pan *et al.*, 1981). If Alu family elements became interspersed among the genome of an ancestor common to rodents and primates, and each human and mouse Alu family member evolved independently, then the genetic distance among members ought to be random. However, the evidence from molecular hybridization (D. Housman, personal communication) and DNA sequencing (Pan *et al.*, 1981) suggests that human Alu family members are more closely related to one another than to mouse Alu family members. Thus, either Alu family members were installed in human and mouse genomes *after* primate and rodent lines diverged or they display "concerted" evolution (Zimmer *et al.*, 1980) in their respective genomes. However,

---

FIG. 1 (a) *Size of fragments produced by KpnI from primate DNA. DNA (10 µg) from selected primates was digested for 4 h at 37°C with 20 units of KpnI (Bethesda Research Labs) then fractionated according to size by electrophoresis in a 1.5% agarose gel and stained with ethidium bromide. From left to right, the DNA sources were: apes = human, gorilla and white-handed gibbon; old world monkeys = royal baboon, vervet guenon and stumptail macaque; new world monkeys = marmoset and capuchin; prosimian = ring-tailed lemur; non-primate = cat.*
(b) *Limited digestion of marmoset DNA by KpnI. Marmoset DNA (100 µg) was mixed with 200 units of KpnI and incubated at 37°C for various lengths of time (from 5 min to 4 h) before electrophoresis. Inreasing time is from right to left. The slots with the most prominent 1100 base-pair band (migrated about 3/4 of the way to the bottom) represents the longest exposures of the DNA to the endonuclease.*

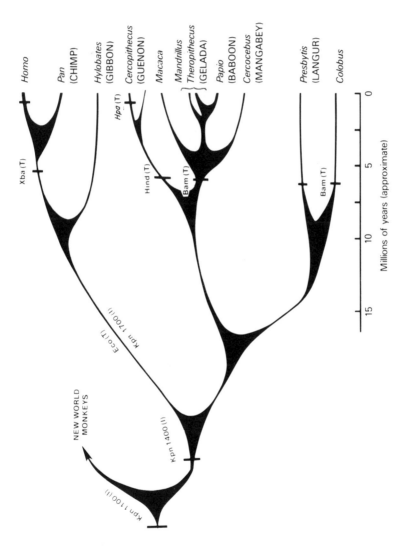

FIG. 2 *Phylogeny of old world primates. The deduced time of appearance of repeated sequences is indicated by hash marks. The family is designated by the restriction endonuclease that cuts it. T, tandemly organized; I, interspersed.*

the unequal crossover mechanism favored for concerted evolution is difficult to apply to interspersed repeated DNA.

2. Long, Interspersed DNA (the "Kpn Family") Exhibits Saltatory Amplification during Primate Evolution

Human DNA treated with the restriction endonuclease *Kpn*I, and
fractionated according to size by electrophoresis through agarose gels displays three bands of DNA fragments about 1100,
1400 and 1600 base-pairs long, respectively (Fig. 1(a)). That
these bands arise from cutting interspersed repeated DNA, as
opposed to tandemly repeated DNA, is indicated by three
experiments. Limited enzyme digestion does not product multimers (Fig. 1(b)), indicating that different Kpn family members
possess different flanking sequences. Second, Kpn family
sequences left after digestion of human DNA with *Hind*III are
found in DNA fragments of heterogeneous size, averaging 6000
base-pairs while tandemly repeated DNA is found in DNA
fragments over 100 000 base-pairs long (not shown). Finally,
about 20% of clones randomly selected from a human genomic
library possess *Kpn*I-related sequences (L. Donehower, unpublished results).

Old world monkey DNA carries the 1100 and 1400 base-pair
Kpn family sequences, DNA from new world monkeys possesses
only the 1100 base-pair Kpn family sequences, while non-
primate DNA lacks Kpn family members entirely, judging by
restriction analysis (Fig. 1(a) and (b)). Human DNA carries
a 1600 base-pair *Kpn* I sequence family in addition to the 1100
and 1400 base-pair families. Thus, the appearance of "new"
Kpn family sequences was sudden (Fig. 2). In addition, the
1100 base-pair *Kpn*I members possess a defined sequence
between and beyond the *Kpn*I restriction sites, in the sense
that cleavage sites for other enzymes can be mapped thereon
(L. Donehower, unpublished results). Finally, cloned human
Alu family probe does not hybridize to the 1100, 1400 or
1600 base-pair bands of human DNA restricted with *Kpn*I (not
shown).

Evolutionary explanations for these results are limited:
(1) Kpn family DNA was amplified as tandem units with spacers
between and with time the spacers accumulated many mutations;
(2) Kpn family DNA was amplified as tandem units, then individuals were interspersed among infrequent sequences; or (3) Kpn
family DNA was amplified as interspersed repeated DNA. Case
(1) demands too high a spacer mutation rate for credibility
and is inconsistent with the cloning results. But cases (2)
and (3) both imply drastic, cataclysmic genomic reorganization
during evolution, regarding the positioning of repeated and
infrequent genomic DNA elements.

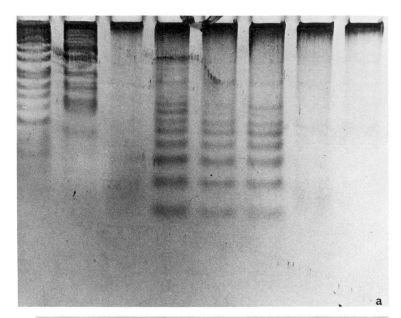

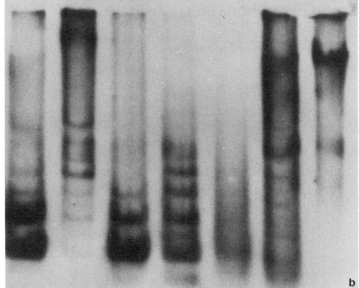

680 bp→

340 bp→

FIG. 3 (a)   *Presence of* HpaII *satellite in guenons.   Conditions are as described for Fig. 1.   From left to right, DNAs are: marker (φX174* HinfI*); marker (φX174* AluI*); Debrazza guenon; patas monkey; mustached guenon; Sykes guenon; Diana*

3. Satellite DNA Amplification in Primates Immediately Preceded Speciation of Karyotypically Coherent Groups

The most recent major amplification of satellite DNA in old world primates immediately preceded active speciation. This fact is clear in the case of the baboon and guenon groups (Gillespie, 1977; Donehower and Gillespie, 1980) and appears to be true in the colobuses, langurs and apes as well (L. Donehower and D. Gillespie, unpublished results).

In the baboons and guenons, the time of amplification can be dated rather precisely (Fig. 2). Each group possesses a distinct alpha satellite (Rosenberg *et al.*, 1978; Donehower and Gillespie, 1980; Donehower *et al.*, 1980). All tested members of each group possess roughly the same amount of the same alpha satellite (Donehower and Gillespie, 1980, and unpublished results); and no genus of the baboon-macaque-mangabey group possesses an additional alpha satellite (Gillespie, 1977). Consequently, the baboon group alpha satellite was amplified six to eight million years ago, during the period after baboons and guenons diverged but before the 20 or more species of baboons, macaques and mangabeys were formed.

Thus, this phylogenetically coherent group of 42-chromosome monkeys appears to be characterized by a fixed quantity of one and only one major satellite DNA. The case is similar for the other major old world primate groups, insofar as they have been studied, but the data are more fragmentary and, in some cases, more complex. The guenon-specific alpha satellite was amplified about the same time as the baboon-specific satellite; i.e. six to eight million years ago, immediately before the formation of the 20 or more guenon species. However, a subgroup of guenon possesses a simple satellite based on a

---

Fig. 3 (continued)
guenon; vervet. Animals not possessing HpaII *satellite DNA, determined in other experiments, are capuchin, langur, colobus, simeng gibbon, agile gibbon, orangutan, chimpanzee, human,* Papio hamadryas, *gelada baboon, rhesus macaque, sooty mangabey, woolly monkey, marmoset, lemur and cat.* **(b)** *Hybridization of* EcoRI *340 base-pair alpha satellite to human DNA cleaved with* XbaI. *Endonuclease-restricted human DNA (10 µg) was fractionated by electrophoresis through a 1% agarose gel, transferred to nitrocellulose after denaturation and hybridized to nick-translated 340 base-pair DNA fragments purified from* EcoRI-*restricted human DNA. From left to right, restriction endonucleases used to cleave human DNA were:* EcoRI, XbaI, HinfI, HaeIII, MboI, AluI, KpnI. *bp, base-pairs.*

ten base-pair repeating unit with a cleavage site for the
restriction endonuclease, HpaII, (Fig. 3(a) and Table I).
These guenons carry quite variable numbers of chromosomes,
while guenons lacking the simple satellite possess chromosome
numbers around 60.   Similarly, apes are characterized by an
alpha satellite with a cleavage site for EcoRI, while great
apes carry, in addition, an alpha satellite with an XbaI cleavage site, and humans possess yet additional satellite DNAs
(summarized in Fig. 2).   The XbaI and EcoRI human alpha satellites are not identical (Fig. 3(b)).   It may be important
that apes and guenons are karyotypically more diverse than
the baboons, macaques and mangabeys, in addition to having
an apparently more complex satellite DNA picture.

Though some cases are more complicated than others, there
are no results known to us that deny the interpretation that
the major satellite DNAs originate in a saltatory amplification event as proposed by Britten and Kohne (1968) before
active speciation and then are retained by subsequent descendants.   In the primates, there is apparent constancy regarding
the amount of the major satellite DNA per genome in individuals of a given group.

It seems reasonable to conclude that the amplification of
alpha satellite DNA immediately preceding active speciation
is causally related to the subsequent emergence of karyologi-
cally coherent phylogenetic groups of animals.   To assume
coincidence is to predict major alpha satellite DNA amplifications in subgroups, e.g. in baboons and not in macaques, an
event so far undiscovered in the baboon group.   Relatively
minor amplifications may occur frequently in nature (see
Kurnit, 1979), and these minor amplifications might reflect
a "selfish" aspect of satellite DNA, but the timing of major
satellite DNA amplifications prompts us to assign a biological role to primate satellite DNA.   The correlation with
karyology leads us to believe that this role involves DNA
sequence organization or reorganization.

4. Spatial Relationships between Repeated and Infrequent DNA

Figure 4(a) presents a "reannealing-rebanding" of baboon and
marmoset DNA.   Randomly fragmented DNA was denatured,
reannealed to a low C_0t value, then fractionated according to
strandedness by centrifugation to equilibrium in NaI gradients.   A smear of DNA is expected if the length ratio of
repeated DNA to flanking infrequent DNA is not conserved
(Fig. 5).   The actual "bands" of DNA obtained suggest a long-
range ordering of repeated and infrequent elements in mammalian genomes.

TABLE I

Distribution of Satellites in Old World Monkeys

	HindIII ($170n$) (base-pairs)	BamHI ($340n$) (base-pairs)	HpaII ($10n$) (base-pairs)	No. chromosomes
Baboons (6 species)	+	+	−	42
Macaques (6 species)	+	+	−	42
Mangabeys (2 species)	+	+	−	42
Guenons				
Group 1:				
Cercopithecus neglectus (Debrazza guenon)	+	−	−	58–62
Cercopithecus diana (Diana guenon)	+	−	−	58–60
Cercopithecus pugerythrus (vervet; African green monkey)	+	−	−	60
Group 2:				
Cercopithecus cephus (mustached guenon)	+	−	+	66
Cercopithecus albogulans (Sykes guenon)	+	−	+	72
Erythrocebus (?) patas (patas monkey)	+	−	+	54

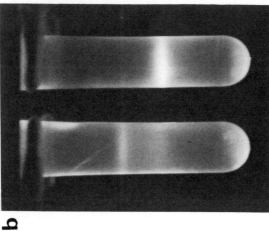

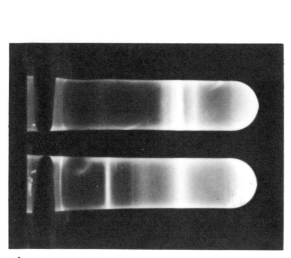

FIG. 4 (a) Annealing-banding of DNA. A solution of DNA in NaI was made by mixing 6.0 ml of saturated NaI, 0.5 ml of 0.1M-Tris, 0.1 M-EDTA (pH 8.0), 0.1 ml of 1 mg ethidium bromide/ml and 3.5 ml of DNA (0.4 mg) in 0.15 M-NaCl. DNA was denatured at 100°C for 20 min. The average single-stranded length of the DNA was 10⁴ bases. The final refractive index of the solution was adjusted to 1.4340, then 10 ml of DNA solution and 2 ml of mineral oil were centrifuged for 3 days at 38°C and 38 000 revs/min in a type 40 rotor, during which time annealing of repeated DNA families occurred. Centrifugation was continued at 23°C for 6 days to achieve maximal separation of annealed complexes. The tubes were photographed under ultraviolet light and approximately 25 fractions were collected from the bottom of the tube.
Left tube, baboon spleen DNA; right tube, 71AP1 transformed marmoset cell line DNA.

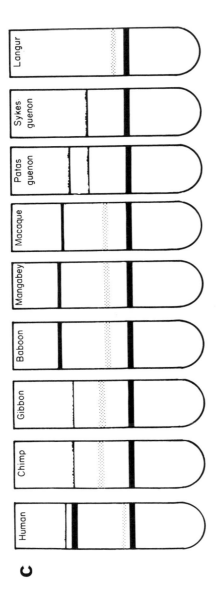

C

Human | Chimp | Gibbon | Baboon | Mangabey | Macaque | Patas guenon | Sykes guenon | Langur

(b) Annealing-banding of human midband DNA. Human DNA was subjected to the annealing banding procedure described in the legend to Fig. 4a. The midband and main band DNAs were collected separately, denatured, reannealed and recentrifuged. The bulk of the DNA reran at the midband position. Left tube, human midband DNA; right tube, human main band DNA. (c) Schematic representation of reannealing-rebanding results using DNA from several different primate species.

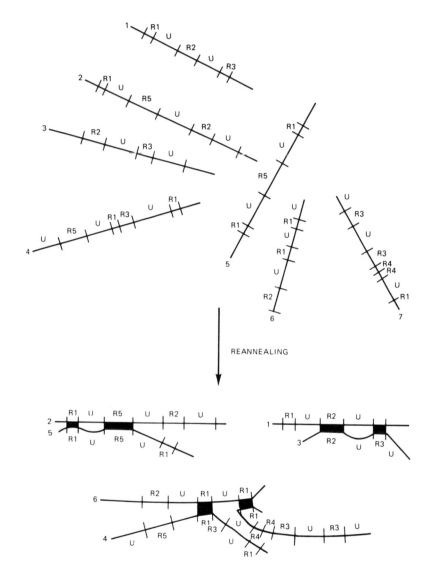

FIG. 5 *Schematic of annealing of repeated DNA segments. R1 to R5, families of repeated DNA; V, unique sequence DNA.*

After reannealing-rebanding, DNA from most animals was recovered in three bands. Tandemly repeated satellite DNA was recovered near the top of the gradient. The DNA in this band was over 80% double-stranded. Up to 25% of the DNA was recovered in the satellite band, depending on the source of DNA. DNA from cultured cells varied widely in the content of

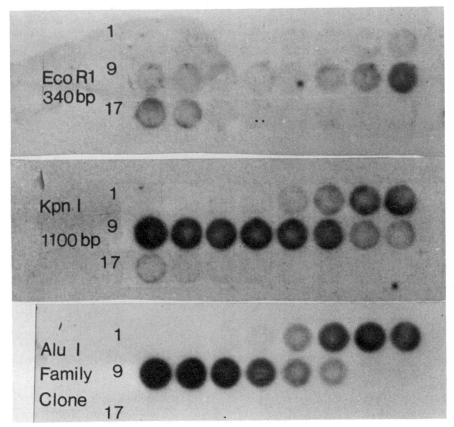

FIG. 6 *Location of Alu and Kpn family repeated DNA in NaI annealing-banding gradients. Human DNA was denatured, annealed and centrifuged in NaI as described in the legend to Fig. 4(a): 1% of each fraction was immobilized on nitrocellulose and hybridized at 65°C in 0.9 M-NaCl, 0.09 M-sodium citrate (pH 7) to the labeled repeated DNA family listed in the Figure. After hybridization, filters were radioautographed and circles containing individual samples were counted by scintillation counting. Only the radioautographic results are presented. Main band, fractions 6 to 9; midband, fractions 10 to 13; satellite bands, fractions 15 to 17.*

satellite band DNA. Half or more of reannealed DNA banded as complexes, which were about 20% double stranded (main band). Few, if any sequences in DNA 3000 nucleotides in length or over band as single-stranded after reannealing to low C_0t values. The remainder of reannealed DNA assumes a position between the top bands and the main band (the midband). The quantity and position of midband DNA varies from animal to animal, as described below.

Many trivial artifacts could be imagined to explain the recovery of discrete bands following rebanding-reannealing. However, annealing-banding patterns were reproducible with DNA from any given animal species, and any given band could be recovered from a NaI gradient, denatured and reannealed, and recentrifuged to its original position in NaI (Fig. 4(b)). It is our conclusion that in any given mammalian DNA there are only a few permitted sequence organizations. This is *not* to say that any selected repeated sequence, e.g. Alu family members, must be spaced regularly in the genome. We *do* contend, however, that in the aggregate the length ratio of repeated and infrequent sequences in a given genome is a regular rather than a random feature.

Rebanding profiles of a given reannealed DNA were not only orderly, they were characteristic (Fig. 4(c)). The reannealing-rebanding profiles of DNA of similar size from humans, baboons and guenons were readily distinguishable, primarily on the basis of the position of midband DNA. Human midband DNA was recovered just above main band DNA. Baboon midband DNA was recovered higher in the gradient, but not as high as guenon midband DNA. These positions were characteristic of the several members of each group tested (n=3, 4 or 5), when experimental conditions are carefully controlled. However, the position of midband DNA was changed by using DNA of different chain lengths or by altering the reannealing conditions. Midband DNA contains Alu and Kpn repeated sequences (Fig. 6).

Thus, DNA from each animal tested has a regular long-range organization. DNA from phylogenetically coherent groups has a characteristic organization, judging from our limited survey. So far, old world primates with a given alpha satellite have a common organization. Thus, humans (one species), baboons-macaques and mangabeys (5 species), guenons (4 species) and langur form four separate groups, each characterized by a distinct alpha satellite and a distinct reannealing-rebanding profile.

Like the Kpn family results, the reannealing-rebanding data suggest saltatory changes in DNA organization during evolution.

ARGUMENTATION

1. Use of Satellite DNA as an Evolutionary Marker

Satellite DNA may represent the only evolutionary marker of a qualitative nature available to us today. In some situations it may be the only marker that is capable of providing an unambiguous branching order in phylogenetic studies. The case of the *Hpa*II satellite DNA in the patas monkey, referred to earlier, is an example of this. Typing with satellite DNA reveals that the patas clearly belongs to a guenon subgroup, while conventional evolutionary markers have been unable to classify this animal.

The use of satellite DNA typing in taxonomy has been challenged on the grounds that these repeats are highly variable from individual to individual. In our experience, this variability is primarily evident in cells maintained in tissue culture and in laboratory animals. Interestingly, the annealing-banding profiles of DNA from the laboratory sources also can be uncharacteristic of the wild animal. We do find minor changes in repeated DNA in some wild primates, e.g. a putative unequal crossover product of alpha satellite DNA in mandrill and geladas (Donehower and Gillespie, 1980). These may turn out to be an advantage for refining the satellite DNA typing system and should not be viewed negatively.

The use of satellite DNA typing in animal systematics has also been criticized on the basis that satellite DNA may have no biological function, and that this classification ignores phenotypic and behavioral aspects of the animals under study. If it is true that satellite DNA amplification is sudden (Britten and Kohne, 1968), and if it is relatively constant once amplified, then all descendants of an animal in which amplification has taken place will possess many (e.g. 10^5) copies of the sequence, while other animals, no matter how closely related, will possess as few as one copy.

While it is appropriate to be concerned about phenotypic traits in evolutionary studies, it is unreasonable to ignore a classification system with the potential power of satellite DNA typing. The most informative approach is to classify animals solely by satellite DNA typing, and then focus evolutionary studies on situations where phenotypic characters would have provided a different classification. The patas monkey is a case in point. On the basis of phenotypic traits, the patas monkey is ordinarily not included in the guenon group; it differs from the remaining guenons in several respects. However, by satellite DNA typing it is clearly a guenon and, indeed, split from the guenon stock rather recently. Consequently, the appropriate question is to ask "why has the patas monkey apparently evolved so rapidly at the phenotypic level?"

Apparently, most or all animals of a species characterized as possessing a particular satellite DNA are homozygous, since in such species no satellite-negative individuals have been found. Homozygosity is not predicted from a strict amplification model, but of course it is extremely useful to the sys-' tematic biologist.

Satellite DNA typing using restriction endonucleases has limitations. There will be situations of satellites that do not have cleavage sites for any tested enzyme. In this instance, molecular hybridization with cloned satellite DNA can be used. The satellite DNA is purified as high molecular weight fragments following digestion of genomic DNA with several restriction endonucleases. There may be animals with no satellite DNA. Such animals have not been reported on but, if they exist, would not be classifiable by the satellite DNA typing system. Finally, and most importantly, the satellite DNA typing system must be interpreted on a tentative basis until the underlying assumptions of sudden amplification and subsequent constancy are verified, or until phylogenetic studies of satellite DNA leave no doubt as to its accuracy.

2. Does Repeated DNA have a Biological Function?

We have argued that the phylogenetic distribution of major satellite DNA species among primates constitutes evidence that the satellite DNAs have a biological role. In the case of the baboon satellite, we can ask whether this really follows or whether satellite DNA must necessarily exhibit phylogenetic coherence, even if it is nothing more than a selfish passenger, simply because of the nature of evolutionary processes. The argument developed below leads us to conclude that the

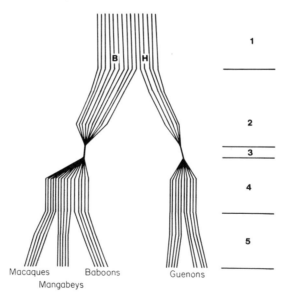

FIG. 7 *Schematic representation of population segregation in 42-chromosome monkeys. See the text for details.*

*Bam*HI satellite-containing baboon ancester most probably was
selected for.

Consider the scheme of Fig. 7. A common interbreeding
population characterized the baboon and guenon ancestors until
about ten million years ago (period 1). Conventional evolutionary theory suggests that two subpopulations segregated
from one another, went through a reduction in number, then
re-expanded to form new genetically separate baboon and
guenon species (periods 2 to 5). Most probably, the *Bam*HI
satellite was amplified during period 1 or 2. Period 3 has
too few individuals. An amplification in period 4 requires
virus-like, germ line transmission among members, a precedented
but unlikely event, and origination of the *Bam*HI alpha satellite in period 5 requires multiple, virtually identical, separate amplification events, since many of the species present
during this period do not interbreed now and presumably did
not interbreed in the past.

However, if amplification took place in period 1 or 2, the
animal carrying the *Bam*HI alpha satellite survived population
reduction; i.e. was selected for. The argument that this happened because satellite amplification was so common is unten-
able, since within species individuals with novel alpha satellites are not detected. It can be argued that satellite
amplification was common at one special period of time, a
period shortly followed by population reduction but, other
than the observation that several alpha satellite amplifications appeared to take place some six to eight million years
ago (Donehower and Gillespie, 1980; and Fig. 2), this idea is
not supported by evidence.

The pattern of distribution of distinct alpha satellites
in the other groups of old world primates and the distribution
of the *Kpn*I interspersed repeated DNA is consistent with the
view that amplification precedes speciation (Fig. 2).

3. The Interspersed Kpn*I* Repeated DNA Family

Phylogenetic surveys of animals possessing interspersed
repeated DNA that can be cut with *Kpn*I suggests that new forms
of these interspersed repeats have "appeared" repeatedly during primate evolution (Fig. 2). We have chosen to call this
phenomenon "genome resetting". As far as we can tell by
examining electropherograms of *Kpn*I-restricted DNA, their
appearance is saltatory, as described in the Britten and Kohne
(1968) model. According to this model, a given sequence is
first multiplied many times, then individual sequences are
distributed throughout the genome. This process is usually
viewed as a two-step process, in which a family of tandemly
arrayed members is first formed through DNA amplification,

then individual family members are excised from the tandem set and interspersed in a single-copy DNA environment.

We have great difficulty with this model if amplification is fast and interspersion is slow, because we cannot detect a tandem form of the *Kpn*I family DNA. Since new interspersed forms have appeared as recently as five million years ago (a 1600 base-pair fragment characterizes humans and apes), we would have expected to find a tandem form of *Kpn*I DNA. To date, we have detected no such tandem form. If there is no tandem form of the *Kpn*I repeat in DNA of extant animals, then either the tandem organization was unstable, amplification to tandem sequences was not detectably more sudden than interspersion, or the amplification products were never tandemly organized.

How the interspersion event is viewed depends on whether it was useful evolutionarily. The *Kpn*I sequences could be viewed as selfish DNA, without biological consequence. But could interspersion at random locations of several thousand new members conceivably have occurred during primate evolution without being lethal? If there exist 10^4 structural genes of about 10^3 nucleotides each, and if an insertion into a structural gene were usually sufficient to select against the recombinant while all other insertions were permissible, the ratio of total nucleotides to structural gene nucleotides ($3 \times 10^9/10^7$) would be significantly less than the number to be inserted (about 10^4). The amplification-insertion process would have to have been frequent to produce at least three separate genome resettings during primate evolution.

There is no such dilemma if insertion is non-random. Non-randomness of insertion is indicated, because the *Kpn*I sequences are concentrated in an atypical DNA organization, judging from results of the annealing-banding methodology (Fig. 6). In the selfish DNA scenario, the non-randomness would be biased toward Kpn family member insertion at evolutionarily permissible DNA sites.

If one takes the view that Kpn family DNA interspersion was a positively selectable event, then the insertion must not only have been non-random, it must have been directed. Amplification-interspersion of these sequences occurred at strategic times during primate evolution; the 1100 base-pair *Kpn*I family was created when primates split from prosimians, the 1400 base-pair family originated when old and new world primates diverged, and 1600 and 1700 base-pair families arose just before the gibbon and ape lineages were formed. However, amplifications could have been more frequent and gone undetected if the amplified sequence lacked the *Kpn*I site.

A system for directing interspersion is likely to be sophisticated, considering the number of elements to be inser-

ted, and assuming that the remainder of the genome is somewhat
sensitive to insertional errors. It is unlikely that such a
system could have arisen simply to confer an evolutionary
selection advantage, by increasing genetic flexibility for
example. More likely, any selective advantage of interspersion confers a physiological advantage to possessor cells,
tissues or individuals. The *Kpn*I sequences are too long to
be candidates for simple regulatory signals, unless the bulk
of the sequence is used for directing interspersion. It is
not known whether they are transcribed or are codogenic, but
we know of no precedent for structural genes present in several thousand copies.

Basically, the sudden appearance of new *Kpn*I interspersed
repeats in primate genomes can be viewed as non-deleterious,
non-random insertions arising from selfish aspects of the
repeated DNA sequence or as directed insertions with biological relevance. If a tandem organization precedes interspersion, it must be transient. A transposition mechanism is consistent with the results, if the transpositions are non-random
or if the accumulation of transpositions is gradual, being
selected one or a few at a time. In the latter case, constraints must be placed upon selection mechanisms to explain
the uniformity of ancestral *Kpn*I repeats (e.g. all old world
monkeys possess a 1400 base-pair *Kpn*I repeat) and the apparent
absence of very recent alternate forms.

4. *Evolution of Long-range Order in DNA.*

The existence of discrete bands after reannealing long,
separated DNA strands and centrifugation of the reannealing
products to equilibrium in NaI density gradients showed that
the positioning of repeated and infrequent elements in DNA is
not random (Fig. 4(a)). The reannealing products of most or
all mammalian DNA is characterized by a main band reflecting
a complex that is about 80% single-stranded and containing
primarily Alu family repeated DNA, a top band reflecting a
complex that is about 80% double-stranded and containing the
satellite type of repeated DNA, and an intermediate "midband",
usually 30 to 50% double-stranded Alu and Kpn carrying interspersed repeated DNA. The position of the midband in NaI
gradients is characteristic of small, phylogenetically coherent groups of animals, and we interpret this as reflecting
DNA organizations that evolved recently.

Like the appearance of satellite DNA and *Kpn*I interspersed
repeated DNA, midband DNA changed suddenly during evolution.
However, the evaluation of midband DNA is more technically
demanding than the restriction and electrophoresis assay of
repeated DNA, and is at present more limited in scope, so we

are not in a position to date the appearance of new midband DNA organizations or to compare particular organizations with the possession of particular *Kpn*I family or alpha satellite DNA sequences. Midband DNA carries structural genes. The evolutionary suddenness of midband sequence reorganization seems established, providing that different locations of midband DNA following centrifugation in NaI reflect different organizations, as we claim.

The maintenance of the midband organization requires some selection toward normalcy. DNA from some tissue culture cells and from some tumors shows annealing-banding patterns that are not comparable to DNA from the same wild species. Therefore, we believe that the organization of repeated and infrequent sequences in midband DNA is important in some function not required of tissue culture and tumor cells. This observation, coupled with the magnitude of genomic reorganization and the need to fix reorganized genomes in the germ line of some animals that are selected for during evolution, lead us to propose that the midband organization is involved in genetic programming of many genes; e.g. those programs used during development and/or differentiation.

It is much too early to say whether genomic reorganization as evidenced by studying reannealed DNA is equivalent to genome resetting, predicted from evaluations of the *Kpn*I family interspersed repeated DNA. Clearly, this conceptual fusion is consistent. Human midband DNA contains over one-third of the *Kpn*I repeated sequences.

In any event, it seems clear that a significant fraction of genomic DNA can undergo reorganization without necessarily creating a large negative selection factor. Whether such reorganizations are biologically advantageous remains to be established but in our opinion the weight of the evidence favors the conclusions that they are.

ACKNOWLEDGEMENTS

This work was supported by grants (GM27270 and HD13860) from and by a donation to the Barry Ashbee Leukemia Research Laboratories by the Philadelphia Flyers. The authors thank M.J. Caranfa, V. Timoteo and A. Apostoli for expert technical assistance, and D. West for preparation of the manuscript.

REFERENCES

Britten, R. and Kohne, D. (1968). Repeated Sequences in DNA. *Science* 161, 529–540.
Donehower, L. and Gillespie, D. (1980). Restriction Site Periodicities in Highly Repetitive DNA of Primates. *J.*

Mol. Biol. **134**, 805–834.

Donehower, L., Furlong, C., Gillespie, D. and Kurnit, D. (1980). DNA Sequence of Baboon Highly Repeated DNA: Evidence for Evolution by Nonrandom Unequal Crossovers. *Proc. Nat. Acad. Sci., U.S.A.* **77**, 2129–2133.

Gillespie, D. (1977). Newly Evolved Repeated DNA Sequences in Primates. *Science* **196**, 889–891.

Kurnit, D. (1979). Satellite DNA and Heterochromatin Variants: The Case for Unequal Mitotic Crossing Over. *Human Genet.* **47**, 169–186.

Pan, J., Elder, J.T., Duncan, L.H. and Weissman, S.M. (1981). Structural Analysis of Interspersed Repetitive Polymerase III Units in Human DNA. *Nucl. Acids Res.* **9**, 1151–1170.

Rosenberg, H., Singer, M. and Rosenberg, M. (1978). Highly Reiterated Sequences in SIMIANSIMIANSIMIANSIMIANSIMAIN. *Science* **200**, 394–399.

Rubin, C.M., Houck, C.M., Deininger, P.L., Friedmann, T. and Schmid, C.W. (1980). Partial Nucleotide Sequence of the 300 NT Interspersed Repeated Human DNA Sequences. *Nature (London)* **284**, 372–374.

Zimmer, E.A., Martin, S.L., Beverley, S.M., Kan, Y.W. and Wilson, A. (1980). Rapid Duplication and Loss of Genes Coding for the α Chains of Hemoglobin. *Proc. Nat. Acad. Sci., U.S.A.* **77**, 2158–2162.

Conserved Sex-associated Repeated DNA Sequences in Vertebrates

K.W. JONES and L. SINGH

Department of Genetics, University of Edinburgh

INTRODUCTION

The biological functions of repeated DNA in most instances remain obscure, despite the fact that such sequences make up much of the bulk of chromosomal DNA (Britten and Kohne, 1968). This state of affairs may, to some extent, be due to the fact that repeated DNA has been investigated for its own sake rather than in a functional context. In this sense, repeated DNA constitutes a solution in search of a problem. However, this situation may alter rapidly now that repeated DNA sequences are being reported in genomic clones of recombinant DNA next to coding sequences of known function (Shen and Maniatis, 1980; Fritsch *et al.*, 1980; Duncan *et al.*, 1979; Kaufman *et al.*, 1980). Satellite DNA, which is highly repetitious, simple sequence DNA isolated in isopycnic gradients of heavy metal salts, constitutes a good example of a type of repeated DNA that has been investigated largely without reference to any functional correlates. This type of DNA was first appreciated in the mouse (Kit, 1961) and, in an attempt to assign some function to it, was first mapped on chromosomes by hybridization *in situ* (Jones, 1970), where it located in constitutive heterochromatin near the centromeres of all but the Y chromosome (Pardue and Gall, 1970). Unfortunately, these demonstrations of its major concentrations have done almost nothing to elucidate its functions and subsequent investigations of satellite DNA sequences (Southern, 1970) and their inter- and intraspecies permutations (Wu and Manuelidis, 1980; Peacock *et al.*, 1973; Fry and Salser, 1977; Pech *et al.*, 1979) have also failed to suggest any useful leads. In fact, the idea that satellite DNA functions in connection with constitutive heterochromatin now seems to be an over-simplification (John and Miklos, 1979), since there are reported instances of such heterochromatin that lacks

highly repetitious DNA, as well as reports that sequences
related to mouse satellite occur interstitially in euchromatic
regions of chromosomes (L. Manuelidis, this volume).

An approach that was designed to overcome some of these
problems was initiated in our laboratory in 1975, when we
decided to investigate highly repeated DNAs in the context of
the evolution and function of sex chromosomes (Singh *et al.*,
1976). We hoped that the isolation of repeated DNAs from
such chromosomes might shed light on some feature that could
be correlated with these aspects and thus provide some more
useful insights than were available at that time. The material that we chose to work with was snake DNA, for the reason
that female snakes were known to carry an entirely constitu-
tively heterochromatic sex-determining chromosome, the W,
partnered by a euchromatic Z chromosome (Ray-Chaudhuri *et al.*,
1970). Males are of the constitution ZZ, and do not exhibit
any very prominent heterochromatin. Thus, since heterochromatin was suggested to contain satellite DNA, it follows that
the female might contain unique satellite DNA originating from
the W chromosome, and that this might be isolated by comparative density-gradient centrifugation. Another potential advantage of snake species was that they contained some families
whose members apparently lack cytologically definable sex
chromosomes, as well as others whose members showed highly
differentiated ZW bivalents. Thus, the process of evolution
of chromosomal sex determination is portrayed in its essential
respects. (Ray-Chaudhuri *et al.*, 1971)

THE ISOLATION AND CYTOLOGICAL MAPPING OF
W-ASSOCIATED SATELLITE DNA

Because of its availability, and because it contains a well-
differentiated W chromosome, we chose to use the Indian banded
krait (*Bungarus fasciatus*) for the isolation of W-related
repeated DNA sequences. DNA extracted from males and females
was compared by analytical ultracentrifugation under various
conditions in isopycnic gradients of cesium salts. Whilst
certain satellite DNA fractions could be seen to be common to
both sexes, one satellite DNA was visible only in female DNA
(Fig. 1), and this was recovered and purified by repeated
centrifugation. This fraction was designated banded krait
minor (Bkm) satellite DNA (Singh *et al.*, 1980). The origin
of this satellite DNA from the W chromosome was confirmed by
using it as a radiolabeled probe in hybridization *in situ*
experiments with male and female chromosome spreads of the
same species. The only chromosome to be extensively and
prominently labeled was the W, in which it could be seen
that Bkm satellite DNA was distributed over the entire length

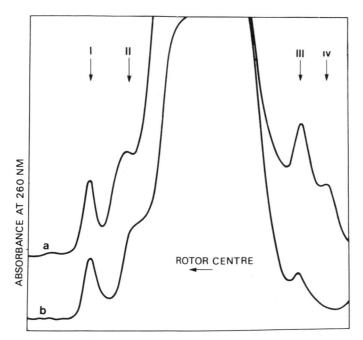

FIG. 1 *Analytical equilibrium density-gradient centrifugation of total DNA of* E. radiata *male and female, respectively, superimposed (male inner trave) to compare the satellite DNA components I, II, III and IV. Satellite III is underrepresented in the male and satellite IV is absent. Both of these satellite DNAs derive from the W chromosome, and satellite IV is homologous with Bkm satellite DNA, which is found in the snake* B. fasciatus, *as described in the text. (Adapted from Singh* et al., *1976.)*

(Fig. 2). Since the contribution of this chromosome to the genome is considerably in excess of the amount of Bkm satellite DNA in gradients of genomal DNA, we assume that the sequences are distributed in an interspersed manner with other W-associated DNA. This suggested that such interspersed non-satellite DNA could be contained in our preparations and isolated together with Bkm satellite DNA. Subsequent experiments showed that some components in our satellite preparation are conserved in evolution, since annealing of Bkm satellite DNA probes to the DNAs of the separate sexes of a very wide selection of different snake species, representing several families, yielded very significant hybridization, which was in all cases in quantitative excess in female DNA when compared to male DNA within a given species (Table I). This effect was confirmed to be due to the persistence of

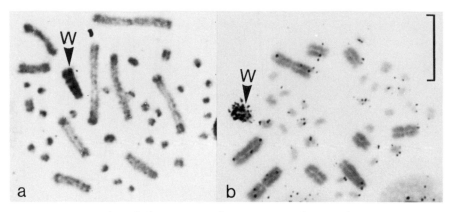

FIG. 2 (a) *C-banded preparation of metaphase chromosomes of*
Naja naja naja *(cobra) family Elapidae. Note the entirely C-
banded W chromosome.* (b) *Autoradiograph of the chromosomes
of the same species as shown in (a) after hybridization with
Bkm satellite DNA, ³H-labeled by nick-translation and exposed
for 20 days. Silver grains are distributed over the entire
length of the W chromosome, with the exception of the minute
short arm. The bar represents 10 μm. (Adapted from Singh
et al., 1980.)*

related sequences on the W chromosomes by subsequent hybridization *in situ* of the same probes to chromosomal spreads
(Singh *et al.*, 1980). It is also notable that significant
hybridization was obtained in the case of male DNA, despite
the fact that no corresponding satellite DNA fraction can be
recovered. Evidently, Bkm satellite DNA is present in both
sexes but has become concentrated on the W chromosome to an
extent that at least equals its total amount in males and in
certain species exceeds this amount. Bkm satellite DNA-
related sequences therefore do not appear to persist in evolution simply because they are located in a privileged site on
the genetically isolated W chromosome. It is worthwhile
pointing out in this context that other snake satellite DNAs
that are common to both sexes do not persist in evolution
between species (unpublished) and the same appears to be true
of some that are located on the W chromosome. Evidently, the
persistence of Bkm satellite DNA is related to some special
structural and/or functional attribute which, at the present
time, remains to be elucidated.
    An important exception to these observations is found in
the case of the Boidae, which are a group of constrictors
lacking cytologically defined sex chromosomes. In the DNAs
of such species, we have not found any evidence of significant

TABLE I

Hybridization of B. fasciatus Minor Nick-translated Satellite DNA with Total Male and Female DNA of Various Species of Snakes Representing Different Evolutionary States of Sex Chromosome Differentiation

Family	Species	Hybridized radioactivity (cts/min)			Sex chromosomes†
		Male	Female	Con.*	
Boidae	Sand boa (*Eryx johni johni*)	717	744	119	−
Colubridae	Rat snake (*Ptyas mucosus*)	1852	4165	116	+(i)
Colubridae	Tree snake (*Elaphe radiata*)	2429	3536	137	+(i)
Colubridae	Water snake (*Natrix piscator*)	2496	5042	116	+(h)
Elapidae	Common Indian krait (*Bungarus caeruleus*)	1861	6681	116	+(h)
Elapidae	Krait (*Bungarus walliwall*)	−	5566	137	+(h)
Elapidae	Banded krait (*Bungarus fasciatus*)	1108	1742	59	+(h)
Elapidae	Tiger snake (*Notechis scutatus*)	2963	10253	158	+(h)
Elapidae	Cobra (*Naja naja oxiana*)	−	2348	55	+(h)
Viperidae	Russell's viper (*Vipera russelli russelli*)	1621	1932	158	+(h)

Details of methods are given by Singh *et al.* (1980), from which source these data have been abstracted.

*Con, Bacterial DNA control

† +, Present; −, absent; (i) isomorphic ZW; (h) heteromorphic ZW.

sex bias in the quantity of Bkm satellite-related DNA (Table
I) and the absolute amount of such DNA may be as much as ten
times less than that found in elapid snakes (Singh *et al.*,
1980). Evidently, there has been a massive increase in the
relative and in the absolute amount of Bkm satellite-related
DNA in the evolutionary period separating these two types of
sex-determining systems. This increase does not relate only
to the elaboration of the heterochromatic W chromosome but has
affected the genomes as a whole. Another relevant observation
is that, whilst most snake W chromosomes from different species show an extensive distribution of Bkm satellite DNA-
related sequences, one of the species that we examined, *Elaphe
radiata*, which has homomorphic Z and W chromosomes, contains a
high concentration of these sequences in a more localized region of the W chromosome. We assume that the homomorphic condition of the sex chromosomes indicates a relatively poorly
differentiated stage in the evolution of such chromosomes,
which reach their ultimate development in species with a much
reduced W chromosome. If this example permits a generalisa-
tion, it would seem that the evolution of the W chromosome has
been paralleled by the spread of Bkm satellite DNA-related
sequences from a localised region to a more general distribution. The significance of this is not clear.

Bkm SATELLITE-RELATED DNA IN OTHER VERTEBRATES

The finding that Bkm satellite-related DNA sequences are to
be found in a wide variety of snake species encouraged us to
examine whether such conservation might extend to birds and
mammals. Accordingly, using conditions of high stringency, we
hybridized Bkm satellite DNA probes to the DNAs of male and
female Japanese quails (*Coturnix coturnix japonica*). Very
significant hybridization compared to control DNA was obtained
which was, as in snakes, biased quantitatively in favour of
female DNA, indicating an association with the W chromosome.
Similar hybridizations were carried out with the filter-bound
DNAs of males and females of species including an amphibian
(*Triturus cristatus carnifex*), a lizard (*Varanus flavescens*)
and mammals (*Homo sapiens*) (Table II). Positive hybridization
was obtained in each case but, unlike snakes and birds, there
was no significant sex-related bias in the amount of hybridization. It was therefore decided to determine whether there
might be an observable difference in sequence arrangement in
the sexes of a mammalian species such as the mouse, in which
we had also shown Bkm satellite-related DNA.

TABLE II

DNA of Different Species Denatured and Filter Bound in Triplicate

| Species | Bkm satellite [³H] DNA (cts/min hybridized per 10⁴ diploid genomes) | | |
	Male	Female	Ratio (f/m)
			—
Insect (*Drosophila melanogaster*)	4		
Sea urchin (*Echinus esculentus*)	40	45	1.13
Amphibian (*Triturus cristatus carnifex*)	2300	2140	0.93
Lizard (*Varanus flavescens*)	294	349	1.19
Snake (*Bungarus caeruleus*)	1660	6630	3.99
Bird (*Coturnix coturnix japonica*)	59	142	2.41
Mammal (*Homo sapiens*)	428	482	1.13
Bacterium (*Micrococcus lysodeikticus*)	0.006		

Values shown are means of 3 determinations and have been normalized for genome size. Conditions of hybridization are detailed by Singh *et al.* (1980). Note higher female/male ratio in species with female hetero-gamety. Hybridization in *Drosophila*, although apparently relatively minor, occurs to DNA located in one region (19F–20AB) at the euchromatic base of the X chromosome (Singh *et al.*, 1981; and this text).

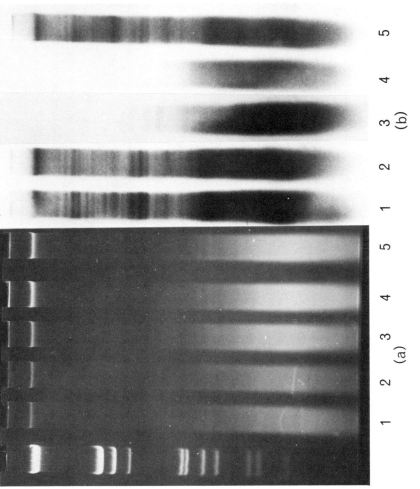

FIG. 3 For legend see opposite.

SEX-SPECIFIC ARRANGEMENT OF Bkm SATELLITE-RELATED DNA

To investigate this question, the genomal DNAs of male and female mice from inbred laboratory strains were digested with different restriction enzymes, blotted onto nitrocellulose filters and hybridized with a ^{32}P-labeled Bkm satellite nick-translated DNA probe under the same stringency conditions used in experiments with Bkm satellite in snake DNA hybridization. With two enzymes, *Hae*III and *Alu*I, striking sex-specific patterns were obtained in the subsequent autoradiographs (e.g. *Alu*I; Fig. 3), which showed high molecular weight Bkm satellite-containing fragments in male DNA that were absent from female DNA (Singh *et al.*, 1981). This finding was further investigated in the case of *Alu*I by extracting DNA from different organs and many different individuals of each sex and running comparative hybridization blots. The patterns obtained showed that there was no detectable differences between individual males or between organs within individual males, but the male-female difference was clearcut and invariant. Controls were devised in which the hybrids with Bkm satellite DNA were melted off the comparative blots, which were subsequently re-hybridized with a ^{32}P-labeled mouse satellite DNA probe (Jones and Singh, 1981). This showed no differences between any of the DNAs, as would be expected for an autosomal DNA sequence, and confirmed that the previous result was not due to some technical factor, such as differences in the digestion conditions or in DNA transfer from gels to nitrocellulose. This sex difference is potentially attributable to three, not necessarily exclusive mechanisms, which have been discussed elsewhere (Singh *et al.*, 1981). Most

---

FIG. 3 (opposite) (a) *DNAs of normal male (1 and 2) normal female (3 and 4) and sex-reversed Sxr XX male (5) digested with AluI, separated on agarose gels by electrophoresis and stained with ethidium bromide for ultraviolet photography. There is no visible difference between the 4 tracks. The extreme left-hand track is a molecular size marker of lambda phage DNA cleaved with restriction enzymes. The DNA in track 4 was from a female of the same genetic background as that in track 5. (b) The DNAs shown in (a) transferred in the same order to a nitrocellulose filter and hybridized with a ^{32}P-labelled Bkm satellite DNA probe. This autoradiograph shows the locations of hybridized fragments. Note that male DNA contains fragments above approximately 2000 base-pairs in size, which hybridize and which appear to be absent from female DNA. However, track 5 (XX Sxr) shows the full range of male-specific hybridization, despite the absence of a cytologically definable Y chromosome. (Adapted from Singh et al., 1981.)*

obviously, the contribution of the Y chromosome will distinguish male from female DNA, and the fragments decorated with
Bkm satellite-related sequences could arise from this source.
It is also possible that some sex-specific modification under-
lies the difference, affecting the digestibility of sex-
related sequences. Finally, it is possible that some developmental restructuring of DNA accompanies the process of sex
determination. The second possibility was investigated by the
use of isochizomers (restriction enzymes recognising the same
sequence), which differ in their sensitivity to DNA methylation, for example *MspI* and *HpaII*, *Cfo* I and *HhaI*; the most pre-
valent modification mechanism. No indication of such a mechanism was found, and enzymes (*HaeIII* and *AluI*) that do not
recognise sequences including CpG, the most frequently modified dinucleotide, do reproduce the sex difference in DNA digestion patterns. We conclude that restriction modification is
not a major factor in this difference. To investigate the
first and most obvious possibility, we resorted to hybridization *in situ*.

Bkm SATELLITE-RELATED DNA SEQUENCES ON THE Y CHROMOSOME

Mouse male and female chromosome spreads hybridized with Bkm
satellite DNA nick-translated with ^{3}H-labeled nucleoside tri-
phosphates show a clear difference in hybridization pattern
in situ (Jones and Singh, 1981 and unpublished results). In
mitotic chromosome preparations, there is a single small chromosome in male spreads, presumptively the Y, which labels in-
intensely in a paracentromeric location (Fig. 4(a)). This is
absent from female spreads. Both karyotypes show a definite,
though relatively diffuse, hybridization to many other chromosomes, and this is most apparent after a longer exposure than
necessary to reveal the presumed Y location. Confirmation of
the Y location was obtained after hybridization to testicular
meiotic preparations, in which the XY bivalent is easily identified (Fig. 4(b)). In these it is very clear that the terminal region of the Y, which is unpaired and thus defined as the
centromeric end, is the region that contains the high concentration of Bkm satellite-hybridizable DNA. This region, perhaps significantly, is the only such region in the mouse karyotype that fails to show a high concentration of mouse satellite DNA by hybridization *in situ* and is thus defined as
qualitatively different from the other paracentromeric locations in this species. We conclude from this investigation
that the male-female difference in blot hybridization with
Bkm satellite DNA probes reflects, to a significant extent,
the contribution of sequences from the proximal region of the
Y chromosome. However, we cannot exclude the possibility

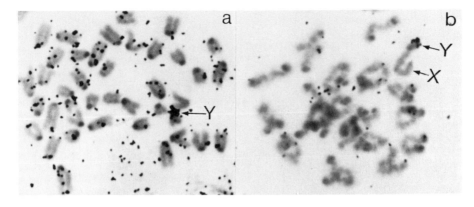

FIG. 4 *Chromosome preparations of normal male mice autoradio-
graphed after hybridization* in situ *with Bkm satellite [³H]
DNA.* **(a)** *Mitotic preparations in which there is a concentration of silver grains over a small chromosome, presumably the
Y. These are located centromerically, denoting the major concentration of Bkm-satellite-related DNA.* **(b)** *Meiotic metaphase in normal male mouse showing the XY bivalent, in which
a cluster of silver grains is seen over the centromeric terminus of the Y chromosome. Such observations support the interpretation that the chromosome labeled in (a) is indeed the Y
chromosome. (Adapted from Jones and Singh, 1981.)*

that the third mechanism, mentioned above, contributes nothing to these observations. The important question arising
from these observations is whether the region of the Y defined
by the Bkm satellite DNA probe has a sex-determining function.
This question was approached by investigating the DNA and
chromosomes of the mouse mutant *Sxr* (see Cattanach, 1975)
which causes XX (chromosomally female) individuals to develop
as phenotypic, but sterile, males.

Bkm SATELLITE-RELATED SEQUENCES IN *Sxr* MICE

The basis of the mutation *Sxr* is not known but, in principle,
it is thought either to reflect the transfer of a sex-
determining Y chromosome region to some other chromosome,
which then effectively becomes a Y chromosome, or to be due
to a constitutive mutation in a sex-determining gene normally
under Y control but located elsehwere (Cattanach, 1975). It
was therefore of considerable interest to examine the DNA of
XX$_{Sxr}$ males using the Bkm satellite DNA probe. This was done
(Fig. 3(b) track 5) and it became clear that this DNA contains
the same fragment pattern that is seen to characterise normal
male DNA (Singh *et al.*, 1981). The only visible difference

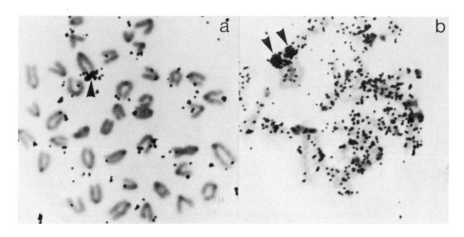

FIG. 5 *Chromosome preparations from* Sxr *mice showing locations of Bkm-satellite-related DNA sequences after hybridization* in situ *and autoradiography.* **(a)** *A mitotic chromosome preparation from an XX* Sxr *mouse showing a major concentration of Bkm-satellite-related DNA near the distal terminus of one autosome and equal in size to the major region of Bkm-satellite DNA-hybridization on the Y chromosome of normal male mice as shown in Fig. 4(a).* **(b)** *A meiotic preparation from an XY male mouse carrying the* Sxr *trait, which is transmitted and expressed in XX progeny, turning them into sterile males. Here there are two major locations of hybridization, one of which is on the Y as in normal mice and the other is located elsewhere, consistent with its attachment to an autosome that transmits the* Sxr *trait. This material is still being analysed and is as yet unpublished.*

is the absence from XX_{Sxr} DNA of one Bkm satellite positive *Hae*III fragment class, which is present in normal male DNA (unpublished results). This finding is consistent with the first of the above explanations; namely, that there has been a transfer of a male-determining chromosome region, although cytological evidence for this has not been found, despite intensive investigations by conventional cytogenetic means (Winsor *et al.*, 1978; Chandley and Fletcher, 1980; Evans *et al.*, 1980). Moreover, attempts to map *Sxr* by genetic means have also failed, despite the fact that all mouse chromosomes are well marked genetically. Nevertheless we examined the chromosomes of XX_{Sxr} male mice by hybridization *in situ* using a probe of Bkm satellite DNA. We found that these chromosomes exhibited an intensely hybridizing location (unpublished results), similar in intensity to the Y proximal region, but located distally on one unidentified but presumed autosome

(Fig. 5(a)). Next, meiotic preparations from XY$_{Sxr}$ carrier males were examined similarly. These exhibited two intensely hybridizing regions (Fig. 5(b)), which were, presumably, the Y chromosome and an autosome, consistent with our observations on mitotic preparations of XX$_{Sxr}$ males. Frequently, these two regions were closely associated with the sex vesicle and with another chromosome, presumptively the X, but this aspect is still under study. Blot hybridization of carrier males yielded identical patterns to those seen in normal males. Females from the *Sxr*-bearing strain show a normal female pattern, lacking the male-specific hybridization bands (unpublished results).

We conclude from these experiments that *Sxr* is caused by the movement of a sex-determining region of the Y chromosome to some other chromosome, which then functions as a Y in sex determination. This would appear to render unnecessary the alternative hypothesis of a gene mutation in an autosomal sex-determining gene. It can also be concluded that Bkm satellite DNA-related sequences are intimately associated with this sex-determining region, although we cannot conclude that they have any direct role in sex determination on present evidence. Nevertheless, it is clear that Bkm satellite DNA can be used to define precisely the DNA sequences that are sex determining, so that its role in the process should become clear. The origin of the *Sxr* condition is of considerable interest, Two aspects are worth considering here. Firstly, as originally pointed out by Cattanach and his co-workers (Cattanach, 1975), *Sxr* could not be due to a *chromosomal* translocation of *the* sex-determining region of a Y chromosome, since this would have resulted in its permanent loss from the original Y chromosome, which would have rendered the transmission of *Sxr* impossible *via* either the male or female line. A translocation model would, therefore, either have to assume a chromatid translocation, or the existence of multiple sex-determining regions on the Y chromosome. The second point is that attempts to map *Sxr*, which should have been expected to succeed, have failed. Evidently, *Sxr* shows no strong linkage and this might indicate that it is unstably associated with one or other autosome. Both of these aspects could be explained if it was assumed that the sex-determining region of the mouse Y chromosome has some of the characteristics of a large transposable element that moves only infrequently, perhaps to a variable location in the genome. Such elements have been described in *Drosophila*, where they are large enough to be easily visible in salivary chromosomes (Ising and Block, 1981). Moreover, these appear to move by some process of replication that leaves an apparently intact copy at the site of origin. In this context, it is important to

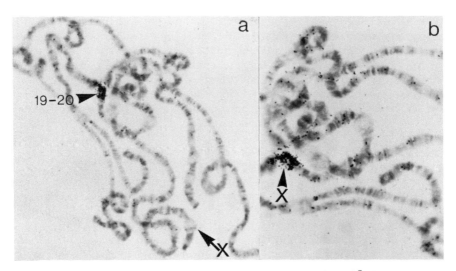

FIG. 6 *Hybridization* in situ *of Bkm satellite [³H]DNA to*
D. melanogaster *salivary gland chromosomes.* (a) *The auto-*
radiograph shows intense hybridization to the base of the X
chromosome only. This region has been mapped as 19F and 20AB
within the euchromatic portion. (b) *Same as (a) but at higher*
magnification. (From Singh et al., 1981.)

recall that human XX males exist in the normal population (de
la Chapelle, 1972) and might be similarly explained, although
this phenomenon remains for future study.

Bkm SATELLITE-RELATED DNA IN INVERTEBRATES

To gain some impression of the extent to which Bkm satellite-
related DNA sequences are to be found in eukaryotes, we have
surveyed a few invertebrate species by quantitative hybridization to genomal DNA, some of which data are shown in Table II.
Our survey included sea urchins (*Echinus esculentus*), insects
(*Drosophila melanogaster*) and yeast (*Saccharomyces cerevis-*
iae). There was detectable hybridization to all of these DNAs
and, to enable some evaluation of its significance, we
examined *Drosophila* salivary gland chromosomes by hybridization *in situ* of Bkm satellite DNA probe labeled with tritium.
This revealed (Singh *et al.*, 1981) that related DNA is concentrated at the base of the X chromosome in a euchromatic
region designated 19F and 20AB, which lies tandem to the
heterochromatic region and the nucleolar organizer (Fig. 6).
Very strong hybridization occurs to this region but to no
other. Consequently, Bkm satellite DNA provides a probe with

which to analyse this developmentally important location.
This result encouraged us to look for homologous sequences in
a "library" of recombinant *Drosophila* DNA in lambda phage,
which yielded several clones, six of which were recovered and
shown to contain sequences homologous to Bkm satellite DNA
tandemly flanked with non-homologous sequences. All of the
hybridization phenomena that we can demonstrate with Bkm
satellite DNA can be reproduced using this *Drosophila* DNA,
illustrating that the homology is not in respect of a minor,
or a different, sequence component of the Bkm satellite DNA.
Analysis of these clones by restriction enzyme digestion
(Singh *et al*., 1981) showed that they were all different in
the nature of the sequences flanking the Bkm satellite-related
insert. Consistent with the idea that the region of 19F and
20AB may be constructed of Bkm satellite-related DNA alternating with other, unrelated DNA (data not shown). Individual
library clones were then hybridized back to salivary chromosomes to confirm their origin in the X base. Whilst each
individual clone did hybridize to this region, surprisingly,
each one also hybridized simultaneously to one other chromosome band located distantly from the X base, and this band was
different in the case of four out of the six clones tested
(Singh *et al*., 1980). The hybridized bands were located on
chromosomes other than the X, as well as on the X chromosome
itself. Sub-cloning of the original library isolates was used
to separate these various specificities. The sub-cloned DNA,
when hybridized *in situ*, identified the distal bands but not
the X base in the two instances tested (unpublished results),
in which the Bkm-satellite DNA-related portion of the clone
had been separated from the non-related portion. The significance of these peculiar findings is unclear but, taken at face
value, they appeared to indicate that DNA fragments that are
tandemly associated in DNA clones established from embryonic
DNA in the original library might have become dissociated in
later embryonic stages and that the Bkm satellite DNA-related
portion has remained at (or been translocated to) the base of
the X chromosome. Several alternative but not necessarily
exclusive explanations, however, are possible for these observations, some of which are less interesting than others. For
example, DNA from genomally distant sites may have become
artefactually linked during the construction, or the manipulation, of the library. However, if this had happened on the
scale necessary to give the result we have described, libraries would be largely useless for examining genome structure,
which they clearly have not been found to be despite certain
limitations (Fritsch *et al*., 1980). On these grounds, we
tend to heavily discount this explanation. Another explanation may be that the non-Bkm satellite-related DNA detected

distally is, in fact, tandem to a stretch of Bkm satellite-
related DNA that is too small to be detected *in situ* at chromosome level but is detectable in library clones. Thus, since
there is a lot of Bkm satellite-related DNA at the X base, it
will be detected *in situ* by the library clone, but not *vice
versa*. In this case, we have a possible technical limitation
arising out of a lack of sensitivity in hybridization *in situ*,
although even very prolonged autoradiographical exposure
failed to reveal additional Bkm satellite DNA locations.
Finally, there is the more interesting possibility that
sequence movement and/or developmental loss of Bkm satellite-
related DNA does in fact occur. To test this, we examined
DNA from different organs of developing fly larvae and from
adult flies by blot hybridization and challenge with Bkm
satellite DNA. Our preliminary results appeared to bear out
the hypothesis, in that radically different banding patterns
were seen in the resulting autoradiographs. However, further
studies (unpublished results) revealed that these differences
were the result of strain differences in the backgrounds of
the giant mutant flies that we used for convenience in dissection. Better control of this aspect showed that in experiments where there is proper control of strain variation, there
appears to be no alteration in Bkm satellite-related DNA, as
determined by blot hybridization, during development of the
flies. However, this remains to be determined in the case of
DNA extracted from salivary glands. It is evident, however,
that such DNA varies strikingly between the genomes of related
strains and, in this respect, Bkm satellite-related DNA resembles other *Drosophila* sequences exemplified by *copia* (Finnegan *et al.*, 1978). Failure to show developmental movement in
Bkm satellite-related DNA itself, however, does not exclude
the possibility that sequences closely associated with it may
move, as suggested by our observations mentioned above on the
cloned fragments recovered with the help of the Bkm satellite
DNA probe from a genomal library. Such movement might not
significantly change the relative size of restriction fragments and might, for this reason, not be readily detected
from a visual examination of genomal DNA restriction blots
after challenge with Bkm satellite DNA. This possibility is
being investigated.

Finally, a quite extraordinary recent observation is
worthy of preliminary description in the context of the present discussion. It was made during an examination of the DNA
of the yeast *S. cerevisiae* in order to establish whether or
not Bkm satellite DNA-related sequences might be present and
thus to establish how far down the scale of structural
elaboration and evolution such sequences might exist. Yeast
is also the organism in which most is currently known about

the genetic control of sexual phenomena and in which there is clear evidence of sequence transposition in the mechanism of mating type control (Hicks *et al.*, 1977; Strathearn *et al.*, 1980; Klar *et al.*, 1980). We argued, therefore, that if Bkm satellite-related DNA sequences were to be found in yeast, there might be considerable advantages in using this organism to establish what their function, if any, might be.

Surprisingly, we found that there were four visible bands of hybridization in an autoradiograph of a total diploid yeast genome DNA *Hin*dIII restriction blot challenged with ^{32}P-labeled Bkm satellite DNA (data not shown). The same blot was then melted to dissociate these hybrids and re-hybridized with a ^{32}P-labeled cloned version of the MAT region of yeast DNA, which contains sequences homologous to both the silent and expressed mating type genes. A subsequent autoradiograph revealed that the same regions that had previously shown hybridization bands with Bkm satellite DNA also hybridized with the MAT clone, but much more strongly. There were also other hybridization locations, which were not seen with Bkm satellite DNA, reflecting other non-MAT sequences in the clone. Comparison of the Bkm satellite DNA-positive bands with those obtained with the MAT clone revealed a precise matching. Molecular weight determinations established that these bands corresponded to those that have been suggested to contain the silent copies (HML and HMR) and the expressed copies (MATa and MATα) of the mating type genes (Nasmyth and Tatchell, 1980). The converse experiment, in which the MAT clone was hybridized to DNA previously shown to contain Bkm satellite-related DNA, however, failed to reveal evidence of hybridization. This suggested that the Bkm satellite-related DNA of yeast is not contained in the MAT clone but lies outside, possibly on the same *Hin*dIII restriction fragments, although this remains to be established. The significance of these remarkable findings are presently obscure but they suggest the exciting possibility that Bkm satellite DNA-like sequences are somehow implicated in sexual phenomena in a simple eukaryote. It should be stressed, however, that such an association might, if it is not fortuitous, reflect the facility with which Bkm satellite-related sequences are able to exploit genetically privileged locations for their own transmission. Alternatively, we may be getting a first glimpse of a mechanism that plays an essential role in the control of gene expression in eukaryotes. Which, if any, of these working hypotheses is correct may become clearer when we have sequenced and compared clones of Bkm satellite-containing DNA from various organisms.

SUMMARY

We have shown that DNA derived from the sex-determining W
chromosome of snakes is related by hybridization under
unreduced criteria to sequences present in the DNAs of an
extremely wide range of eukaryotic species. At present, it
appears that this DNA has persisted most obviously in connection with sex-determining regions of chromosomes. However, it
is possible that this reflects the fact that we have sought it
in such a connection, rather than that it has a specific function in sex determination only. It may thus be that if other
analyses are applied to control systems unrelated to sex
determination, that Bkm satellite DNA-related sequences might
also be detected. It seems, nevertheless, that the Bkm satellite class of sequences is quantitatively dominant on sex
chromosomes, and this might either reflect the fact that such
chromosomes are genetically isolated and thus suitable for the
accumulation of DNA that does not have a function necessarily
related to the phenotype. Alternatively, since sex determination involves a very basic switch in cell behaviour and properties leading to the development of one or other mating type
in single-celled eukaryotes and alternative gonads in multi-
cellular eukaryotes, particular nucleotide sequences may have
been preferred in connection with the underlying control at
the DNA level, surprising as this might seem. The finding of
Bkm satellite-related DNA apparently only in association with
the restriction fragments containing the mating type genes in
yeast may offer the strongest immediate hope of testing this
idea, and this observation certainly suggests a significant,
as opposed to a trivial, explanation of these sequences. The
outcome of future studies in this system might well be to establish that sex determination in eukaryotes has some common
features right across the board, if only in terms of the type
of parasitic, selfish or passenger, sequences, which it har-
bors. Finally, in the practical context at least, we must
emphasise the point that, whatever the functional significance
of the Bkm satellite class of DNA, it affords most valuable
tools with which to probe for and recover genes from important
chromosomal domains, including those that undoubtedly do control the genes involved in specifying the sexual phenotype.

REFERENCES

Britten, R.J. and Kohne, D.E. (1968). Repeated sequences in
    DNA, *Science* **161**, 529-540.
Cattanach, B.M. (1975). Sex reversal in the mouse and other
    mammals. *Brit. Soc. Dev. Biol. Symp.* 2, 305-317.
Chandley, A.C. and Fletcher, J.M. (1980). Meiosis in *Sxr*

male mice   1. Does a Y-autosome rearrangement exist in sex-reversed (*Sxr*) mice? *Chromosoma* **81**, 7-17.

Chapelle, de la, A. (1972). Analytical review: Nature and origin of males with XX chromosomes. *Amer. J. Human Genet.* 24. 71-105.

Duncan, C., Biro, P.A., Choudary, P.V., Elder, J.T., Wang, R.R.C., Forget, B.G., BeRiel, J.K. and Weissmann, S.M. (1979). *Proc. Nat. Acad. Sci., U.S.A.* **76**, 5095-5099.

Evans, E.P., Burtenshaw, M.D. and Brown, B.B. (1980). Meiosis in *Sxr* male mice   11. Further absence of cytological evidence for a Y autosome rearrangement in sex-reversed (*Sxr*) mice. *Chromosoma* **81**, 19-26.

Finnegan, D.J., Rubin, G.M., Young, M.W. and Hogness, D.S. (1978).   Repeated gene families in *Drosophila melanogaster*. *Cold Spring Harbor Symp. Quant. Biol.* **42**, 1053-1063.

Fritsch, E.F., Lawn, R.M. and Maniatis, T. (1980).   Molecular cloning and characterisation of the human β-like globin gene cluster. *Cell* **19**, 959-972.

Fry, K. and Salser, W. (1977).   Nucleotide sequences of HS satellite DNA from kangaroo rat *Dipodomys ordii* and characterization in other rodents. *Cell* **12**, 1069-1084.

Hicks, J.B., Strathearn, J.N. and Herskowitz, I. (1977).   The cassette model of mating type interconversion. *In* "DNA insertion elements Plasmids and Episomes" (Bukhari, A., Shapiro, J. and Adhya, S., eds), pp. 457-462, Cold Spring Harbor Laboratory Press, Cold Spring Harbor.

Ising, G. and Block, K. (1981).   Derivation dependent distribution of insertion sites for a *Drosophila* transposon. *Cold Spring Harbor Symp. Quant. Biol.* **45**, 527-544.

John, B. and Miklos, G.L.G. (1979).   Functional aspects of satellite DNA and heterochromatin. *Int. Rev. Cytol.* 1-14.

Jones, K.W. (1970).   Chromosomal and nuclear location of mouse satellite DNA in individual cells. *Nature (London)* **225**, 912-915.

Jones, K.W. and Singh, L. (1981).   Conserved repeated DNA sequences in vertebrate sex chromosomes. *Human Genet.* 58, 46-53.

Kaufman, R.E., Fretschmer, P.J., Adams, J.W., Coon, H.C., Anderson, W.F. and Nienhuis, A.W. (1980).   Cloning and characterization of DNA sequences surrounding the human γ,δ,β-globin genes. *Proc. Nat. Acad. Sci., U.S.A.* **77**, 4229-4233.

Kit, S. (1961).   Equilibrium sedimentation in density gradients of DNA preparations from animal tissues. *J. Mol. Biol.* 3, 711-716.

Klar, A.J.S., McIndoo, J., Strathearn, J.N. and Hicks, J.B. (1980).   Evidence for a physical interaction between the transposed and the substituted sequences during mating type gene transposition in yeast. *Cell* 22, 291-298.

Nasmyth, K.A. and Tatchell, K. (1980). The structure of transposable yeast mating type loci. *Cell* 19, 753-764.

Pardue, M.L. and Gall, J.G. (1970). Chromosomal localization
of mouse satellite DNA. *Science* 168, 1356-1358.

Pech, M., Streeck, R.E. and Zachau, H.G. (1979). Patchwork
structure of a bovine satellite DNA. *Cell* 18, 883-893.

Ray-Chaudhuri, S.P., Singh, L. and Sharma, T. (1970). Sexual
dimorphism in somatic interphase nuclei of snakes. *Cytogenetics* 91, 410-423.

Ray-Chaudhuri, S.P., Singh, L. and Sharma, T. (1971). Evolution of sex chromosomes and formation of W chromatin in
snakes. *Chromosoma (Berlin)* 33, 239-251.

Shen, C.-K.J. and Maniatis, T. (1980). The organization of
repetitive sequences in a cluster of rabbit β-like globin
genes. *Cell* 19, 379-391.

Singh, L., Purdom, I.F. and Jones, K.W. (1976). Satellite DNA
and evolution of sex chromosomes. *Chromosoma (Berlin)* 59,
43-62.

Singh, L., Purdon, I.F. and Jones, K.W. (1980). Sex chromosome associated satellite DNA: evolution and conservation.
Chromosoma (Berlin) 79, 137-157.

Singh, L., Purdom, I.F. and Jones, K.W. (1981). Conserved
sex chromosome associated nucleotide sequences in eukaryotes. *Cold Spring Harbor Symp. Quant. Genet.* 45,
805-813.

Southern, E.M. (1970). Base sequence and evolution of guinea
pig satellite DNA. *Nature (London)* 227, 794-798.

Strathearn, J.N., Spatola, E., McGill, C. and Hicks, J.B.
(1980). Structure and organization of transposable mating
type cassettes in *Saccharomyces* yeasts. *Proc. Nat. Acad.
Sci., U.S.A.* 77, 2839-2843.

Peacock, W.J., Appels, R., Dunsmuir, P., Lohe, A.R. and Gerlach, W.L. (1973). The organization of highly repeated DNA
sequences: chromosomal localization and evolutionary conservatism. *Int. Cell Biol.* 496-506.

Winsor, E.J.T., Ferguson-Smith, M.A. and Shire, J.G.M. (1978).
Meiotic studies in mice carrying the sex reversal (*Sxr*)
factor. *Cytogenet. Cell Genet.* 21, 11-18.

Wu, J.C. and Manuelidis, L. (1980). Sequence definition and
organization of a human repeated DNA. *J. Mol. Biol.* 142,
363-386.

PART II
EVOLUTION OF GENE FAMILIES

Evolution of Globin Genes

A.J. JEFFREYS

Genetics Department, University of Leicester, England

INTRODUCTION

Globin genes are an ideal system for studying the molecular evolution of genes and multigene families. Globins are widespread in nature and include the tetrameric haemoglobins of higher vertebrates, monomeric haemoglobins of protochordates, a variety of invertebrate globins characterised in molluscs, the midge larva and the bloodworm, monomeric myoglobins, and the monomeric leghaemoglobins found in the root nodules of nitrogen-fixing plants (Hunt *et al.*, 1978). The amino acid sequences of many of these globins have been determined, and sequence homologies point to a common evolutionary origin of most or all globins. Detailed sequence comparisons of different species' globins have enabled phylogenetic trees of globins to be constructed (Dayhoff, 1972; Hunt *et al.*, 1978), and have shown that globin sequences tend to diverge in evolution at a constant rate, independent of the lineage being studied. This evolutionary molecular clock is typical of many protein sequences, and has proved an invaluable aid in determining the time of species divergence or gene duplication (see Wilson *et al.*, 1977).

Higher vertebrates code for a variety of globins; for example, man has eight active globin genes specifying haemoglobin polypeptides, which can be divided by sequence homology into two families. The α-globin related family consists of a ζ-globin gene expressed during early embryogenesis and two almost-identical α-globin genes expressed in the foetus and adult. The β-related family contains a single embryonic ε-globin gene, two very similar foetal globin genes ($^{G}\gamma$ and $^{A}\gamma$), a minor adult δ-globin gene and the major adult β-globin gene. In addition, there are an unknown number of myoglobin genes. All of these gene products show significant amino acid sequence homology and, by using the molecular clock rate of globin sequence divergence, it is possible to deduce the timing of the gene duplications that gave rise to these families

(Dayhoff, 1972; Efstratiadis *et al.*, 1980). The most ancient
duplication, about 1100 million years ago, gave rise to the
ancestors of haemoglobin and myoglobin genes. The αβ-globin
gene duplication occurred about 500 million years ago, early
in the evolution of vertebrates. The β-globin gene family
evolved more recently, with a foetal-adult duplication about
200 million years ago, an ε-γ duplication 100 million years
ago and a δ-β duplication 40 million years ago.

Over the last four years, vertebrate haemoglobin genes have
been investigated intensively using recombinant DNA techniques. This analysis has been facilitated by the isolation
and purification *via* complementary DNA (cDNA) cloning of α-
and β-globin messenger RNA sequences. Globin cDNA clones have
been used to detect corresponding globin genes in Southern
blot analyses of total genomic DNA cleaved with restriction
endonucleases, and to isolate these genes from recombinant
λ bacteriophage libraries. To date, α- and β-related globin
genes have been studied in detail in animals ranging from
amphibians to man and, so far, complete DNA sequences of some
15 different genes have been reported.

These studies at the DNA level have further shown the suit-
ability of globin genes for evolutionary studies. Globin
genes, in common with many other genes in higher eukaryotes,
contain intervening sequences, the evolutionary history of
which is beginning to be traced by comparative molecular studies. The α- and β-globin gene families have been found to
be arranged in gene clusters that have probably arisen by tandem gene duplication. Evidence for genetic interchange
between members of a gene cluster is accumulating. Additional
inactive pseudogenes, unsuspected from protein studies, have
been found.

This paper reviews some of the recent advances in our
understanding of gene evolution that have resulted from comparative molecular studies of globin genes.

EVOLUTION OF GLOBIN CODING SEQUENCES

By comparing the DNA sequences of the coding regions (exons)
of globin genes, we can learn something of the rates and modes
of exon evolution. Many such comparisons, both of duplicated,
diverged genes within a single species, and of homologous
genes in different species, have shown that nucleotide substitutions are not scattered at random throughout exons, but show
a marked clustering at third codon positions, where they cause
synonymous codon changes (Kafatos *et al.*, 1977, Efstratiadis
et al., 1980; Perler *et al.*, 1980). Clearly, selection has
eliminated the bulk of substitutions that caused amino acid
replacements (replacement site substitutions), whereas many

synonymous or silent site substitutions probably have little
or no effect on gene function and can be fixed in evolution.
A similar phenomenon has been found in many other gene systems
(see Jukes and King, 1979; Jukes, 1980).

There have been several attempts to quantify the rates of
replacement site and silent site substitution (Efstratiadis
et al., 1980; Miyata *et al.*, 1980; Perler *et al.*, 1980; Kimura, 1981; Miyata and Yasunaga, 1981). Replacement sites
diverge in a clock-like fashion at about 0.1% per million
years; this monotonous change of sequence is of course expected, in view of the protein evolutionary clock hypothesis plus
the fact that most divergence times used by Perler *et al.*
(1980) and Efstratiadis *et al.* (1980) were estimated from protein sequence data, including globin sequences. Replacement
site divergences can be used interchangeably with amino acid
sequence divergences to construct molecular phylogenies (Efstratiadis *et al.*, 1980). The clock rates for these divergences vary from protein to protein, and presumably reflect
the proportion of amino acid sites that can be altered without
causing loss of protein function. Invariant sites are likely
to be important for protein function and in the globins
include, in particular, residues involved in haem binding.

Silent sites in globin exons tend to diverge much more
rapidly, at an initial rate of about 1% per million years (cf.
0.1% million years for replacement sites: Efstratiadis *et
al.*, 1980; Miyata *et al.*, 1980; Perler *et al.*, 1980). It is
not known whether this rate is constant in evolution and independent of lineage, since few closely related genes have been
sequenced. There is tentative evidence from other gene comparisons to suggest that silent sites, unlike replacement
sites, might evolve at a constant rate in all genes (Miyata
et al., 1980; Perler *et al.*, 1980; Kimura, 1981). If so, then
this silent site clock would be of great use for constructing
phylogenies of closely related species.

If a silent site clock running at the same rate for all
genes does exist, then this would support the idea, repeatedly
proposed, that those silent site changes that are fixed in
evolution are selectively neutral (Jukes and King, 1979;
Jukes, 1980; Kimura, 1981). However, not all possible silent
site replacements are necessarily neutral. In a comparison of
human and rabbit β-globin mRNA sequences, Kafatos *et al.*
(1977) noted that silent changes were not scattered at random
along the mRNA but were clustered into regions that tended
also to be rich in replacement site substitutions. Similarly,
Miyata and Yasunaga (1981) find that silent site changes seem
to accumulate less rapidly than nucleotide changes in possibly
functionless pseudogenes (see below). Thus, some silent
changes are eliminated in evolution, perhaps as a result of

interfering with transcription or mRNA/precursor mRNA structure, processing or export. Alternatively, a silent substitution might generate a synonymous codon for which no abundant
transfer RNA exists.

Perler *et al.* (1980) and Efstratiadis *et al.* (1980) find
that the rate of silent site substitution apparently slows
down after about 100 million years of divergence and there-
after proceeds at a rate similar to replacement site substitutions. It is still not certain whether this shift in rate is
real or an artifact caused by analysing highly divergent
sequences, or by the methods used for correcting these divergences for multiple substitutions (see Jeffreys, 1981; Kimura,
1981). If the shift is real, then it suggests that there are
(at least) two classes of silent site substitution. One
occurs exceedingly rapidly, at about 1.5% per million years
for globin, and might represent neutral sites that become
fully randomised within 100 million years of divergence. The
second set changes slowly (about 0.1%/million years) and might
represent the gradual appearance of adaptive changes in gene
expression, or might be driven by a slow shift in the codon
utilisation pattern of the genome (Grantham *et al.*, 1981).

Perler *et al.* (1980) have proposed that the replacement
site substitution clock, and therefore amino acid divergence,
is driven primarily by selection. They argue that, since
replacement sites diverge at about 10% of the rate of silent
changes, 90% of replacement site substitutions are eliminated
by selection. Thus, only 10% of possible replacement changes
could be neutral, which suggests that replacement site divergence should saturate at 10%; this does not occur. This argument has not been accepted by Kimura (1981).

EVOLUTION OF INTERVENING SEQUENCES IN GLOBIN GENES

A wide variety of α- and β-related globin genes have been analysed in detail in species including man, rabbit, mouse,
chicken and *Xenopus laevis* (see Jeffreys, 1981). All active
vertebrate globin genes studied contain two intervening
sequences interrupting the protein coding sequence. In every
case, the intervening sequences occur at precisely homologous
positions in the genes. As noted by Leder *et al.* (1978), this
establishes that the discontinuous structure must have been
in existence at least 500 million years ago, before the αβ-
globin gene duplication arose, and that no intervening
sequence has subsequently been gained or lost by active vertebrate globin genes. This structure is also preserved in
the ψα1 pseudogene in man (Proudfoot and Maniatis, 1980), in
the rabbit ψβ2 sequence (Lacy and Maniatis, 1980) and in the
goat β^X pseudogene (Cleary *et al.*, 1980).

A remarkable exception to this rule occurs in the mouse
ψα3 pseudogene, in which both intervening sequences have been
precisely removed recently in evolution (Nishioka *et al.*, 1980;
1980; Vanin *et al.*, 1980). In addition, this pseudogene has
accumulated a number of deletions, insertions and frameshifts
sufficient to render the gene non-functional. It is not known
whether the removal of the intervening sequences was initially
responsible for silencing the gene. Precise intron loss has
also been noted in the rat preproinsulin I gene, a functional
gene that lacks one of the two introns seen in the closely
related rat preproinsulin II gene as well as in human and
chicken preproinsulin genes (Lomedico *et al.*, 1979; Bell *et
al.*, 1980; Perler *et al.*, 1980).

Jensen *et al.* (1981) have recently analysed the structure
of a leghaemoglobin (Lb) gene coded by soybean DNA. This gene
has three intervening sequences, not two, but the first and
third introns appear to be at positions homologous to the two
vertebrate introns. The central intron in the Lb gene inter-
rupts what is a continuous coding sequence in animal genes.
The extraordinary thing is that there is any similarity at all
between plant and animal globin genes. It has often been assumed that animal globins and Lb are the products of convergent
evolution. However, the similar gene structure, as well as
(limited) sequence homology, points to a common evolutionary
origin of these globins. Since the vast majority of plants
do not produce Lb, it is difficult to see how this relationship can be traced back to the common ancestor of plants and
animals. Instead, it seems possible that horizontal gene
transfer might have occurred between animals and plants,
although sequence comparisons between vertebrate and invertebrate haemoglobins, myoglobins and Lb give no clue as to the
source of the Lb gene (Hunt *et al.*, 1978). It will be very
interesting to see whether invertebrate globin genes, or vertebrate myoglobin genes, also contain three introns. If so,
then it would seem likely that the central intron was eliminated in the lineage leading to the vertebrates at some time
before the αβ-globin gene duplication.

How are introns removed so precisely in germ cell DNA during evolution? A variety of mechanisms has been postulated,
including: recombination of a cDNA copy of globin mRNA into a
split gene; annealing of mRNA to the coding strand during DNA
replication, followed by excision of the displaced single-
stranded intron DNA loops; direct action of the RNA splicing
system on the anti-coding DNA strand at the replication fork;
inclusion of a gene within a proretrovirus, followed by the
production of spliced retroviral RNA and subsequent retroviral reintegration to produce an intron-less gene (Goff *et
al.*, 1980; Nishioka *et al.*, 1980; Vanin *et al.*, 1980).

Comparisons of intron sequences in homologous globin genes have shown that, in contrast to coding sequences, intervening sequences evolve rapidly by base substitution and by small deletions and insertions (Van Den Berg *et al.*, 1978; Konkel *et al.*, 1979; Van Ooyen *et al.*, 1979). These microdeletions/ insertions might have been generated by slipped mispairing during DNA replication (Efstratiadis *et al.*, 1980), and would not generally be tolerated in exons. No accurate estimate of the rate of intron divergence has yet been reported, nor have any conserved intron sequences been noted, except for the consensus sequences at splice junctions (Breathnach and Chambon, 1981).

Despite the high frequency of microdeletions/insertions in globin genes during evolution, the lengths of introns have remained surprisingly constant; for example, the length of the first intervening sequence in mammalian α- and β-globin genes lies within the range of 116 to 130 base-pairs, despite the extreme age of the αβ-globin gene duplication (see Jeffreys, 1981). Van Den Berg *et al.* (1978) have suggested that intron length, rather than sequence, might be important in some way for globin gene function. However, rapid evolutionary changes in intron lengths have occurred in other genes, such as preproinsulin genes (Perler *et al.*, 1980), vitellogenin genes (Wahli *et al.*, 1980), δ-crystallin genes (Jones *et al.*, 1980) and ovalbumin-related genes (Heilig *et al.*, 1980).

The exons in vertebrate globin genes appear to correlate with domains in haemoglobin, as first proposed in general for split genes by Blake (1979). The central exon codes for the haem-binding domain of globin, although the other two exon products are required to maintain a stable haem-protein complex (Craik *et al.*, 1980,1981). The distribution of intersubunit contacts in haemoglobin appears to be non-random with respect to the three exons (Eaton, 1980). At first glance, the finding of an extra intron in the haem-binding exon of the leghaemoglobin gene (Jensen *et al.*, 1981) seems incompatible with the exon-domain correlation. However, Gō (1981) has analysed the folding pattern of the human β-globin polypeptide, and discerns four domains. He suggests that globin genes were originally composed of four exons, and predicts an additional intron in the haem-binding exon at almost precisely the position of the extra intron found in the Lb gene. Unfortunately, other approaches for detecting globular domains in haemoglobin do not reveal these exon-related domains (Rashin, 1981). Thus, the correlation between exons and structural or functional domains in globins and other proteins (see Jeffreys, 1981) remains highly suggestive but not completely proven.

The likely relation between exons and stable domains in proteins lends support to the idea that exons were originally "mini-genes" that evolved to specify some simple protein function such as haem-binding (Gilbert, 1978; Darnell, 1978; Reanney, 1979). A more sophisticated globin polypeptide could then be evolved by the rearrangement of unrelated exons and their inclusion within a common transcriptional unit; the final fusion of the exon-specified polypeptides would be achieved by RNA splicing. Similarly mutations could open up new pathways of RNA splicing in a transcriptional unit, resulting in the appearance of new combinations of exons in mature mRNA. These mechanisms for shuffling exons could provide a major source of novel genetic functions during evolution. Doolittle (1978) has argued that this evolution by exon shuffling is ancient, and probably predates the divergence of prokaryotes and eukaryotes.

Darnell (1978) and Crick (1979) have suggested that intervening sequences might have arisen by insertion of transposable and spliceable elements into originally continuous genes. Presumably, these insertions would only be tolerated at locations where a regional disturbance of the amino acid sequence would not affect protein function; these locations might tend to be at domain boundaries and could give rise to the observed correlation between exons and protein domains. However, no example of intron gain by a gene during evolution has been documented.

There are instances where at least part of the discontinuous nature of a gene has resulted from tandem duplication of a smaller less-split gene (Ohno, 1980). Immunoglobulin C_H genes appear to have evolved by tandem duplication of a single C exon (Early *et al.*, 1979; Sakano, *et al.*, 1979a,b), and ovalbumin, ovomucoid and collagen genes also show clear signs of internal reduplications (Cochet *et al.*, 1979; Stein *et al.*, 1980; Ohkubo *et al.*, 1980). However, there is no evidence of any structural or functional homology between globin exons, and it seems more likely that the split gene has arisen by shuffling of unrelated exons, or by insertion of transposable elements into a once-continuous gene. In either case, introns might be regarded as non-functional DNA, and have repeatedly been included within the discussions of "selfish" or "junk" DNA (see Doolittle and Sapienza, 1980; Orgel and Crick, 1980; Ohno, 1980; and contributions within this volume).

EVOLUTION OF GLOBIN GENE CLUSTERS

The arrangement of human globin genes has been studied intensively both by restriction endonuclease analysis of human DNA, and by analysis of cloned DNA fragments containing these

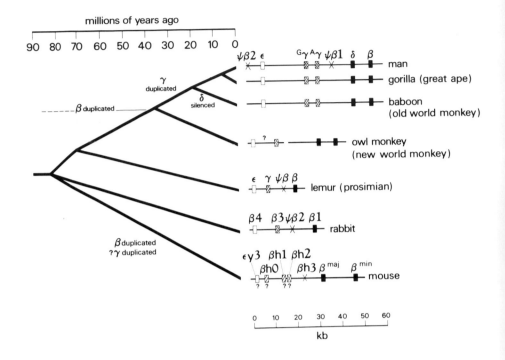

FIG. 1 *Phylogeny of the β-globin gene cluster in mammals. The arrangement of the human cluster was determined by Fritsch et al. (1980), the primate clusters by Barrie et al. (1981), the rabbit cluster by Lacy et al. (1979) and the mouse cluster by Jahn et al. (1980). These maps show genes that are probably expressed (boxes) and known pseudogenes (crosses). All genes are transcribed from left to right, and intervening sequences within these genes are not shown. ε-Related genes are shown by open boxes, γ-related genes by hatched boxes and adult β-globin genes by filled boxes (Barrie et al., 1981). Orthologies with the mouse embryonic genes have not been determined, and the assignment of εy3 to ε and the closely related βh0 and βh1 to γ is completely speculative. Mouse βh2 and βh3 are highly divergent, and βh3, at least, is a pseudogene. Genes orthologous to human ψβ1 and ψβ2 are probably present at equivalent positions in the gorilla and baboon. The linkage of owl monkey ε- and γ-globin genes has not been firmly established, and the linkage of ε,γ to δ,β is unknown. These maps are arranged phylogenetically, using divergence times cited by Dayhoff (1972), Romero-Herrera et al. (1973) and Sarich and Cronin (1977). (Reprinted from Jeffreys, 1981.) kb, × 10³ base-pairs.*

genes. At the moment, these genes, and their homologues in other vertebrates, provide the most detailed account of the evolutionary history of a multigene family.

Human globins are coded by two unlinked clusters of genes (see Efstratiadis *et al.*, 1980; Proudfoot *et al.*, 1980). The α-globin gene cluster on human chromosome 16 contains two ζ- and two α-globin genes scattered over 25×10^3 base-pairs and arranged in the order 5'-ζ2-ζ1-α2-α1-3'. All genes are orientated in the same direction and are separated by substantial tracts of intergenic DNA. The β-globin gene cluster on human chromosome 11 contains five active genes spread over 45×10^3 base-pairs and arranged 5'-ε-^{G}γ-^{A}γ-δ-β-3' (see Fig. 1). Both clusters have clearly evolved by a series of tandem globin gene duplications and have at some stage become unlinked to give the separate α- and β-globin gene clusters. After duplication, various genes must have diverged both in sequence and in developmental expression to give the current organisation of developmentally regulated genes.

Only about 8% of the DNA in these clusters is used to code globin mRNA. An additional 8% makes up the globin introns, and therefore appears in precursor mRNAs. The role of the remaining 84% of cluster DNA found between genes is a complete mystery. This intergenic DNA consists of a complex mixture of single copy DNA and elements repeated both within the cluster and elsewhere in the genome (Adams *et al.*, 1980; Baralle *et al.*, 1980; Coggins *et al.*, 1980; Fritsch *et al.*, 1980). The small homologous sequence found near the unlinked α- and β-globin genes in mouse (Leder *et al.*, 1978) might be an example of such a dispersed repetitive element. The function of virtually all of this intergenic DNA is a complete mystery. Close to genes, one can discern small conserved elements such as the TATA box considered to be the putative promoter for RNA polymerase II (see Breathnach and Chambon, 1981). Control functions further out from globin genes are tentatively suggested by the phenotypes of various deletion mutants in the human β-globin gene cluster (Fritsch *et al.*, 1979; Van Der Ploeg *et al.*, 1980). The apparent scarcity of coding sequences in the α- and β-globin gene clusters, and indeed in other mammalian gene clusters, seems to reflect the DNA excess in higher eukaryotic genomes. Indeed, Orgel and Crick (1980) and Ohno (1980) have suggested that these extensive intergenic regions might have no function, but instead represent "junk" DNA. If so, then one would predict that they should evolve rapidly (at about 1 to 2%/million years) and should show rapid changes in length as a result of deletions and regional duplications driven in particular by recombination between repetitive elements in these intergenic regions.

Both of these junk predictions seem not to be borne out for

the β-globin gene cluster. The δ- and β-globin gene arrangement in man was shown to be conserved in great apes and Old
World monkeys (Martin *et al*., 1980; Zimmer *et al*., 1980;
Jeffreys and Barrie, 1981). More remarkably, the arrangement
of the entire β-globin gene cluster was found to be indistinguishable in man, gorilla and baboon, indicating that the
arrangement of the human cluster had been fully established
20 to 40 million years ago and faithfully preserved since
(Barrie *et al*., 1981; Fig. 1). Furthermore, by comparing
restriction endonuclease cleavage maps of intergenic DNA regions in these different primate species, it is possible to
estimate rates of DNA sequence divergence in this region
(Barrie *et al*., 1981). Surprisingly, intergenic regions
evolve slowly at about 0.2% per million years (compared with
0.1%/million years for replacement site substitutions and 1
to 2% per million years predicted for junk DNA). This conservation of the arrangement and sequence of intergenic DNA
strongly suggests that these sequences have been substantially
constrained by selection, at least during recent primate evolution. This seems to be incompatible with the notion that
intergenic DNA is junk. Instead, it might be more meaningful
to regard the entire cluster (and possibly regions beyond) as
a single co-adapted functional supergene rather than an
assembly of largely autonomous globin genes loosely arranged
by evolutionary accident into a gene cluster. One physical
basis for such a large functional unit might be the existence
of extensive chromatin domains including such a region; regulation of gene activity might then be seen as a consequence
of modulating the packing conformation of such domains (Van
Der Ploeg *et al*., 1980). While such ideas account for the
evolutionary stability of cluster arrangement, they do not
readily account for the apparently strong conservation of DNA
sequence.

In contrast, New World monkeys and prosimians have radically different arrangements of the β-globin gene cluster,
including altered numbers of genes and major shifts in the
lengths of intergenic regions (Barrie *et al*., 1981; see Fig.
1). However, when two diverged species of lemur were compared, the arrangement, though not sequence, of the entire
cluster was found to have been completely preserved. The
impression one gains is that the arrangement of the β-globin
gene cluster does not evolve smoothly, but instead proceeds
in major jumps interspersed with long periods of stasis and
selective constraint. It is still not certain whether such
critical jumps have occurred and, if so, whether these bursts
of change might be related to shifts in the physiology of
erythropoiesis in primates.

Phylogenetic analyses of gene clusters, such as that

shown in Fig. 1, provide a new method for timing gene duplication events. The duplicated $\delta\beta$-globin gene seems to have arisen about 40 to 70 million years ago, consistent with the estimate of 40 million years by comparing δ- and β-globin gene sequences (Efstratiadis *et al.*, 1980). Curiously, Old World monkeys do not produce detectable δ-globin, and Martin *et al.* (1980) have suggested that the gene has recently become silenced in Old World monkeys. In sharp contrast, the γ-globin gene duplicated 20 to 40 million years ago, yet the duplicates have remained almost identical in sequence in man (Slightom *et al.*, 1980). This failure to diverge after duplication might be an example of concerted evolution (see below).

The β-globin gene cluster in the lemur (a prosimian) is the shortest so far found in mammals (Barrie *et al.*, 1981) and is similar in arrangement to the rabbit cluster (Lacy *et al.*, 1979; see Fig. 1). This suggests that a simple cluster like that of the rabbit and lemur was established at least 85 million years ago, before the radiation of the mammals. The corresponding antiquity of the ϵ- , γ- and β-globin genes is consistent with their times of divergence deduced from DNA sequences (Efstratiadis *et al.*, 1980).

The β-globin gene cluster has also been characterised in the mouse (Jahn *et al.*, 1980; see Fig. 1). A complex cluster was found, with three to four active embryonic/foetal globin genes and duplicated adult β^{maj}- and β^{min}-globin genes. Although orthologies between the mouse non-adult genes and human $\epsilon\gamma$-globin genes have not been established, it seems likely that the mouse cluster evolved from a simple ancestral cluster by β and γ duplications independent of those seen in the lineage leading to man (Fig. 1). This model readily accounts for the reversed order of major and minor adult β-globin genes in man and mouse.

By comparison, there is little information on the evolution of the α-globin gene cluster. Zimmer *et al.* (1980) compared the $\alpha2$-$\alpha1$-globin gene arrangement in man and apes, and found identical arrangements apart from the occasional small deletion or insertion.

Unlinked clusters of α- and β-globin genes also exist in the chicken (Hughes *et al.*, 1979; Engel and Dodgson, 1980). However, the β-globin gene cluster contains an embryonic gene on the 3' side of the adult gene, an arrangement not seen in mammals (Dodgson *et al.*, 1979). The relation between these genes and mammalian β-related globin genes is not known: birds and mammals diverged about 270 million years ago, perhaps before the emergence of mammalian ϵ- , γ- and β-globin genes. It is entirely possible that the avian cluster arose by an entirely independent series of duplications of a β-related globin gene.

In sharp contrast, the major adult α1- and β1-globin genes
in the amphibian X. laevis are closely linked in the order
5'-α1-β1-3' (Jeffreys et al., 1980; Patient et al., 1980).
This arrangement strongly suggests that the initial αβ-globin
gene duplication that occurred about 500 million years ago
was a tandem duplication, and that the tandem duplicates have
since remained closely linked in amphibia.   In contrast, these
genes probably became unlinked in the reptilian ancestors of
birds and mammals, perhaps about 300 million years ago.   This
unlinking could have occurred by several mechanisms: inclusion
of one or other globin gene within a transposable element,
followed by transposition; a translocation between the α- and
β-globin genes; chromosome duplication to give two unlinked
αβ clusters, followed by silencing of linked β- or α-globin
genes.   The last model would predict the possible existence
of homologous genetic functions (conserved DNA sequences)
shared by regions neighbouring the mammalian α- and β-clus-
ters; such elements should be detectable by DNA analysis.

   X. laevis possesses a second αβ-globin gene cluster that
codes for minor adult globins (Jeffreys et al., 1980).   This
cluster appears to have arisen by tetraploidisation in an
ancestor of X. laevis, and a contemporary equivalent of this
ancestor, X. tropicalis, has a single αβ cluster as expected.
Thus chromosome duplication has generated globin diversity in
X. laevis.   The importance of chromosome duplication and poly-
ploidisation in vertebrate evolution has been emphasised
repeatedly (Ohno, 1970,1973; Lalley et al., 1978; Lundin,
1979).

   Human globin gene clusters still retain the capacity for
expanding and contracting their numbers of genes by unequal
crossing over.   The fused δβ-globin polypeptide in haemoglobin
Lepore has been shown, by direct DNA analysis, to be the product of unequal crossing-over between δ- and β-globin genes
(Flavell et al., 1978).   Similarly, haemoglobin Kenya probably
results from unequal crossing-over between $^A\gamma$- and β-globin
genes (see Jones et al., 1981).   Chromosomes carrying three
α-globin genes or a single α-globin gene have been detected
in man (Embury et al., 1979; Goossens et al., 1980; Higgs et
al., 1980) and the chimpanzee (Zimmer et al., 1980).   These
three-gene and one-gene chromosomes appear to be the reciprocal products of unequal crossing-over between duplicated α-
loci.   The point of crossover need not necessarily be within
an α-globin gene, since each α-globin gene resides within an
extensive (4 × 10^3 base-pair) tandem repeat (Lauer et al.,
1980; Proudfoot and Maniatis, 1980).   Thus, the duplicated
α-locus probably arose by tandem duplication of a large region
containing an α-globin gene.   The mechanism for this duplication is unknown, although unequal crossing-over between

sequences repeated throughout globin gene clusters (Coggins
et al., 1980; Fritsch et al., 1980) could have been responsible. By this argument, the initial spacing of gene duplicates would be dictated largely by the location of dispersed
repetitive sequences relative to globin genes.

PSEUDOGENES

Recent analyses of globin gene families have revealed the
existence of additional gene sequences that have accumulated
mutations and no longer function as active globin genes.
These additional pseudogenes have all been detected by cross-
hybridisation with active globin gene sequences.

The human β-globin gene cluster contains two $\psi\beta$ sequences,
one at the 5' end of the cluster and one between the $^A\gamma$- and
δ-globin genes (Fritsch et al., 1980; Efstratiadis et al.,
1980; see Fig. 1). The sequences of these $\psi\beta$ genes have not
been published, and thus their true pseudogene status is not
clear. These $\psi\beta$ sequences are probably also present in apes
and Old World monkeys (Barrie et al., 1981). The rabbit β-
globin gene cluster contains a $\psi\beta2$ pseudogene between the $\beta3$
(γ-like) and $\beta1$(β-like) globin genes (Lacy and Maniatis,
1980; see Fig. 1). Pseudogenes are also seen (at probably
similar positions) in the β-globin gene cluster of the mouse
(Jahn et al., 1980), lemur (Barrie et al., 1981) and goat
(Cleary et al., 1980). An inactive α-globin pseudogene
($\psi\alpha1$) has been found between the embryonic and adult α-globin
genes in man (Proudfoot and Maniatis, 1980). The extraordin-
ary mouse α-globin pseudogene, which has lost both intervening
sequences (Nishioka et al., 1980; Vanin et al., 1980), has
already been described, although its linkage arrangement to
functional α-globin genes has not been reported.

Several globin pseudogenes have been sequenced completely.
The rabbit $\psi\beta2$ gene shows substantial divergence from the $\beta1$
gene, including frameshift mutations, premature termination
codons and disruption of normal intron/exon junction sequences
sufficient to render the gene incapable of coding globin (Lacy
and Maniatis, 1980). The human $\psi\alpha1$ gene contains a similar
range of abnormalities, plus an initiation codon mutation
(Proudfoot and Maniatis, 1980). The mouse pseudogene $\beta h3$
(Fig. 1) shows substantial divergence from the adult β-globin
gene only at the 5' end of the gene (Jahn et al., 1980). The
lemur $\psi\beta$ gene appears to contain the 3' end of a β-globin
gene preceded by sequences related to the 5' end of an ϵ-
globin gene, although sequence data are not available (Barrie
et al., 1981).

Globin pseudogenes can be regarded as supernumerary sequences generated by gene duplication and silenced by divergence.

There is evidence to indicate that these duplications might
have initially given rise to active globin genes, which became
inactivated later in evolution. For example, Lacy and Maniatis (1980) compared the rabbit ψβ2 and β1 globin genes and
found that replacement sites had diverged less than silent
sites, to an extent that suggested that in ψβ2 the former
sites (at least) were under selective constraint for some time
after duplication. By using the approximate rates of replacement site and silent site divergence in active globin genes,
they estimate that the duplication arose at least 50 million
years ago, and that ψβ2 was eventually silenced about 30 million years ago. In a more detailed analysis of the mouse ψα3
gene, which has lost its intervening sequences, Miyata and
Yasunaga (1981) calculate that the α-ψα3 duplication arose at
least 24 million years ago and that the ψα3 gene was silenced
17 million years ago, after which it diverged at nearly twice
the rate of silent site substitutions in active globin genes.
    This apparently rapid rate of pseudogene divergence is con-
sistent with the notion that these sequences are junk and free
from selective constraint. In this respect, it will be important to carry out a direct phylogenetic comparison of homologous pseudogenes; the ψβ1 and ψβ2 genes of man, apes and Old
World monkeys would provide ideal test cases. This rapid divergence also predicts the existence of many more globin pseudogenes that are too diverged to be detected by hybridisation.
In contrast, the conservation of intergenic DNA in the human
β-globin gene cluster, and the consistent appearance of pseudogenes between mammalian γ- and β-globin genes, might point
to some functional role. Vanin et al. (1980) have suggested
that pseudogenes might be involved in the control of gene
expression, perhaps by diverting transcription into non-
productive pathways or by encoding control RNA species.
Alternatively, pseudogenes might mimic active genes in maintaining the architecture of chromatin domains involved in the
regulation of gene activity. If pseudogenes and active genes
can interchange sequences by recombination or gene conversion
(see below), then they could be of use as generators of diversity that could ultimately appear within active genes.

CONCERTED EVOLUTION

When a gene duplicates, each duplicate locus does not necessarily diverge independently. Instead, the duplicates sometimes appear to interchange sequences by some mechanism that
maintains a close sequence homology between the two loci.
Zimmer et al. (1980) have called this process "concerted evolution".
    The human ^{G}γ- and ^{A}γ-globin genes provide a clear example.

The γ-globin gene duplication arose about 20 to 40 million years ago (Barrie *et al.*, 1981; see Fig. 1). In contrast, the $^G\gamma$- and $^A\gamma$-globin genes isolated from a single human chromosome show extreme homology, particularly over the 5' region of the genes (Slightom *et al.*, 1980). Furthermore, it is likely that a duplicated restriction endonuclease cleavage site polymorphism common to both γ-globin genes has arisen by sequence interchange between these loci (Jeffreys, 1979). Several mechanisms could permit such interchanges; all involve mispairing of $^G\gamma$- and $^A\gamma$-globin genes at meiosis. Recombination could then lead to the appearance of three-γ and one-γ chromosomes, or to the interchange of sequence blocks between the paired genes, or to correction of one sequence to the other by gene conversion (see Slightom *et al.*, 1980; Zimmer *et al.*, 1980; Jeffreys, 1981). In all cases, homogenisation of sequence differences could ensue, leading to concerted evolution at the duplicated γ-locus.

A similar phenomenon appears to be occurring at the duplicated $\alpha2$-$\alpha1$ locus. The duplication arose at least 8 million years ago (Zimmer *et al.*, 1980), yet the $\alpha2$- and $\alpha1$-globin genes are almost identical in sequence (Liebhaber *et al.*, 1981). Again, homology is most marked in a localised area of these genes, suggesting that the most recent round of "correction" was confined to only part of the α-globin gene sequence. The ability of α-globin genes to mispair and recombine at meiosis is strongly supported by the existence of three-α and one-α loci in man (see above).

Concerted evolution by gene conversion or multiple unequal crossing-over can also be invoked to account for the lack of divergence of the first intervening sequence in the mouse β^{maj}- and β^{min}-globin genes (Konkel *et al.*, 1979), the relative similarity of the 5' ends of the human β- and δ-globin genes (Efstratiadis *et al.*, 1980), the remarkable homology of 5' flanking regions near the goat γ- , β^c- and β^A-globin genes (Haynes *et al.*, 1980), and what appears to be an ϵ-β hybrid structure of the lemur $\psi\beta$ gene (Barrie *et al.*, 1981). Clearly, concerted evolution is not a rare phenomenon, and seems to occur between even relatively distantly related genes and between active genes and pseudogenes. One wonders about the level of selective constraint required to prevent this type of process from completely homogenising a globin gene cluster.

Concerted evolution is not restricted to globin genes, and has been documented in a variety of other gene families and in repetitive DNA sequences (see Dover and Coen, 1981; Jeffreys, 1981).

CONCLUDING REMARKS

The application of recombinant DNA methods to the study of
gene evolution, especially of globin genes, is rapidly producing a mass of data concerning rates and modes of gene evolution, and the types of informational interchange and selective
constraints that might operate on gene families and clusters.
While much more information is required on functional gene
sequences, the time is ripe for focusing attention on clusters
and, particularly, intergenic DNA. How much of this intergenic DNA is occupied by divergent pseudogenes? Are they
really functionless? Can one discern large conserved domains
encompassing entire clusters? Do clusters evolve in a discontinuous fashion? Do transposable elements exist in higher
eukaryotes and what influence would they have on the appearance and dispersal of multigene families?

As yet, these studies give few clues as to key events responsible for karyotypic evolution, speciation, and morphological and behavioural shifts in evolution. Yet they are important for understanding the basic rules that govern DNA evolution, and time alone should enable us to bridge the gap
between these fine structural studies of molecular evolution
and the effects of these molecular changes on phenotype and
their relation to macro-evolutionary phenomena.

REFERENCES

Adams, J.W., Kaufman, R.E., Kretschmer, P.J., Harrison, M.
     and Nienhuis, A.W. (1980). *Nucl. Acids Res.* **8**, 6113-6128.
Baralle, F.E., Shoulders, C.C., Goodbourn, S., Jeffreys, A.
     and Proudfoot, N.J. (1980). *Nucl. Acids Res.* **8**, 4393-4404.
Barrie, P.A., Jeffreys, A.J. and Scott, A.F. (1981). *J. Mol.
     Biol.* **149**, 319-336.
Bell, G.I., Pictet, R.L., Rutter, W.J., Cordell, B., Tischer,
     E. and Goodman, H.M. (1980). *Nature (London)* **284**, 26-32.
Blake, C.C.F. (1979). *Nature (London)* **277**, 598.
Breathnach, R. and Chambon, P. (1981). *Annu. Rev. Biochem.*
     50, in the press.
Cleary, M.L., Haynes, J.R., Schon, E.A. and Lingrel, J.B.
     (1980). *Nucl. Acids Res.* **8**, 4791-4802.
Cochet, M., Gannon, F., Hen, R., Maroteaux, L., Perrin, F.
     and Chambon, P. (1979). *Nature (London)* **282**, 567-574.
Coggins, L.W., Grindlay, G.J., Vass, J.K., Slater, A.A.,
     Montague, P., Stinson, M.A. and Paul, J. (1980). *Nucl.
     Acids Res.* **8**, 3319-3333.
Craik, C.S., Buchman, S.R. and Beychok, S. (1980). *Proc.
     Nat. Acad. Sci., U.S.A.* **77**, 1384-1388.
Craik, C.S., Buchman, S.R. and Beychok, S. (1981). *Nature*

(London) **291**, 87-90.

Crick, F. (1979). *Science* **204**, 264-271.

Darnell, J.E. (1978). *Science* **202**, 1257-1260.

Dayhoff, M.O. (1972). Editor of *Atlas of protein sequence and structure*, vol. 5, Natl. Biomed. Res. Found, Silver Spring, Md.

Dodgson, J.B., Strommer, J. and Engel, J.D. (1979). *Cell* **17**, 879-887.

Doolittle, W.F. (1978). *Nature (London)* **272**, 581-582.

Doolittle, W.F. and Sapienza, C. (1980). *Nature (London)* **284**, 601-603.

Dover, G. and Coen, E. (1981). *Nature (London)* **290**, 731-732.

Early, P.W., Davis, M.M., Kaback, D.B., Davidson, N. and Hood, L. (1979). *Proc. Nat. Acad. Sci., U.S.A.* **76**, 857-861.

Eaton, W.A. (1980). *Nature (London)* **284**, 183-185.

Efstratiadis, A., Posakony, J.W., Maniatis, T., Lawn, R.M., O'Connell, C., Spritz, R.A., DeRiel, J.K., Forget, B.G., Weissman, S.M., Slightom, J.L., Blechl, A.E., Smithies, O., Baralle, F.E., Shoulders, C.C. and Proudfoot, N.J. (1980). *Cell* **21**, 653-668.

Embury, S.H., Miller, J., Chan, V., Todd, D., Dozy, A.M. and Kan, Y.W. (1979). *Blood* **54** (suppl.), 53a.

Engel, J.D. and Dodgson, J.B. (1980). *Proc. Nat. Acad. Sci., U.S.A.* **77**, 2596-2600.

Flavell, R.A., Kooter, J.M., De Boer, E., Little, P.F.R. and Williamson, R. (1978). *Cell* **15**, 25-41.

Fritsch, E.F., Lawn, R.M. and Maniatis, T. (1979). *Nature (London)* **279**, 598-603.

Fritsch, E.F., Lawn, R.M. and Maniatis, T. (1980). *Cell* **19**, 959-972.

Gilbert, W. (1978). *Nature (London)* **271**, 501.

Gō, M. (1981). *Nature (London)* **291**, 90-92.

Goff, S.P., Gilboa, E., Witte, O.N. and Baltimore, D. (1980). *Cell* **22**, 777-785.

Goossens, M., Dozy, A.M., Embury, S.H., Zachariades, Z., Hadjiminas, M.G., Stamatoyannopoulos, G. and Kan, Y.T. (1980). *Proc. Nat. Acad. Sci., U.S.A.* **77**, 518-521.

Grantham, R., Gautier, C., Gouy, M., Jacobzone, M. and Mercier, R. (1981). *Nucl. Acids Res.* **9**, r43-r74.

Haynes, J.R., Rosteck, P. and Lingrel, J.B. (1980). *Proc. Nat. Acad. Sci., U.S.A.* **77**, 7127-7131.

Heilig, R., Perrin, F., Gannon, F., Mandel, J.L. and Chambon, P. (1980). *Cell* 625-637.

Higgs, D.R., Old, J.M., Pressley, L., Clegg, J.B. and Weatherall, D.J. (1980). *Nature (London)* **284**, 632-635.

Hughes, S.H., Stubblefield, E., Payvar, F., Engel, J.D., Dodgson, J.B., Spector, D., Cordell, B., Schimke, R.T. and Varmus, H.E. (1979). *Proc. Nat. Acad. Sci., U.S.A.* **76**, 1348-1352.

174   A.J. JEFFREYS

Hunt, T.L., Hurst-Calderone, S. and Dayhoff, M.O. (1978). *In* "Atlas of protein sequence and structure" vol. 5 suppl. 3,229-251.

Jahn, C.L., Hutchison, C.A., Phillips, S.J., Weaver, S., Haigwood, N.L., Voliva, C.F. and Edgell, M.H. (1980). *Cell* 21, 159-168.

Jeffreys, A.J. (1979). *Cell* 18, 1-10.

Jeffreys, A.J. (1981). *In* "Genetic Engineering". (Williamson, R., ed.), vol. 2, Academic Press, London, New York.

Jeffreys, A.J. and Barrie, P.A. (1981). *Phil. Trans. Roy. Soc. Ser. B* 292, 133-142.

Jeffreys, A.J., Wilson, V., Wood, D., Simons, J.P., Kay, R.M. and Williams, J.G. (1980). *Cell* 21, 555-564.

Jensen, E.O., Paludan, K., Hyldig-Nielsen, J.J., Jørgensen, P. and Marcker, K.A. (1981). *Nature (London)* 291, 677-679.

Jones, R.E., Bhat, S.P., Sullivan, M.A. and Piatigorsky, J. (1980). *Proc. Nat. Acad. Sci., U.S.A.* 77, 5879-5883.

Jones, R.W., Old, J.M., Trent, R.J., Clegg, J.B. and Weatherall, D.J. (1981). *Nature (London)* 291, 39-44.

Jukes, T.H. (1980). *Science* 210, 973-978.

Jukes, T.H. and King, J.L. (1979). *Nature (London)* 281, 605-606.

Kafatos, F.C., Efstratiadis, A., Forget, B.G. and Weissman, S.M. (1977). *Proc. Nat. Acad. Sci., U.S.A.* 74, 5618-5622.

Kimura, M. (1981). *Proc. Nat. Acad. Sci., U.S.A.* 78, 454-458.

Konkel, D.A., Maizel, J.V. and Leder, P. (1979). *Cell* 18, 865-873.

Lacy, E. and Maniatis, T. (1980). *Cell* 21, 545-553.

Lacy, E., Hardison, R.C., Quon, D. and Maniatis, T. (1979). *Cell* 18, 1273-1283.

Lalley, P.A., Minna, J.D. and Francke, U. (1978). *Nature (London)* 274, 160-162.

Lauer, J., Shen, C.K.J. and Maniatis, T. (1980). *Cell* 20, 119-130.

Leder, A., Miller, H.I., Hamer, D.H., Seidman, J.G., Norman, B., Sullivan, M. and Leder, P. (1978). *Proc. Nat. Acad. Sci., U.S.A.* 75, 6187-6191.

Liebhaber, S.A., Goossens, M. and Kan, Y.W. (1981). *Nature (London)* 290, 26-29.

Lomedico, P., Rosenthal, N., Efstratiadis, A., Gilbert, W., Kolodner, R. and Tizard, R. (1979). *Cell* 18, 545-558.

Lundin, L.-G. (1979). *Clin. Genet.* 16, 72-81.

Martin, S.L., Zimmer, E.A., Kan, Y.W. and Wilson, A.C. (1980). *Proc. Nat. Acad. Sci., U.S.A.* 77, 3563-3566.

Miyata, T. and Yasunaga, T. (1981). *Proc. Nat. Acad. Sci., U.S.A.* 78, 450-453.

Miyata, T., Yasunaga, T. and Nishida, T. (1980). *Proc. Nat. Acad. Sci., U.S.A.* 77, 7328-7332.

Nishioka, Y., Leder, A. and Leder, P. (1980). *Proc. Nat. Acad. Sci., U.S.A.* **77**, 2806-2809.

Ohkubo, H., Vogeli, G., Mudryj, M., Avvedimento, V.E., Sullivan, M., Pastan, I. and De Crombrugghe, B. (1980). *Proc. Nat. Acad. Sci., U.S.A.* **77**, 7059-7063.

Ohno, S. (1980). "Evolution by gene duplication". Springer-Verlag, Berlin, Heidelberg and New York.

Ohno, S. (1973). *Nature (London)* **244**, 259-262.

Ohno, S. (1980). *Rev. Brasil. Genet. III*, 2, 99-114.

Orgel, L.E. and Crick, F.H.C. (1980). *Nature (London)* **284**, 604-606.

Patient, R.K., Elkington, J.A., Kay, R.M. and Williams, J.G. (1980). *Cell* **21**, 565-573.

Perler, F., Efstratiadis, A., Lomedico, P., Gilbert, W., Kolodner, R. and Dodgson, J. (1980). *Cell* **20**, 555-566.

Proudfoot, N.J. and Maniatis, T. (1980). *Cell* **21**, 537-545.

Proudfoot, N.J., Shander, M.H.M., Manley, J.L., Gefter, M.L. and Maniatis, T. (1980). *Science* **209**, 1329-1336.

Rashin, A.A. (1981). *Nature (London)* **291**, 85-87.

Reanney, D. (1979). *Nature (London)* **277**, 598-600.

Romero-Herrera, A.E., Lehmann, H., Joysey, K.A. and Friday, A.E. (1973). *Nature (London)* **246**, 389-395.

Sakano, H., Rogers, J.H., Hüppi, K., Brack, C., Trannecker, A., Maki, R., Wall, R. and Tonegawa, S. (1979a). *Nature, (London)* **277**, 627-633.

Sakano, H., Hüppi, K., Heinrich, G. and Tonegawa, S. (1979b). *Nature (London)* **280**, 288-294.

Sarich, V.M. and Cronin, J.E. (1977). *Nature (London)* **269**, 354.

Slightom, J.L., Blechl, A.E. and Smithies, O. (1980). *Cell* **21**, 627-638.

Stein, J.P., Catterall, J.F., Kristo, P., Means, A.R. and O'Malley, B.W. (1980). *Cell* **21**, 681-687.

Van Den Berg, J., Van Ooyen, A., Mantei, N., Schambück, A., Grosveld, G., Flavell, R.A. and Weissmann, C. (1978). *Nature (London)* **276**, 37-44.

Van Der Ploeg, L.H.T., Konings, A., Oort, M., Roos, D., Bernini, L. and Flavell, R.A. (1980). *Nature (London)* **283**, 637-642.

Van Ooyen, A., Van Den Berg, J., Mantei, N. and Weissmann, C. (1979). *Science* **206**, 337-344.

Vanin, E.F., Goldberg, G.I., Tucker, P.W. and Smithies, O. (1980). *Nature (London)* **286**, 222-226.

Wahli, W., Dawid, I.B., Wyler, T., Weber, R. and Ryffel, G.U. (1980). *Cell* **20**, 107-117.

Wilson, A.C., Carlson, S.S. and White, T.J. (1977). *Annu. Rev. Biochem.* **46**, 573-639.

Zimmer, E.A., Martin, S.L., Beverley, S.M., Kan, Y.W. and

Wilson, A.C. (1980). *Proc. Nat. Acad. Sci., U.S.A.* **77**, 2158–2162.

The Sea Urchin Actin Genes, and a Speculation on the Evolutionary Significance of Small Gene Families

ERIC H. DAVIDSON, TERRY L. THOMAS,
RICHARD H. SCHELLER* and ROY J. BRITTEN

Division of Biology, California Institute of Technology, Pasadena, Calif., U.S.A.

Animal genomes contain multiple and diverse actin genes, differing in structure and in developmental expression. The organization of this gene family has been reported from studies on actin genes isolated from recombinant DNA libraries of *Dictyostelium* (Kindle and Firtel, 1978; McKeown *et al.*, McKeown and Firtel, 1981), of the nematode *Caenorhabditis* (D. Hirsh and J. Files, personal communication), of *Drosophila* (Fyrberg *et al.*, 1980,1981; Tobin *et al.*, 1980), of the sea urchin *Strongylocentrotus purpuratus* (Durica *et al.*, 1980; Schuler and Keller, 1981; Scheller *et al.*, 1981), of the chicken (C. Ordahl, personal communication), and of the rat (U. Nudel, personal communcation). At least six diverse actin proteins, some tissue-specific, are present in vertebrates (Vandekerckhove and Weber, 1978a), and several cloned vertebrate actin messenger RNAs differing in sequence content and tissue representation have also been described (Ordahl *et al.*, 1980; Cleveland *et al.*, 1980). Though knowledge of this complex gene family is still not extensive, comparisons of the diverse actin genes both within and among species have raised several fascinating evolutionary issues. In this paper we briefly review some recently discovered characteristics of the sea urchin actin genes. We then focus on some evolutionary problems that are illuminated by this interesting example of a small ubiquitous gene family.

1. *The sea urchin actin gene family*

The following abbreviated description of the actin genes of

---

*Present address: Institute of Cancer Research, 701 West 168th Street, New York, New York 10068, USA.

S. purpuratus derives mainly from the work of Scheller *et al*.
(1981), and from more recent unpublished data. Figure 1 summarizes relevant aspects of our knowledge of the 11 actin
genes so far identified. As shown there, seven of these genes
are known to be linked to other actin genes at distances of
several kilobases (kb). The remaining four genes occurred
singly on the recombinant DNA fragments on which they were
isolated. However, it is not unlikely that some of these
genes, or some of the separate actin gene clusters indicated
in Fig. 1, are also linked in the genome, but at greater distances than have yet been examined. Nor can it be excluded
that several additional actin genes exist in the DNA of *S.
purpuratus*, though we are confident that Fig. 1 includes the
majority of the actin genes in this species (Durica *et al*.,
1980; Scheller *et al*., 1981).

The actin genes are of several diverse types. As described
by Scheller *et al*. (1981), they differ in their 5' and 3'
flanking sequences, and also in their intervening sequences.
Gene *J* has four intervening sequences, while genes *A*, *B*, *C*,
F, *G*, *H* and *I* have two. The intervening sequences of genes
D, *E* and *K* are not known. Sequence data (Durica *et al*.,
1980; Scheller *et al*., 1981) show that genes *C* and *J* both
possess intervening sequences at positions 121 and 203 in the
amino acid sequence, and that the additional two intervening
sequences in gene *J* are located at positions 41 and 267 (see
Table 1). It is probable from heteroduplex and restriction
map data that genes *A*, *B*, *F*, *G*, *H* and *I* have intervening
sequences at the same positions as found in gene *C*. However,
these intervening sequences themselves are often quite divergent. Thus, when certain of the actin genes are renatured
with each other and the heteroduplexes are observed in the
electron microscope, the intervening sequences appear as
single-stranded loops separated by the double-stranded coding
regions of the repetitive genes. The intervening sequences
of other pairs of actin genes are sufficiently similar to
renature under the same conditions.

The various actin genes are also distinguished by non-
homologous sequences, several hundred nucleotides in length,
that are represented in the 3' untranslated regions of the
actin mRNA transcripts. As shown in Fig. 1, the 11 actin
genes fall into at least three distinct categories with respect to these 3' untranslated sequences. Genes *A*, *B*, *C*, *D*, *E*
and J share the most common of the 3' untranslated sequences,
which is referred to below as the 3' *C* sequence, after gene *C*,
the example for which the primary sequence has been published
(Scheller *et al*., 1981). At least two of these five genes are
active during oogenesis and in early sea urchin embryos, when
they produce a characteristic 2.2 kb actin mRNA transcript

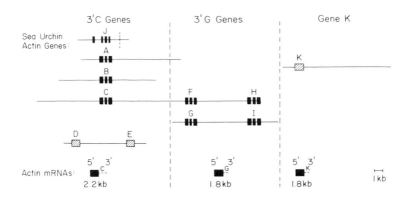

FIG. 1 *Organization and expression of different sea urchin actin gene types. Sea urchin genomic DNA sequences containing actin genes, exclusive of gene J, were isolated from a bacteriophage λ library (Scheller et al., 1981). Gene J was isolated from a subgenomic library cloned into pBR322 (Durica et al., 1980). See Scheller et al. (1981) for a description of these recombinants. Individual non-allelic actin genes were identified by restriction site analysis and are designated by capital letters. Shaded regions indicate the position of nucleotides coding for actin proteins. Intervening sequences are shown by thin lines interrupting the solid regions. The internal structure of genes D, E and K is not known. These genes are indicated by hatched boxes. Intervening sequences in gene J are located at amino acid positions 41, 121, 203 and 267. Intervening sequences in genes A to C and F to I are located at positions 121 and 203. Two different actin genes occur on the same phage λ recombinant in the cases of genes C and F, F and H, G and I, and D and E. Genes C, F and H occur on overlapping phage λ recombinants and are shown here as a single continuous 28 kb DNA element. The broken vertical lines separate the actin genes into 3 types based on the sequence homologies of the 3' untranslated regions of their respective mRNAs. These homologies were determined by sequencing of the 3' untranslated regions of different actin complementary DNA recombinants and hybridizing them with the various genomic actin genes (Scheller et al., 1981; and our unpublished results). The 3 types of 3' untranslated regions are represented in embryonic poly(A)$^+$ RNAs by 3 different transcript types. These were detected by hybridizing probes representing the 3' untranslated regions to RNA gel blots. Each type of transcript is shown with a different 3' terminus, indicated by broken lines. The dashed line in J indicates that the exact location of the last coding region is not yet known.*

(see Fig. 1).   Genes *F*, *G*, *H* and *I* share a second 3'untranslated sequence, referred to here as the 3'*G* sequence, after
gene *G*.   Beginning 13 nucleotides (nt) beyond the translation
stop codons, the 3'*C* and 3'*G* sequences become totally non-
homologous.   In RNA gel blots, a probe containing a sequence
element (from gene *I*) homologous with the 3'*G* sequence reacts
only with a 1.8 kb actin mRNA transcript found in embryos, and
not with the 2.2 kb actin mRNA.   The prevalence of 1.8 kb
actin transcripts rises very rapidly during the blastula stage
of embryogenesis (Crain *et al.*, 1981; Scheller *et al.*, 1981).
Recent measurements suggest that most of these new embryo
transcripts may derive from gene *K*.   This actin gene contains
a third 3' untranslated sequence which, though not yet determined, displays no detectable homology with either the 3'*C* or
the 3'*G* untranslated sequence in hybridization reactions.   In
contrast the 1.8 kb transcripts found in embryos are not
represented in adult cell types such as coelomocytes or intes-
tine.   The prominent actin transcripts in these cells are 2.2
kb in length and are likely to contain the 3'*C* untranslated
sequence.

The general import of these findings for our present purposes is that the actin genes used at different times in
development are distinguished by different 3' untranslated
sequences.   It is thus possible to determine which actin genes
are expressed in given stages or cell types, and where the
specific genes are located with respect to each other.   A
major future object of this analysis, as discussed below, is
to discover whether the actin genes expressed in different
states of differentiation are linked, or are located in separate regions of the sea urchin genome.

2. *Are the 3' non-coding sequences of the actin genes functional?*

Among the prominent themes of this symposium is the justifica-
tion, or lack thereof, for considering *a priori* that DNA
sequences not specifying protein sequences are essentially
functionless, that is, in physiological terms.   The tissue
and stage-specific 3' untranslated sequences of the various
actin mRNAs provide an interesting example of a set of non-
coding sequences associated with known genes.   There are at
least three different arguments to suggest that in fact these
sequences might perform some physiological role, though current knowledge is insufficient to more than hint at what this
role might be.

(a) The 3' non-coding sequences of at least some actin genes show more similarity to each other than do other parts of the same genes. For example, Scheller *et al.* (1981) showed that in heteroduplexes formed between genes *G* and *I* (see Fig. 1) both pairs of intervening sequences remain single-stranded at a 65% (v/v) formamide, room temperature criterion, whereas some regions of the 3' non-coding portions and all of the coding regions are completely duplexed. In 50% formamide at room temperature, one pair of intervening sequences in these same genes is still seen as a single-stranded loop, while all of the 3' untranslated region is now observed as a duplex. The 3' untranslated probe from gene *I* also reacts intensely with the equivalent region of gene *G* in blot hybridizations, as indicated above. Whether the rate of divergence of the intervening sequences in genes *G* and *I* is the maximum or "free substitution" rate that might be expected of sequences not under any selective pressure (Britten and Davidson, 1976) is of course unknown, since the time of the duplication event giving rise to these linked actin genes is indeterminate. However, in relative terms, the 3' untranslated sequences have clearly changed less than have the intervening sequences of the same transcription units, implying the effects of selection, and hence that the 3' non-coding sequences are indeed functionally significant.

(b) Non-coding actin 3' mRNA sequences hundreds of nucleotides long may be a ubiquitous feature of the actin gene family. In other organisms as well as sea urchin these sequences are specifically associated with particular actin mRNAs found in different cell types. Thus, in chick skeletal muscle α-actin mRNA, such a sequence has been reported (Ordahl *et al.*, 1980), and other specific 3' non-coding tail sequences are found in the messages for the β and γ "cytoplasmic" actins of chicken brain (Cleveland *et al.*, 1980). In *Drosophila* there are six actin genes specifying at least five different actin proteins, and these are also characterized by long, non-homologous 3' mRNA sequences following the termination codons (Fyrberg *et al.*, 1981; Zulauf *et al.*, 1981). Distinct 3' terminal sequences have been seen, in addition, among the 17 actin genes of *Dictyostelium* (McKeown and Firtel, 1981). The 3' non-coding mRNA sequences are not conserved over great phylogenetic distances, e.g. between chicken, human and sea urchin (Cleveland *et al.*, 1980) as are the actin coding sequences themselves. However, neither are they unique to each gene, as might be the case if they were merely the result of "readthrough" into the diverse flanking sequences. As noted above, in the sea urchin the 3'*C* and 3'*G* sequences are

subclass markers, occurring in five and four different actin
genes, respectively. Similarly, Cleveland *et al.* (1980)
reported that four different chicken actin genes share the 3'
non-coding sequence of the cytoplasmic β–actin message. The
correlation between actin gene expression and the subfamilies
defined by the specific 3' non-coding regions could be
explained in each case as the result of a complex series of
gene duplications and sequence divergences. However, it may
also mean that, in general, the 3' non-coding sequences of
the actin mRNAs are utilized in different ways in different
animal cell types.

*(c) Some evidence suggests that in sea urchin embryos the 3'
untranslated sequences may affect actin mRNA prevalence by determining mRNA turnover rate.* A direct measurement (C.V. Cab-
rera and J.J. Lee, this laboratory) indicates that the 2.2 kb
embryo transcripts that contain the 3'*C* sequence are very
stable, while the 1.8 kb transcripts, containing mainly the
3'*K* sequence, turn over with about a three-hour half-life in
the same embryos. These mRNAs may contain distinct 5' non-
coding leaders as well, however, and it is not known which
sequence element is responsible for the measured turnover rate
difference.

  There is obviously no direct evidence yet that the 3' non-
coding sequences of the actin genes are physiologically functional. However, there are reasonable grounds to suspect that
they are, and their significance is to be sought in the realm
of regulation of actin gene expression, probably at the mRNA
level.

*3. Phylogeny of the actin genes and the location of
   intervening sequences*

Table I lists the available data regarding the positions at
which actin genes are interrupted by intervening sequences in
various organisms. It is immediately evident that the actin
gene family does not display the striking conservation of
intervening sequence positions observed for several other
gene families. In the cases of the globin genes (Efstradiatis *et al.*, 1980), the ovalbumin genes and their homologues
(Royal *et al.*, 1979; Heilig *et al.*, 1980), the vitellogenin
genes (Wahli *et al.*, 1980), and the immunoglobulin genes
(Ellison *et al.*, 1981; Yamawaki-Kataoka *et al.*, 1981) for
example, the members of the gene family present in a given
genome all possess intervening sequences at exactly the same
locations. In contrast, the positions of intervening sequences vary among the *Drosophila* actin genes (Fyrberg *et al.*,

TABLE I

Positions of Intervening Sequences in Actin Genes with Coding Regions

Organism	Actin gene	Intervening sequence locations*	Reference
Yeast	There is a single actin gene	4	Ng and Abelson (1980)
Dictyostelium			
Oxytricha			
Sea urchin	Gene C, probably ≥ 6 others	None	Firtel et al. (1979)
		None	Kaine and Spear (1981)
		121, 203	Durica et al. (1980)
	Gene J	41, 121, 203, 267	Scheller et al. (1981)
Chicken	Skeletal muscle α-actin	41, 150, 203, 267, 327	C. Ordahl (personal communication)
Rat	Skeletal muscle α-actin	41, 150, 203, 267, 327	U. Nudel (personal communication)
Caenorhabditis elegans	Cytoplasmic β-actin[†] Genes I and III[†]	121, 267	D. Hirsh and J. Files (personal communication)
		63	
Drosophila	DmA6 and DmA1	307	Fyrberg et al (1981)
	DmA4	13	

*Location is indicated as the number of the actin codon within which or at the 3' boundary of which the intervening sequence occurs, using the rat skeletal muscle α-actin sequence (Elzinger and Lu, 1976; Vandekerckhove and Wever, 1978b) as the reference sequence.
[†]Other intervening sequences, whose locations are not known, may also exist in these genes.

1981), and the same is true of the sea urchin, rat and chicken actin genes. In considering the actin genes of these different organisms, we note that the amount of phylogenetic alteration in the intervening sequence positions (Table I) cannot be compared directly with any other example. No other highly conserved protein coding gene family has been studied over the same enormous phylogenetic distances, except the histone genes, which lack intervening sequences. The nearest case is the globin gene family, in which the locations of intervening sequences have remained constant at least throughout vertebrate evolution (Patient *et al*., 1980; Barrie *et al*., 1981). Thus, despite the generally accepted view that intervening sequence positions are evolutionarily stable, a different result may be obtained when genes are compared across greater biological distances. Perhaps change in intervening sequence positions between deuterostomial and protostomial lines of evolution, or between deuterostomial invertebrates and vertebrates, will turn out to be the rule rather than the exception.

On the basis of the smaller amount of data then available, Fyrberg *et al*. (1980) suggested that the "ancestral" metazoan actin gene(s) included all the intervening sequences found in more highly evolved forms, and that in the various branches of evolution these genes underwent selective deletion of different intervening sequences. According to Table I, this argument would require that the ancestral actin gene possessed at least 11 intervening sequences, if the yeast example is included. While this possibility cannot be excluded, it is not suggested by the structure of the actin genes in any of the three primitive eukaryotes included in Table I. The actin genes in two of these organisms possess no intervening sequences at all, and the third has one intervening sequence so far not found in metazoa. We propose, on the basis of the phylogenetic distribution of intervening sequence locations shown in Table I, exactly the opposite process. It seems likely to us that *insertion* of new intervening sequences has occurred in different locations in the gene at various stages of evolution, followed often by a low-level multiplication of the new actin gene form. Thus, for example, the intervening sequences at positions 121, 203 and 267 might have been inserted early in deuterstome evolution, since they occur both in echinoderm and in vertebrate actin genes, but not in *Drosophila*, a protosomial creature, or in the lower eukaryotes. The intervening sequence at position 41 of the *S. purpuratus* actin *J* gene could be an echinoderm invention, while those at position 150 and 327 of the rat muscle α-actin gene could be particular to vertebrates. No doubt, some deletion of pre-existing intervening sequences has also occurred along the way. In any case the phylogenetic rate of change in actin gene intervening sequence position

seems to have been particularly fortuitous for the examination
of broad evolutionary relationships. Further study may reveal
which invertebrate deuterostome actin genes, for example, are
most closely related to the skeletal muscle actin genes utilized in vertebrates.

Some interesting consequences follow from the proposition
that the actin gene family has evolved by a series of independent intervening sequence insertions (and deletions) followed
by gene multiplications. It would be extremely unlikely, in
this case, that the positions of the intervening sequences in
current actin genes demarcate the primitive functional domains
of the protein, or its original evolutionary building blocks.
This has been claimed for several other genes (Tonegawa *et
al.*, 1978; Craik *et al.*, 1980; Stein *et al.*, 1980; Sargent *et
al.*, 1981). Instead, the implication is that intervening
sequences may behave during evolution like mobile transposable
elements, which can insert in many if not all sequence locations, and can also delete precisely at their termini. Lomedico *et al.* (1979) have reported a clear example of a precise
intervening sequence deletion in one of the rat insulin genes,
and another such deletion is known to have occurred in a
silent mouse α-globin gene (Nishioka *et al.*, 1980). We would
predict that many other cases of both evolutionary insertion
and deletion of intervening sequences will come to light as
more genes are examined over large phylogenetic distances.

4. Location of gene family members and the evolution of novel biological properties

In *Drosophila*, the six actin genes are all located distantly
from one another in the genome (Tobin *et al.*, 1980; Fyrberg
et al., 1980,1981). As noted above, it is not known whether
the sea urchin actin genes reside in different regions of the
genome or are all present in a single cluster, though some of
them are clearly linked (Fig. 1). One obvious implication
of gene linkage is that the contiguous genes are regulated
together; that is, they are used at the same time and place
in the life-cycle. The other side of the coin is that non-
linked, distant members of the same gene family might be
expressed in different regulatory modes, at other times or in
other cell types. One pertinent example is provided by the
histone genes of the sea urchin. The 5 "early" histone genes
are utilized in pregastrular embryos and are linked in tandem
array (Cohn *et al.*, 1976; Schaffner *et al.*, 1976), while the
"late" histone genes expressed in more advanced embryos are
found isolated in other genomic regions (R. Maxson and L.
Kedes, personal communication). A second example is found in
the chorion genes of the silk moth, where several distinct

clusters of genes are known that are utilized at separate stages of choriogenesis (Jones and Kafatos, 1980; Iatrou *et al.*, 1980). The general inference of such observations is that gene activity is specified by the regions of the genome in which the members of these gene families find themselves embedded. However, it is not necessary to invoke very large extents of DNA sequence when considering the genomic information required to specify induction of gene activity. Thus, recent DNA transformation experiments have convincingly shown for several selected cases that functional regulatory sequences exist in the immediate vicinity of structural genes. Examples include the *his3* gene of yeast (Struhl, 1981), the late $\alpha_{2\mu}$ globulin gene (Kurtz, 1981), the human growth hormone gene (D. Robins, P. Seeburg and R. Axel, personal communication), the Tk gene of Herpes virus (McKnight *et al.*, 1981) and the heat shock genes of *Drosophila* (Corces *et al.*, 1981). Induction of activity depends, for each of these genes, on specific natural effectors or stimuli, and 100 to <2000 nucleotides of 5' leading DNA sequence are found to contain the necessary and sufficient genomic regulatory information to obtain a striking induction of the experimentally introduced gene. These examples show that the extent of DNA sequence needed to ensure programmed regulation of gene activity may not be very large. Other cases, such as the extensive histone gene clusters in the sea urchin, may imply that regional regulation is an essential feature of genomic organization for some genes. In any case, however, the activity of a given structural gene may be said to depend in part on its location in the genome; that is, on the genomic regulatory sequences in which it is embedded.

One approach to the problem of how novel biological structure arose in evolution is to consider changes in the regulation of structural genes involved in morphogenesis, or in determining the properties of organs and tissues. It is easy to imagine translocational events that would result in bringing such a structural gene (or genes) into action in a new regulatory context; that is, by adding it to the functional battery of genes specifying some biological structure. Either the regulatory sequences could be inserted appropriately next to the structural gene(s) in question, or the gene could itself be duplicated, translocated and inserted appropriately in a set of pre-existing regulatory sequences. Processes of the first type must be invoked for genes that remain single copy but at some point in evolution became utilized in new and additional regulatory contexts in association with other genes. This kind of evolutionary mechanism implies changes in a dispersed regulatory sequence network, as discussed in detail by Britten and Davidson (1971). The unexpectedly high incidence

of small, dispersed gene families, possibly including many
cases of genes coding for isozymes, focuses attention on
alternative possibilities, though these two classes of concept
are by no means to be regarded as mutually exclusive.

A general evolutionary hypothesis for small dispersed gene
families is that they arose, and were fixed, as family members that were created by duplication, and were transposed to
genomic regulatory modules specifying biological structure in
which the new gene could play a useful role (Scheller *et al.*,
1981; Davidson, 1981).   Such events could have occurred during
the evolution of each major taxon, since phylogenetic groups
are distinguished by the different properties of their biological structures.   To take the actin gene family as an example,
we would argue that different sets of actin genes are apparent
in different groups of organisms because in the evolution of
each group new copies of actin genes were introduced into the
regulatory units specifying the specific actin-containing morphological structures of that taxon.

A fascinating and relevant insight into the evolution of
deuterostome and protostome actins has recently become available from the sequence analyses that have been carried out on
Drosophila and *Caenorhabditis elegans* actin genes.   (For purposes of discussion we here consider the nematodes as pseudo-
coelomate animals with protostomial affinities, as indicated
by e.g. Barnes (1980)).   The sequence data reported by Fyrberg
et al. (1981) show that all the *Drosophila* actin genes code
for proteins that are very similar to the "cytoplasmic" actins
rather than to the muscle actins of mammals.   However, some of
these *Drosophila* genes code for the actin proteins actually
utilized in striated muscle.   Similarly, D. Hirsh and J. Files
(personal communication) showed that the DNA sequence of an
actin gene expressed in the body wall musculature of *Caenorhabditis* indicates that this gene codes for a protein similar
to vertebrate cytoskeletal actins rather than to vertebrate
muscle actins.   Thus, it is reasonable to suppose that in
these protostomes the actin genes functioning to produce
muscle proteins do so because they belong to a muscle onto-
genic regulatory module, rather than because the gene coding
sequence specifies a muscle and not a cytoskeletal protein.
Similar actin coding sequences are used in the deuterstomes to
produce cytoskeletal actins.   The intervening sequence distributions reviewed above indicate that new actin gene subfamilies arose during deuterostome evolution, and some of these
evidently provide actin muscle fibrils in echinoderm tissues.
It is likely (Vandekerckhove and Weber, 1978a) that some new
actin isoforms again arose during vertebrate evolution.

There is clearly insufficient comparative information on
either the sequence of actin proteins or the structure of

actin genes to draw a detailed phylogenetic tree for this gene
family. We suppose that when it can be constructed, such a
tree would resemble the taxonomic phylogenetic tree. In our
view this would be no accident, since the creation and dispersal of individual actin genes may have been part of the same
process by which the structures of major taxonomic groups
arose.

ACKNOWLEDGEMENTS

We thank Dr Norman Davidson for his critical and helpful
review of this manuscript. This study was supported by
National Institutes of Health grant GM-20927.

REFERENCES

Barnes, R.D. (1980). "Invertebrate Zoology". Saunders
   College, Philadelphia.
Barrie, P.A., Jeffreys, A.J. and Scott, A.F. (1981). Evolu-
   tion of the β-globin gene cluster in man and the primates.
   J. Mol. Biol. **149**, 319-336.
Britten, R.J. and Davidson, E.H. (1971). Repetitive and non-
   repetitive DNA sequences and a speculation on the origins
   of evolutionary novelty. *Quart. Rev. Biol.* 46, 111-138.
Britten, R.J. and Davidson, E.H. (1976). DNA sequence
   arrangement and preliminary evidence on its evolution.
   Fed. Proc. Fed. Amer. Soc. Exp. Biol. 35, 2151-2157.
Cleveland, D.W., Lopata, M.A., Macdonald, R.J., Cowan, N.J.,
   Rutter, W.J. and Kirschner, M.W. (1980). Number and evolu-
   tionary conservation of α- and β-tubulin and cytoplasmic
   β- and γ-actin genes using specific cloned cDNA probes.
   Cell 20, 95-105.
Cohn, R.H., Lowry, J.C. and Kedes, L.H. (1976). Histone genes
   of the sea urchin (*S. purpuratus*) cloned in *E. coli*: order,
   polarity, and strandedness of the five histone-coding and
   spacer regions. *Cell* **9**, 147-161.
Corces, V., Pellicer, A., Axel, R. and Meselson, M. (1981).
   Integration, transcription and control of a *Drosophila* heat
   shock gene in mouse cells. *Proc. Nat. Acad. Sci., U.S.A.*
   78, 7038-7042.
Craik, C.S., Buchman, S.R. and Beychok, S. (1980). Characterization of globin domains: heme binding to the central exon
   product. *Proc. Nat. Acad. Sci., U.S.A.* 77, 1384-1388.
Crain, W.R., Durica, D.S. and Van Doren, K. (1981). Actin
   gene expression in developing sea urchin embryos. *Mol.
   Cell. Biol.* 1, 711-720.
Davidson, E.H. (1981). Evolutionary change in genomic regulatory organization: speculations on the origins of novel

biological structure. *In* "Evolution and Development" (Bonner, J.T., ed.), Springer-Verlag, Heidelberg, in the press.

Durica, D.S., Schloss, J.A. and Crain, W.R. (1980). Organization of actin gene sequences in the sea urchin: molecular cloning of an intron-containing DNA sequence coding for a cytoplasmic actin. *Proc. Nat. Acad. Sci., U.S.A.* **77**, 5683–5687.

Efstradiatis, A., Posakony, J.W., Maniatis, T., Lawn, R.M., O'Connell, C., Spritz, R.A., DeRiel, J.K., Forget, B.G., Weissman, S.M., Slightom, J.L., Blechl, A.E., Smithies, O., O., Baralle, F.E., Shoulders, C.C. and Proudfoot, N.J. (1980). The structure and evolution of the human β-globin gene family. *Cell* **21**, 653–668.

Ellison, J., Buxbaum, J. and Hood, L. (1981). The nucleotide sequence of a human immunoglobulin $C_{\gamma 4}$ gene. *Recombinant DNA*, in the press.

Elzinger, M. and Lu, R.C. (1976). Comparative amino acid sequence studies on actins. *In* "Contractile Systems in Non-muscle Tissues" (Perry, S.V., Margreth, A. and Adelstein, R.S., (eds), pp. 29–37, North-Holland, Amsterdam.

Firtel, R.A., Timm, R., Kimmel, A.R. and McKeown, M. (1979). Unusual nucleotide sequences at the 5' end of actin genes in *Dictyostelium discoideum*. *Proc. Nat. Acad. Sci., U.S.A.* **76**, 6206–6210.

Fyrberg, E.A., Kindle, K.L., Davidson, N. and Sodja, A. (1980). The actin genes of *Drosophila*: a dispersed multigene family. *Cell* **19**, 365–378.

Fyrberg, E.A., Bond, B.J., Hershey, N.D., Mixter, K.S. and Davidson, N. (1981). The actin genes of *Drosophila*: protein coding regions are highly conserved but intron positions are not. *Cell* **24**, 107–116.

Heilig, R., Perrin, F., Gannon, F., Mandel, J.L. and Chambon, P. (1980). The ovalbumin gene family: structure of the X gene and evolution of duplicated split genes. *Cell* **20**, 625–637.

Iatrou, K., Tsitilou, S.G., Goldsmith, M.R. and Kafatos, F.C. (1980). Molecular analysis of the Gr[B] mutation in *Bombyx mori* through the use of a chorion cDNA library. *Cell* **20**, 659–669.

Jones, C.W. and Kafatos, F.C. (1980). Coordinately expressed members of two chorion multigene families are clustered, alternating and divergently orientated. *Nature (London)* **284**, 635–638.

Kaine, B.P. and Spear, B.B. (1981). Nucleotide sequence of a macronuclear gene for actin in *Oxytricha fallax*. *Nature (London)*, in the press.

Kindle, K.L. and Firtel, R.A. (1978). Identification and

analysis of *Dictyostelium* actin genes, a family of moderately repeated genes. *Cell* 15, 763–778.

Kurtz, D.T. (1981). Hormonal inducibility of rat α_{2u} globulin genes in transfected mouse cells. *Nature (London)* 291, 629–631.

Lomedico, P., Rosenthal, N., Efstratiadis, A., Gilbert, W., Kolodner, R. and Tizard, R. (1979). The structure and evolution of the two nonallelic rat preinsulin genes. *Cell* 18, 545–558.

McKeown, M. and Firtel, R.A. (1981). Evidence for subfamilies of actin genes in *Dictyostelium* as determined by comparisons of 3' end sequences. *Cell*, in the press.

McKeown, M., Taylor, W.C., Kindle, K.L., Firtel, R.A., Bender, W. and Davidson, N. (1978). Multiple, heterogeneous actin genes in *Dictyostelium*. *Cell* 15, 789–800.

McKnight, S.L., Gavis, E.R., Kingsbury, R. and Axel, R. (1981). Analysis of transcriptional regulatory signals of the HSV thymidine kinase gene: identification of an upstream control region. *Cell* 25, 385–398.

Nishioka. Y., Leder, A. and Leder, P. (1980). Unusual α-globin-like gene that has cleanly lost both globin intervening sequences. *Proc. Nat. Acad. Sci., U.S.A.* 77, 2806–2809.

Ng, R. and Abelson, J. (1980). Isolation and sequence of the gene for actin in *Sacchromyces cerevisiae*. *Proc. Nat. Acad. Sci., U.S.A.* 77, 3912–3916.

Ordahl, C.P., Tilghman, S.M., Ovitt, C., Fornwald, J. and Largen, M.T. (1980). Structure and developmental expression of the chick α-actin gene. *Nucl. Acids Res.* 8, 4988–5005.

Patient, R.K., Elkington, J.A., Kay, R.M. and Williams, J.G. (1980). Internal organization of the major adult α- and β-globin genes of *X. laevis*. *Cell* 21, 565–573.

Royal, A., Garapin, A., Cami, B., Perrin, F., Mandel, J.L., LeMeur, M., Bregegegre, F., Gannon, F., LePennec, J.P., Chambon, P. and Kourilsky, P. (1979). The ovalbumin gene region: common features in the organization of three genes expressed in chicken oviduct under hormonal control. *Nature (London)* 279, 125–132.

Sargent, T.D., Jagodzinski, L.L., Yang, M. and Bonner, J. (1981). Fine structure and evolution of the rat serum albumin gene. *Mol. Cell. Biol*, 1, 871–883.

Schaffner, W., Gross, K., Telford, J. and Birnstiel, M. (1976). Molecular analysis of the histone gene cluster in *Psammechinus miliaris*: II. The arrangement of the five histone coding and spacer sequences. *Cell* 8, 471–478.

Scheller, R.H., McAllister, L.B., Crain, W.R., Durica, D.S., Posakony, J.W., Thomas, T.L., Britten, R.J. and Davidson,

E.H. (1981). Organization and expression of multiple actin genes in the sea urchin. *Mol. Cell. Biol.* 1, 609–628.

Schuler, M.A. and Keller, E.B. (1981). The chromosomal arrangement of two linked actin genes in the sea urchin *S. purpuratus*. *Nucl. Acids Res.* **9**, 591–604.

Stein, J.P., Caterall, J.F., Kristo, P., Means, A.R. and O'Malley, B.W. (1980). Ovomucoid intervening sequences specify functional domains and generate protein polymorphism. *Cell* 21, 681–687.

Struhl, K. (1981). Deletion mapping a eukaryotic promoter. *Proc. Nat. Acad. Sci., U.S.A.* **78**, 4461–4465.

Tobin, S.L., Zulauf, E., Sanchez, F., Craig, E.A. and McCarthy, B.J. (1980). Multiple actin-related sequences in the *Drosophila melanogaster* genome. *Cell* **19**, 121–131.

Tonegawa, S., Maxam, A.M., Tizard, R., Bernard, O. and Gilbert, W. (1978). Sequence of a mouse germline gene for a variable region of an immunoglobulin light chain. *Proc. Nat. Acad. Sci., U.S.A.* **75**, 1485–1489.

Vanderkerckhove, J. and Weber, K. (1978a). At least six different actins are expressed in a higher mammal: an analysis based on the amino acid sequence of the amino-terminal tryptic peptide. *J. Mol. Biol.* **126**, 783–802.

Vandekerckhove, J. and Weber, K. (1978b). Actin amino-acid sequences. Comparison of actins from calf thymus, bovine brain, and SV40-transformed mouse 3TC cells with rabbit skeletal muscle actin. *Eur. J. Biochem.* **90**, 451–462.

Wahli, W., Dawid, I.B., Wyler, T., Weber, R. and Ryffel, G.U. (1980). Comparative analysis of the structural organization of two closely related vitellogenin genes in *X. laevis*. *Cell* **20**, 107–117.

Yamawaki-Kataoka, Y., Miyata, T. and Honjo, T. (1981). The complete nucleotide sequence of mouse immunoglobulin γ2a gene and evolution of heavy chain genes: further evidence for intervening sequence mediated gene transfer. *Nucl. Acids Res.* **9**, 1365–1381.

Zulauf, E., Sanchez, F., Tobin, S.L., Rdest, U. and McCarthy, B.J. (1981). Developmental expression of a *Drosophila* actin gene encoding actin I. *Nature (London)* **292**, 556–558.

On the Generation of Antibody Diversity and On Computer Aided Analysis of V Kappa Gene Sequences

H.G. ZACHAU, J. HÖCHTL, P.S. NEUMAIER, M. PECH
and H. SCHNELL

Institut für Physiologische Chemie, Physikalische Biochemie und Zellbiologie der Universität München, Goethestrasse 33, 8000 München 2, GFR

The generation of antibody diversity has been a central problem in immunology for many years (for summaries, see Tonegawa *et al.*, 1977; Seidman *et al.*, 1978; Robertson, 1981). The antibody repertoire has been estimated to exceed 10^6 different immunoglobulin molecules (Williamson, 1976); several mechanisms contribute to its generation. If the combinatorial association of heavy and light chains takes place without restrictions, the immunocompetent cell population should have the potential to produce roughly 10^3 different heavy chains and 10^3 different light chains. There is no need, however, for the existence of that many variable (V) genes in the germ line, since there are other elements of generation of diversity. Somatic recombination occurs between germ line V gene segments and J gene segments (Sakano *et al.*, 1979; Max *et al.*, 1979) and, in heavy chain genes, between V, D and J gene segments (Early *et al.*, 1980). Antibody diversity is further generated by variation in the exact site of V-J joining (Sakano *et al.*, 1979; Max *et al.*, 1979; Rudikoff *et al.*, 1980). A schematic presentation of the rearrangement leading to kappa light chain genes is shown in Fig. 1.

If one does not invoke somatic point mutations as an additional mechanism of generation of diversity, a few hundred germ line V gene segments would be required for both heavy and light chains. Two lines of evidence indicate that in mouse such numbers of V gene segments do exist: classification of immunoglobulin sequences in groups and subgroups (Potter, 1977) and hybridization experiments with V-gene containing fragments (Seidman *et al.*, 1978; Rabbitts *et al.*, 1980a,b; Steinmetz *et al.*, 1980). One estimates that there are about 50 groups of protein sequences. In blot

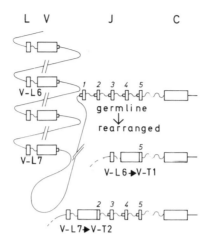

FIG. 1 *Schematic presentation of the rearrangement of V gene segments. For designations, see the text and Fig. 2.*

hybridization experiments with germ line DNA digests, on the other hand, V gene probes were found to hybridize with several, for instance six, V gene segments each. This leads to the definition of groups of V genes according to cross-hybridization. The resulting number of $50 \times 6 = 300$ germ line V_K gene segments (Valbuena *et al.*, 1978; similarly Seidman *et al.*, (1978) for light chains is, of course, only a rough estimate. The number of protein sequence groups is an extrapolation from the number of groups hitherto identified. It is still an assumption that the groups defined by protein sequences correspond to the groups defined by DNA hybridization experiments. Moreover, the size of the groups, as defined by the latter type of experiment, has been determined in only a small number of cases. Cross-hybridization with V gene segments from other groups is likely. It should also be mentioned that DNA hybridization cannot distinguish between potentially functional gene segments and pseudo-genes as they have been found in several gene systems including the immunoglobulin gene system (e.g. see Bentley and Rabbitts, 1980). These reservations concerning the above-mentioned hypothetical number of germ line V gene segments have been listed to show that the actual number may be quite a bit larger or smaller. It is generally accepted that the germ line repertoire of V gene segments and their combinatorial joining are principal sources of antibody diversity. The exact size of the germ line repertoire is not known, however; and, with that, it is still an open question whether somatic point mutations have to contribute significantly to the generation of antibody diversity.

An answer to some aspects of this question should come from sequence comparisons between rearranged V genes and their germ line counterparts. Are they identical or have somatic mutations occurred in the process of rearrangement? In P. Leder's laboratory, one rearranged mouse V_κ gene segment has been found to be identical in sequence to its germ line counterpart (Seidman et al., 1979). In another pair of germ line and rearranged V_κ gene segments, the situation is similar (Max et al., 1980) except that a number of sequence assignments have been left uncertain. In the genes for a λ_I light chain, on the other hand, and in the surrounding regions three bases were found to be different from the germ line sequence (Bernard et al., 1978). A number of sequence differences was also found between a rearranged heavy chain V gene segment and its germ line counterpart (Sakano et al., 1980); differences were also found in the adjacent sequences. Some sequence differences were also observed in another pair of V_H gene segments (Early et al., 1980) but there it was not fully established that the germ line sequence was the counterpart of the rearranged V_H gene segment.

A recent publication from our laboratory (Pech et al., 1981) is pertinent to the question of somatic mutations. The sequence data should not be shown here but discussion of a few general points seems appropriate. The variable regions of two rearranged κ light chain genes from the same mouse tumor, called myeloma T (Steinmetz and Zachau, 1980), have been sequenced (Altenburger et al., 1980). The two gene segments that had been generated by joining to the J5 and J2 gene segments were designated V-T1 and V-T2, respectively (Steinmetz and Zachau, 1980). While V-T1 represents the expressed allele, V-T2 is non-functional because of a one-nucleotide deletion at the V-J junction. Presumptive germ line counterparts were cloned from liver DNA of BALB/c mice (Steinmetz et al., 1980; Höchtl, 1980) and were designated V-L6 and V-L7, respectively. Other fragments hybridizing with V-T1 or V-T2 can be excluded because of their weak hybridization signals and, in some cases, also because of their restriction patterns (Steinmetz et al., 1980; Höchtl, 1980). The general picture is shown in Fig. 1, and a summary of the cloned fragments is given in Fig. 2.

The presumptive germ line gene segments V-L6 and V-L7 and the adjacent regions were sequenced (Pech et al., 1981). It turned out that both differed from their respective rearranged gene segments, V-T1 and V-T2, in six positions, while the adjacent regions were identical in the germ line and the rearranged genes. The results are summarized in Fig. 3. Apparently, somatic mutations had occurred in the process of rearrangement. Some of the sequence differences are located in

H.G. ZACHAU *ET AL.*

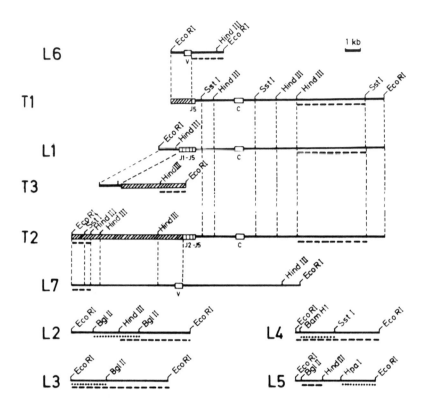

FIG. 2 *Comparison of cloned kappa light chain gene segments and their flanking regions from tumor T DNA (fragments T1 to T3) with the germ line J and C_K gene segments (fragment L1) and with some germ line V gene segments (fragments L2 to L7). The V genes V-L6 and V-L7 are the germ line counterparts of V-T1 and V-T2, whereas L2 to L5 contain V genes cross-hybridizing with V-T1. Dotted lines, fragments containing a V gene; broken lines, fragments containing middle repetitive sequences. DNA segments recombined to a J gene segment or the flank of a J gene segment in tumour T are hatched (taken from Meitinger, 1981); kb, 10^3 base-pairs.*

hypervariable regions, others are in framework sequences (Fig. 4).

 The conclusion that somatic point mutations had occurred rests of course, on the argument that V–L6 and V–L7 are the

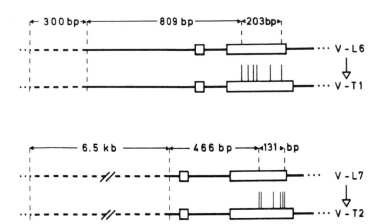

FIG. 3 *Schematic presentation of sequence differences (vertical lines) between V-L6 and V-T1 and between V-L7 and V-T2. Regions represented by unbroken lines and boxes (L and V) have been sequenced. Broken lines show regions identical in restriction mapping (Pech et al., 1981); bp, base-pairs; kb, 10^3 base-pairs.*

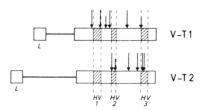

FIG. 4 *Distribution of somatic mutation (vertical arrows) within the V gene segments V-T1 and V-T2. The hypervariable regions are hatched.*

true germ line counterparts of V-T1 and V-T2, respectively. We cannot exclude a number of alternative but, as we think, unlikely possibilities for the explanation of the sequence differences: pseudo-allelism (Hilschmann *et al.*, 1978), residual allele heterogeneity or polymorphism among individual BALB/c mice (Early *et al.*, 1980; Bernard *et al*., 1978) or even the possibility that myeloma T, although it is propagated on BALB/c mice, is derived from another mouse strain. These possibilities have been discussed at some length (Pech *et al.*, 1981). But considering our blot hybridizations, restriction mapping experiments, the sequencing of several independent clones, etc., we concluded that it is the simplest and rather likely interpretation that we are dealing actually with the rearranged and non-rearranged counterparts.

The differences between the germ line and the rearranged V_K genes could have been generated by a somatic mutation mechanism acting preferentially on the coding sequences (perhaps concomitantly with the V-J rearrangement and possibly employing a recombination repair mechanism, as first proposed by Brenner and Milstein, 1966) or, alternatively, by the absence in those sequences of an elsewhere operating surveillance process. That such a mechanism is not acting on all V_K genes is clear from the work of P. Leder's group, who found the rearranged V_K gene sequences of MOPC-41 (Seidman *et al.*, 1979) and MOPC-173B (Max *et al.*, 1980) to be identical with germ line sequences. Our results add to the known instances of somatic mutations investigated at the DNA level (Early *et al.*, 1980; Bernard *et al.*, 1978; Sakano *et al.*, 1980) and at the protein level (Gearhart *et al.*, 1981), a clear case which, because of the clustering of several sequence differences in a small region and the absence of any differences in the adjoining regions, argues in favor of a localized mutation mechanism. Further work is needed to show to what extent such a mutation mechanism contributes to the generation of a functional antibody repertoire.

While a few years ago most discussions on the generation of antibody diversity were on the basis of protein sequences (see, for instance, the Cold Spring Harbor Symposium, 1976), more recently the nucleotide sequences of the gene segments and their flanking regions became available and turned out to be very informative. The use of computers facilitates the analysis of the accumulating nucleotide sequence data. The programs include comparisons of sequence homologies (Hieter *et al.*, 1980; Steinmetz *et al.*, 1981), the analysis of codon usage (Grantham *et al.*, 1980), of dinucleotide frequencies (Altenburger *et al.*, 1981), and of the base distribution along the sequence. Two examples of our work will be mentioned.

A dot matrix comparison (Hieter *et al.*, 1980) uses the axes of a co-ordinate system to represent the sequences to be compared. It generates dots at the co-ordinates that correspond to the positions of homologies in the sequences. Thus, large homologies will be visible as diagonal lines.

FIG. 5 *Dot matrix comparisons for homologies between the V gene regions of (a) MOPC-173B and L6, (b) L7 and L6. Coding regions are boxed (L, leader segment; V, variable region gene segment). An homology of 4 nucleotides scores a dot. In (a) a shift in the broken diagonal line indicates a deletion in the intron of L6 or an insertion in MOPC-173B.*

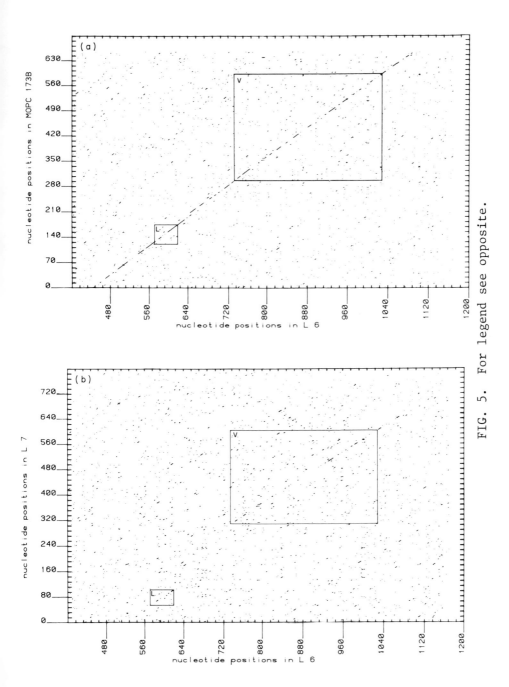

FIG. 5. For legend see opposite.

Interruptions and parallel shifts of the lines are due to non-
homologous parts and insertions or deletions in the sequences,
respectively. In order to diminish the background, it is necessary to define a minimum degree of homology needed for a
score.

In Fig. 5(a) an example of two closely related sequences is
shown and in Fig. 5(b) one in which there is a low level of
homology that is restricted to the framework sequences. The
homologies in the flanking and coding regions of the MOPC-173B
and L6 sequences are obvious, although the respective kappa
chain proteins belong to different groups. There is a 65.8%
homology at the protein level. The sequence homologies extend
also into the introns and hypervariable regions (Fig. 5(a)).
Sequence homologies between L7 and L6, on the other hand, are
hardly discernible within the background of dots. The amino
acid sequence homology of the proteins coded for by these two
gene segments is 47.9%. Compared to the former case, the
homology between L7 and L6 is especially low in the introns
and the flanking regions (Fig. 5(b)). Such sequence

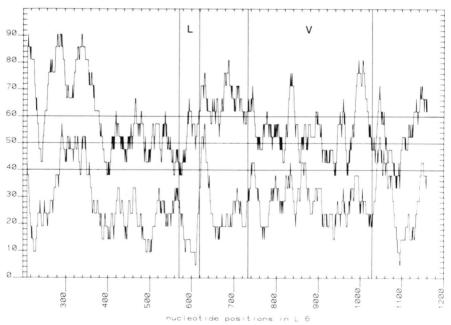

FIG. 6 *Base distribution analysis of a part of the L6 sequence. The* y *axis represents the A+T (upper graph) and A
(lower graph) percentages of 21-nucleotide long segments cen-
tered around every base position (represented by the* x *axis).
Vertical lines show the borders of the leader and the variable region coding segments.*

comparisons are pertinent to the classification of immunoglobulin genes and also to the discussion of their evolution.

Base distribution analyses show the frequencies of certain
bases (e.g. A or A+T) along a given sequence, and the transitions between the regions of different base abundance. They
were used to discriminate A+T-rich from G+C-rich regions
(Török and Karch, 1980). When used to compare the base distribution patterns of functionally related sequences, they can
help to find common features that may, but need not, reflect
sequence similarities.

A comparison of the base distribution diagrams (see Fig. 6
for an example) of seven variable gene regions (L6, L7 (Altenburger *et al.*, 1980), MOPC-41 (Seidman *et al.*, 1979), MOPC-
173B (Max *et al.*, 1980), K2 (Nishioka and Leder, 1980), HK 101
and HK 102 (Bentley and Rabbitts, 1980)) allows the following
conclusions. In all seven leader sequences analyzed, the A
content falls to values of less than 5%, while the T frequency
increases. The A deficiency in the codons -14 to -6 cannot be
explained by the amino acid composition alone; it is due, in
part, to a discrimination against A in the third position of
the codons; an A residue appears in only two of the 47 positions where it could occur according to the amino acid desig-
nation. There may be a selection pressure at the RNA level
influencing the third positions. Immediately downstream of
the leader sequence, the relative A content rises steeply
(except in L7) to between 42 and 62%. This can be discussed
in terms of an hypothesis proposed by Bina *et al.* (1980) on
RNA splicing. In all cases, the introns between the L and V
gene segments show two peaks in the A+T graphs; the second
peak near the end of the intron is broad and/or split. A
further peak in the A+T content is found about 15 nucleotides
in front of the third hypervariable region.

The high A+T content 200 to 300 base-pairs upstream of the
leader segment (Fig. 6; similarly in those V gene regions that
have been sequenced that far upstream) may be of functional
significance in RNA polymerase initiation.

In conclusion, it may be said that the problem of the
generation of antibody diversity, which was the subject of
heated discussions in the sixties and seventies, is not solved
yet. But great progress has been made and the problem can be
discussed now on a more rational basis. Among several approaches, DNA sequence comparisons of germ line and rearranged
gene segments (with and without the help of a computer) will
contribute to the understanding of the underlying mechanisms.

ACKNOWLEDGMENTS

Our work mentioned in this article was supported by Bundes-
ministerium für Forschung und Technologie.

REFERENCES

Altenburger, W., Steinmetz, M. and Zachau, H.G. (1980). Functional and non-functional joining in immunoglobulin light chain genes of a mouse myeloma. *Nature (London)* **287**, 603-607.

Altenburger, W., Neumaier, P.S., Steinmetz, M. and Zachau, H.G. (1981). DNA sequence of the constant gene region of the mouse immunoglobulin kappa chain. *Nucl. Acids Res.* **9**, 971-981.

Bentley, D.L. and Rabbitts, T.H. (1980). Human immunoglobulin variable region genes - DNA sequences of two V_K genes and a pseudogene. *Nature (London)* **288**, 730-733.

Bina, M., Feldmann, R.J. and Deeley, R.G. (1980). Could poly (A) align the splicing sites of messenger RNA precursors? *Proc. Nat. Acad. Sci., U.S.A.* **77**, 1278-1282.

Bernard, O., Hozumi, N. and Tonegawa, S. (1978). Sequences of mouse immunoglobulin light chain genes before and after somatic changes. *Cell* **15**, 1133-1144.

Brenner, S. and Milstein, C. (1966). Origin of antibody variation. *Nature (London)* **211**, 242-243.

Early, P., Huang, H., Davis, M., Calame, K. and Hood, L. (1980). An immunoglobulin heavy chain variable region gene is generated from three segments of DNA: V, D and J. *Cell* **19**, 981-992.

Gearhart, P.J., Johnson, N.D., Douglas, R. and Hood L. (1981). IgG antibodies to phosphorylcholine exhibit more diversity than their IgM counterparts. *Nature (London)* **291**, 29-34.

Grantham, R., Gautier, C. and Gouy, M. (1980). Codon frequencies in 119 individual genes confirm consistent choices of degenerate bases according to genome type. *Nucl. Acids Res.* **8**, 1893-1912.

Hieter, P.A., Max, E.E., Seidman, J.G., Maizel, J.V. Jr and Leder, P. (1980). Cloned human and mouse kappa immunoglobulin constant and J region genes conserve homology in functional segments. *Cell* **22**, 197-207.

Hilschmann, N., Barnikol, H.U., Kratzin, H., Altevogt, P., Engelhard, M. and Barnikol-Watanabe, S. (1978). Genetic determination of antibody specificity. *Naturwissenschaften* **65**, 616-639.

Höchtl, J. (1980). Isolierung und Charakterisierung von V-Gensegmenten aus Leber-DNA von Balb/c-Mäusen und Charakterisierung der Umlagerung eines V-Gensegmentes in Myelom T. Diploma work, Universität München.

Max, E.E., Seidman, J.G. and Leder, P. (1979). Sequences of five potential recombination sites encoded close to an immunoglobulin κ constant region gene. *Proc. Nat. Acad. Sci., U.S.A.* **76**, 3450-3454.

Max, E.E., Seidman, J.G., Miller, H. and Leder, P. (1980),
    Variation in the crossover point of kappa immunoglobulin
    gene V-J recombination: Evidence from a cryptic gene. *Cell*
    21, 793–799.

Meitinger, T. (1981). Charakterisierung repetitiver DNA in
    der Nähe von Immunglobulin-Gensegmenten der Maus. Diploma
    work, Universität München.

Nishioka, Y. and Leder, P. (1980). Organization and complete
    sequence of identical embryonic and plasmacytoma κ V-region
    genes. *J. Biol. Chem.* 255, 3691–3694.

Pech, M., Höchtl, J., Schnell, H. and Zachau, H.G. (1981).
    Differences between germ line and rearranged immunoglobulin
    V_K coding sequences suggest a localized mutation mechanism.
    Nature (London). 291, 668–670.

Potter, M. (1977). Antigen-binding myeloma proteins of mice.
    Advan. Immunol. 25, 141–211.

Rabbitts, T.H., Matthyssens, G. and Hamlyn, P.H. (1980a).
    Contribution of immunoglobulin heavy-chain variable-region
    genes to antibody diversity. *Nature (London)* 284, 238–243.

Rabbitts, T.H., Hamlyn, P.H., Matthyssens, G. and Roe, B.A.
    (1980b). The variability, arrangement, and rearrangement
    of immunoglobulin genes. *Canad. J. Biochem.* 58, 176–187.

Robertson, M. (1981). Genes of lymphocytes I: Diverse means
    to antibody diversity. *Nature (London)* 290, 625–627.

Rudikoff, S., Rao, D.N., Glaudemans, C.P.J. and Potter, M.
    (1980). κ chain joining segments and structural diversity
    of antibody combining site. *Proc. Nat. Acad. Sci., U.S.A.*
    77, 4270–4274.

Sakano, H., Hüppi, K., Heinrich, G. and Tonegawa, S. (1979).
    Sequences at the somatic recombination sites of immuno-
    globulin light-chain genes. *Nature (London)* 280, 288–294.

Sakano, H., Maki, R., Kurosawa, Y., Roeder, W. and Tonegawa,
    S. (1980). Two types of somatic recombination are neces-
    sary for the generation of complete immunoglobulin heavy-
    chain genes. *Nature (London)* 286, 676–683.

Seidman, J.G., Leder, A., Nau, M., Norman, B. and Leder, P.
    (1978). Antibody diversity. *Science* 202, 11–17.

Seidman, J.G., Max, E.E. and Leder, P. (1979). A κ-immuno-
    globulin gene is formed by site-specific recombination
    without further somatic mutation. *Nature (London)* 280,
    370–375.

Steinmetz, M. and Zachau, H.G. (1980). Two rearranged immuno-
    globulin kappa light chain genes in one mouse myeloma.
    Nucl. Acids. Res. 8, 1693–1707.

Steinmetz, M., Höchtl, J., Schnell, H., Gebhard, W. and
    Zachau, H.G. (1980). Cloning of V region fragments from
    mouse liver DNA and localization of repetitive DNA sequen-
    ces in the vicinity of immunoglobulin gene segments. *Nucl.*

Acids Res. 8, 1721-1729.

Steinmetz, M., Frelinger, J.G., Fisher, D., Hunkapiller, T., Pereira, D., Weissman, S.M., Uehara, H., Nathenson, S. and Hood, L. (1981). Three cDNA clones encoding mouse transplantation antigens: Homology to immunoglobulin genes. *Cell* 24, 125-134.

Tonegawa, S., Hozumi, N., Brack, C. and Schuller, R. (1977). Arrangement and rearrangement of immunoglobulin genes. *In* "Immune System: Genetics and regulation" (Sercarz, E.E., Herzenberg, L.A. and Fox, C.F., eds), pp. 43-55, Academic Press, New York.

Török, I. and Karch, F. (1980). Nucleotide sequences of heat shock activated genes in *Drosophila melanogaster*. I Sequences in the regions of the 5' and 3' end of the *hsp* 70 gene in the hybrid plasmid 56HB. *Nucl. Acids Res.* 8, 3105-3123.

Valbuena, O., Marcu, K.B., Weigert, M. and Perry, R.P. (1978). Mouse κ chains with implications for the generation of immunoglobulin diversity. *Nature (London)* 276, 780-784.

Williamson, A.R. (1976). The biological origin of antibody diversity. *Ann. Rev. Biochem.* 45, 467-500.

Human Antibody Genes: Evolutionary Comparisons as a Guide to Function and the Mechanisms of DNA Rearrangement

T.H. RABBITTS, D.L. BENTLEY, A. FORSTER,
CELIA P. MILSTEIN* and G. MATTHYSSENS†

*Medical Research Council Laboratory of Molecular Biology,
Cambridge, England*

Antibody molecules are composed of pairs of heavy (H) and light (L) chains, which are the products of three unlinked sets of autosomal genes. There are two types of L chains (κ or λ chains), which are defined by distinct amino acid sequences within the carboxy-terminal half of the molecule (the constant or C region). The amino-terminal half of the molecule (approximately 100 amino acids) has a variable amino acid sequence (the so-called variable or V region), which forms the antigen binding site. The H chains are also made up of V and C regions; the V_H region is approximately the same size as the V_L region but the C_H region is larger, comprising three or four regions of homology called domains. Amino acid sequence data of this C region has defined classes (μ, δ, γ, ε and α) and subclasses (γ_1, γ_2, γ_3, γ_4, α_1 and α_2) of H chains. Immunoglobulins are designated according to the class of H chain they carry: IgM(μ), IgD(δ), IgG(γ), IgE(ε) and IgA(α).

Studies of the cellular expression of the immunoglobulin molecules have revealed complex patterns of dual and sequential production of different types of immunoglobulins. The first chains produced by the pre-B lymphocytes are intracellular μ chains (comprising a V_H segment and the $C\mu$ segment), which subsequently appear at the cell surface in conjunction with L chains as IgM. These cells later differentiate to

---

*Permanent address: Institute of Animal Physiology, Babraham, Cambridge, England.
†Present address: Free University of Brussels, Institute of Molecular Biology, B-1640 St Genesius-Rhode, Belgium.

co-produce IgM and IgD: a particularly important feature of this co-expression is that the V_H region joined to both C_H regions is identical within one cell. A further differentiation step allows cells to switch from IgM + IgD to IgG, IgA or IgE production (the H chain class switch) and, again, the same V_H region sequence is apparently switched from between C_H classes. Finally, cells can change from the production of surface to secreted antibody.

The expression of the genes for these immunoglobulins clearly represents one of the most versatile systems thus far discovered, and this versatility is achieved by procedures that involve both rearrangement and deletion of chromosomal segments as well as alternative RNA splicing patterns.

It was first proposed by Dreyer and Bennett (1965) that the DNA sequences encoding the V and C regions were separate in the genome, and that the DNA rearrangement in lymphocytes fused the V and C region genes to form the active antibody gene. Subsequent molecular studies of the immunoglobulin genes of mice have established broadly the validity of this hypothesis. In fact, the situation is even more complex. In the light chains there are three DNA segments involved, called V, J and C: V gene integration being the joining of V and J segments (which then remain separate from the C_K gene by a large intervening sequence (IVS)). In heavy chains, a further segment (D) is involved in V gene integration.

Recently, we have begun to study the antibody gene system in man, which has allowed an evolutionary comparison of sequences that are functionally important in the processes of expression of these genes. In this paper, we describe the results from some of these studies (for reviews of much of the work on mouse and human genes, see *Immunological Reviews* (1981) vol. 59).

HUMAN V GENES

We have studied in detail the structure of human V genes of the V_K and V_H types. Figure 1 is a diagram of the structural features of an example of each type of gene. The arrangement of the sequences parallels the analogous genes in mouse. At the 5' end of both human V genes, we observed a region coding for a predicted signal peptide (19 codons for the V_H3 and 22 codons for the V_K1 genes). As with mouse V genes, an intervening sequence (around 100 residues) splits this signal peptide between codons for residues -5 and -4. The remainder of the V-gene segment is uninterrupted, and codes for framework (FR) and complementarity determining regions (CDR) of the protein in the order FR1, CDR1, FR2, CDR2 and FR3. An important difference exists between V_H and V_L gene segments in both

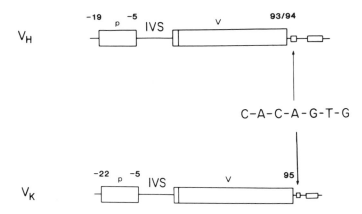

FIG. 1 *The structure of isolated human V genes from foetal liver DNA. P, precursor or leader sequence; IVS, intervening sequences; V, V-gene coding segment.*

human and mouse with respect to the CDR3. In the $V_\kappa 1$ genes of man, almost all of CDR3 occurs in the V segment itself. The last codon of the gene is for proline 95, whilst the remainder of the V region is encoded with one of four separate J segments (Hieter *et al.*, 1980). This parallels the situation in mouse DNA and implies that, as in mouse, sequence variability within CDR3 of human $V_\kappa 1$ genes (in addition to combinatorial joining of different V genes with four J segments) occurs at residue 96 by a misalignment mechanism of V − J joining (Max *et al.*, 1979; Sakano *et al.*, 1979). In human $V_H 3$ genes, CDR3 occurs almost entirely outside of the V segment, the final codon of the V gene being for amino acid 93/94. The 5'-terminal codon of the Cμ gene (the first C_H gene to be expressed) is that for glycine 114. So in the human DNA, the joining segments must encode the missing residues. It has been shown in mouse that the V_H region is created by the joining of a V segment to D and J segments (Early *et al.*, 1980a,b). The results of the human H chain genes implies a similar situation, in which variability is achieved by V − D − J joining and any associated misalignments that might occur.

What sequence homology can we observe near the ends of the V segments that might give clues to the nature of the V gene integration process? It was noted in mouse V_L genes that, immediately downstream of the coding region, two boxes of sequences occur (consensus C-A-C-A-G-T-G and A-C-A-A-A-A-A-C-C), and that similar sequences occur in an inverted orientation upstream of the JL segments (Max *et al.*, 1979; Sakano *et al.*, 1979). It was postulated that these sets of sequences could base-pair with one another to form the stem of a loop,

TABLE I

Sequences Adjacent to the 3' end of Human V_H and V_K clones

$V_H 26$	C–A–C–A–G–T–G	A–G–G–A–A–G–T–C–A–T–G–T–C–A–G–C–C–C–A–G	A–C–A–C–A–A–A–C–C
$V_H 52$	C–A–C–A–G–T–G	A–G–G–A–A–G–T–C–A–G–T–G–T–G–A–G–C–C–C–A–G	A–C–A–C–A–A–A–C–C
$V_H 32$	C–A–C–A–G–T–G	A–G–G–A–A–G–T–C–A–G–T–G–A–G–A ------	
$H_K 101$	C–A–C–A–G–T–G	T–T–A–C–A–C–A–C–C–C–A–A	A–C–A–T–A–A–A–C–C
$H_K 102$	C–A–C–A–G–T–G	T–T–A–C–A–C–A–C–C–C–G–A	A–C–A–T–A–A–A–C–C

V_H clones taken from Matthyssens and Rabbitts (1980 b) and H_K clones from Bentley and Rabbitts (1980)

Sequences within boxed areas are homology regions thought to be involved in V gene joining, and the consensus sequences of homology boxes from H and L chains of human and mouse are C–A–C–A–G–T–G and A–C–A–A–A–A–C–C

which is subsequently excised from the chromosome with the consequence that V and J segments become continuous. When we examined the human V_K and V_H genes, we observed very similar sequences in the homologous downstream positions. Table I shows the sequences adjacent to the 3' end of 2 human V_K and 3 V_H genes. The sequence C-A-C-A-G-T-G is fully conserved between human and mouse in all cases, and the second boxed sequence shows a maximum of one base change from the consensus. This conservation of these sequences between species and between L and H chains is itself highly suggestive evidence that they are functionally important in the process of V - J joining. It is also clear that some conservation of the sequences between and beyond the boxes occurs. The human HK101 and mouse K2 sequences are 74% homologous over a 90-base region immediately downstream of the V_K gene (Bentley and Rabbitts, 1980). Two human V_K sequences differ from each other by only one residue in the 11 positions between the boxes, whilst a mouse V_K sequence differs at four of the 11 residues. However, the sequence of human V_H26, for example, differs at only two residues within the whole region, shown in Table I compared to V_H107, a mouse V_H counterpart (Early *et al.*, 1980a). On the basis of such comparisons, it seems likely that the V gene integration process is indeed facilitated by the conserved sequences bordering V segments: this joining represents one type of DNA rearrangement process utilised by B-cells.

In the course of studies on the isolated human V_K genes, we observed that not all detectable "V_K" sequences were capable of encoding functional V region polypeptides (Bentley and Rabbitts, 1980). The non-functional V_K genes represent pseudo-genes analogous to those found in the globin gene system. The sequence of a V_K pseudo-gene is shown in Fig. 2. This gene was detected by a V_K gene probe and the sequence is recognisable as V_K, except that small insertions and deletions have created a reading frame shift resulting in the occurrence of in-phase chain termination codons, which would prematurely curtail any translation product of the gene. In addition, this pseudo-gene has a G·C triplet at the 5' splicing site of the short intervening sequence; since G·T is invariably found at this site (Lerner *et al.*, 1980; Breathnach *et al.*, 1978), the G·C probably represents a defective splicing signal. The pseudo-gene may in fact also be defective in V - J joining, since the putative V - J joining signals previously discussed are quite divergent from the consensus in the second homology box. The pseudo-gene is apparently widespread in the human population (Bentley and Rabbitts, 1980); therefore, it may represent a DNA sequence that has lost its ability to be expressed and is thus able to drift more rapidly than coding sequences.

FIG. 2 The sequence of a human Vκ pseudo-gene. The DNA sequence of a genomic clone from human DNA, together with the in-phase reading frame (indicated by arrows). Asterisks represent chain termination codons. The reading frames corresponding to normal immunoglobulin protein are in heavy boxes with amino acid positions marked according to standard numbering of normal kappa light chains. Circled amino acids indicate positions of highly conserved amino acids replaced by apparent base-pair substitutions. The abnormal 5' splicing site of the intervening sequence is starred and the putative V − J joining blocks are in light boxes. Data from Bentley and Rabbitts (1980).

HUMAN C$_H$ GENES

We have studied in detail the genomic locus encoding the human Cμ and Cδ genes (Rabbitts *et al.*, 1981) and will describe how closely the general arrangement and sequences parallel those present in the mouse. The human Cμ gene consists of four domains at the DNA level (analogous to those domains of the protein), separated by short intervening sequences (about 150 to 200 bases). Table II enumerates the domain structure of

TABLE II

Domain Structure of the Human Cμ$_s$ gene

GENETIC DOMAIN	SIZE OF IVS	FIRST AMINO ACID	LAST AMINO ACID
Cμ1	130	Gly114	Pro216
Cμ2	242	Val217	Pro328
Cμ3	150	Asp329	Lys433
Cμ4 + tp		Gly434	Tyr565

The size of the intervening sequence (IVS) is deduced from restriction mapping and nucleotide sequencing. The first codon of each domain is created by RNA splicing. The numbers refer to the amino acid number according to the sequence GAL (Watanabe *et al.*, 1973).

this gene. This genetic domain structure is analogous to that of mouse Cμ and appears to be the general structure of mouse and human C$_H$ genes. The four-domain structure of Cμ represents genetic information used for the production of secreted IgM. The production of membrane IgM (μm) in the mouse results from the presence of a different carboxy-terminal portion, attached to Cμ4, which is encoded by two down-stream coding segments (Alt *et al.*, 1980; Early *et al.*, 1980b). We have identified similar coding segments down-stream of the human Cμ gene. Figure 3 gives a map of the relevant region of human DNA. In both human and mouse DNA, the μm coding segments are about 2×10^3 bases down from the Cμ4 domain. The nucleotide sequence analysis of the first μm coding segment in human (Fig. 4) illustrates again a very high conservation between species. The putative translation product of the human μm segment is identical to that of the mouse, despite a 7% base divergence. The RNA splicing positions at both ends are also conserved, implying the existence of a second μm coding segment in the DNA of man as in mouse. Furthermore, the intra-Cμ4 splicing of both species (also compared in Fig. 4) is also conserved. The general similarity

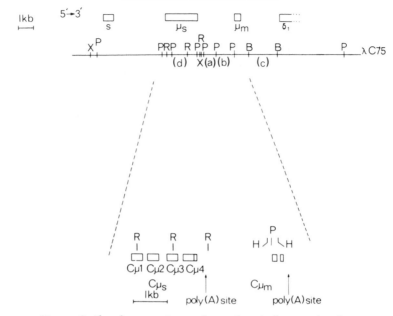

Structure of human Cμ & Cδ locus

FIG. 3 *Map of the human C_μ and C_δ loci determined from cloned DNA. The direction of transcription is shown as $5' \rightarrow 3'$. R, EcoRI; P, PstI; X, XbaI; B, BamHI; and H, HpaII. S refers to tandemly repeated sequences related to the H chain switch. At the bottom of the Figure is an expanded picture of the μs and μm coding segments. Data from Rabbitts* et al. *(1981) 1 kb = 10^3 bases.*

in the μs-μm region in the two species indicates that similar mechanisms exist for changing from m to μs expression in man as in the mouse. Figure 5 diagrams the probable pathway in man. Simply, the production of μs results from splicing the intervening sequences between V-Cμ1, Cμ1-Cμ2, Cμ2-Cμ3 and Cμ3-Cμ4. The μm form results from this splicing of these intervening sequences, but a further splicing site within Cμ4 is joined to the beginning of the first μm coding segment and the intervening sequence between the two μm segments is also spliced out. An extra poly(A)addition site near μm would then be utilised to generate mature μm messenger RNA. The concordance of the μm sequence in mouse and man again indicates strong selective pressures to maintain both a general hydrophobic sequence and the specific protein sequence. It is conceivable, therefore, as suggested by Early *et al.* (1980b), that the hydrophobic sequence anchors surface IgM in the membrane. In addition, however, it is possible that, since the amino acid sequence is itself completely conserved between

```
       E   G   E   V  N  A  E  E  G  F  E  N  L  W  T  T  A  S  T  F  I  V  L
Hu  GTGTCTCCTGCGAGAGGGGGAGGTGAACGCCGAAGAGGGGCTTTGAAACCTGTGGACCACGGCTCCAACCTTGATCGTCCTC
Mo                T     T     G        A           G                  T
```

```
    F   L   L   S   L  F  Y  S  T  T  V  T  L  F  K
Hu  TTCCTCCTGAGCCTCTTCTACAGTACCACCGTCACCTTGTTCAAGGTAGCACGGCTG
Mo                C                         C
```

FIG. 4 *The nucleotide sequence of the first μm coding segment in human (Hu) DNA compared to mouse (Mo). Arrows show RNA splicing positions and the human protein sequence was derived by comparison to that of mouse.*

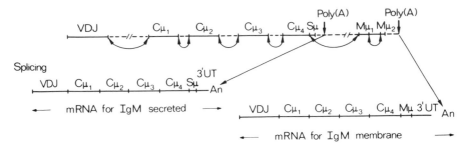

FIG. 5 *Alternative RNA splicing routes for production of human μs and μm. This scheme is by analogy to that postulated for the mouse (Early et al., 1980a,b).*

species, this segment may play an additional role, for example in specific interaction with other membrane components involved in triggering B-lymphocyte differentiation subsequent to antigen binding.

After IgM production, lymphocytes switch antibody synthesis to IgD, IgG and IgA, etc., whilst expressing the same V_H region gene. This process is known as the H chain class switch. In the mouse, the switch between Cμ and Cγ or Cα occurs at a variety of sites within the 5' flanking regions of Cμ and the various other C_H genes (Davis *et al.*, 1980; Katakoa, *et al.*, 1981; Sakano *et al.*, 1980). During this process, the DNA between the two switch sites is deleted from the chromosome (Honjo and Katakoa, 1978; Rabbitts *et al.*, 1980): therefore, like V gene integration, the class-switch involves DNA rearrangements and incurs DNA deletion. The explanation of cells that co-synthesise IgM and IgD, however, does not readily fit into a model involving DNA deletion, and we may need to invoke different mechanisms to explain the later results.

With regard to the mechanism of the μ to γ switch in the mouse, tandemly repeated sequences have been observed in these areas and their presence has been taken to imply a functional role in the class-switch (Dunnick *et al.*, 1980). Similar sequences have also been observed adjacent to human Cμ and Cγ genes. When we compared these repetitive sequences upstream of human Cμ with that near mouse Cμ (Fig. 6), it was again striking that significant homologies had been maintained. The basic repeat unit in the mouse is G-G-G-G-T (G-A-G-C-T)$_n$, whilst in the human sequence we found G-G-G-C-T(G-A-G-C-T)$_n$. Although we have been able to detect related sequences near a Cγ gene, we could find no evidence of any related sequences near the Cδ gene. The occurrence of conserved tandem repeats in the two species, within a region

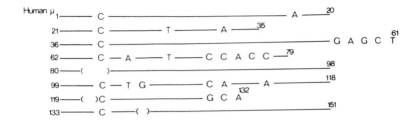

FIG. 6 *Comparison of the nucleotide sequence of tandemly repeated units upstream of the human and mouse Cμ genes. The continuous line implies sequence identity and parentheses represent base deletions.*

known to be important in the H chain switch, argues on evolutionary grounds that the sequences are functionally important in this switch. Furthermore, it supports the hypothesis that such sequences facilitate unequal alignment of homologous regions adjacent to C_H genes before recombination by sister chromatid exchange (Rabbitts *et al.*, 1980) or intrachromosomal looping out.

The most closely linked C_H gene to the human Cμ is the Cδ gene. The 5' end of the Cδ1 gene is located only about 5 × 10^3 bases downstream of the Cμ4 domain, whereas the next C_H gene appears to be more than 25 × 10^3 bases away. The analogous Cδ gene in mouse occurs at about the same distance from Cμ4 (Liu *et al.*, 1980). Interestingly, although there is a great similarity in the maps of the human and mouse Cμ and Cδ region, the homology within the protein sequence of Cδ1 in these species is only 28%. We have been unable to detect (by hybridisation) sequences related to the tandem repeats (which we postulate to be important in the H chain switch) up-stream of the Cδ gene. This observation, taken with the facts that Cμ and Cδ genes are so closely linked and that only the Cμ gene possesses adjacent J_H segments, suggests that Cμ and Cδ expression could result from co-transcription of the two genes and RNA splicing strategies that need not necessarily involve further DNA rearrangement and deletions. If this hypothesis is correct, it means that μs, μm and δ (probably δm and δs, since both membrane and secreted chains are known) can be made from the same transcription unit. This complex transcript, however, would need to circumvent the linear order of RNA splice sites in order to join the V_H segment to the $C\delta$.

CONCLUSION

The comparative sequence studies of the immunoglobulin genes of humans and mice described in this paper help to identify regions conserved in evolution, which have possible functions in the DNA rearrangements characteristic of antibody genes. The types of sequences and mechanism involved in the main DNA rearrangements (i.e. V gene integration and the class switch) are contrasted in Table III. The process of V gene integration (both H and L) seems to be confined to limited areas

TABLE III

Comparison of V gene Joining and the H Chain Class Switch

	V GENE INTEGRATION	CLASS SWITCH
Site	Specific	Variable
Probable sequences involved	Inverted repeats (C-A-C-A-G-T-G)	Tandem repeats (G-A-G-C-T)
Probable mechanism	Looping out	Homologous recombination

bordering V genes, and J or D segments utilising inverted repeat sequences. The H chain class switch, on the other hand, can occur over a wide area of sequence occurring in intervening sequences or flanking segments of C_H genes, and probably occurs by homologous recombination mediated by tandemly repeated sequences.

REFERENCES

Alt, F.W., Bothwell, A.L.M., Knapp, M., Siden, E., Mather, E., Koshland, M. and Baltimore, D. (1980). Synthesis of secreted and membrane-bound immunoglobulin Mu heavy chains is directed by mRNAs that differ at their 3' ends. *Cell* 20, 306-311.

Bentley, D.L. and Rabbitts, T.H. (1980). Human immunoglobulin variable region genes - DNA sequences of two V_K genes and a pseudogene. *Nature (London)* **288**, 730-733.

Breathnach, R., Benoist, C., O'Hare, K., Gannon, F. and Chambon, P. (1978). Ovalbumin gene: evidence for a leader sequence in mRNA and DNA sequences at the exon-intron boundaries. *Proc. Nat. Acad. Sci., U.S.A.* **75**, 4853-4858.

Davis, M.M., Calame, K., Early, P.W., Livant, D.L., Joho, R., Weissman, I.L. and Hood, L. (1980). An immunoglobulin heavy chain gene is formed by at least two recombinational events. *Nature (London)* **283**, 733-739.

Dreyer, W.J. and Bennett, J.C. (1965). The two genes - one polypeptide hypothesis. *Proc. Nat. Acad. Sci., U.S.A.* **54**, 864-869.

Dunnick, W., Rabbitts, T.H. and Milstein, C. (1980). An immunoglobulin deletion mutant with implications for the heavy chain switch and RNA splicing. *Nature (London)* **286**, 669-675.

Early, P., Huang, H., Davis, M., Calame, K. and Hood, L. (1980a). An immunoglobulin heavy chain variable gene is generated from three segments of DNA. *Cell* **19**, 981-998.

Early, P., Rogers, J., Davis, M., Calame, K., Bond, W., Wall, R. and Hood, L. (1980b). Two mRNAs can be produced from a single immunoglobulin μ gene by alternative RNA processing pathways. *Cell* **20**, 313-319.

Hieter, P.A., Max, E.E., Seidman, J.G., Maizel, J.V. and Leder, P. (1980). Cloned human and mouse kappa immunoglobulin constant and J region genes conserves homology in functional segments. *Cell* **22**, 197-207.

Honjo, T. and Katoaka, T. (1978). Organisation of immunoglobulin heavy chain genes and allelic deletion model. *Proc. Nat. Acad. Sci., U.S.A.* **75**, 2140-2144.

Katoaka, T., Kawakami, T., Takahashi, N. and Honjo, T. (1980). Rearrangement of immunoglobulin γ_1,-chain gene and mechanism for heavy-chain class switch. *Proc. Nat. Acad. Sci., U.S.A.* **77**, 919-923.

Lerner, M.R., Boyle, J.A., Moust, S.M., Wolin, S.L. and Steitz, J.A. (1980). Are snRNPs involved in splicing? *Nature (London)* **283**, 220-224.

Liu, C.P., Tucker, P.W., Mushinski, J.F. and Blattner, F. (1980). Mapping of heavy chain genes for mouse immunoglobulins M and D. *Science* **209**, 1348-1353.

Matthyssens, G. and Rabbitts, T.H. (1980). Structure and multiplicity of human immunoglobulin heavy chain variable region genes. *Proc. Nat. Acad. Sci., U.S.A.* **77**, 6561-6565.

Max, E.E., Seidman, J.G. and Leder, P. (1979). Sequences of five potential recombination sites encoded close to an immunoglobulin κ constant region gene. *Proc. Nat. Acad. Sci., U.S.A.* **76**, 3450-3454.

Rabbitts, T.H., Hamlyn, P.H., Matthyssens, G. and Roe, B.A. (1980). The variability, arrangement and rearrangement of immunoglobulin genes. *Canad. J. Biochem.* **58**, 176–187.

Rabbitts, T.H., Forster, A. and Milstein, C.P. (1981). Human immunoglobulin heavy chain genes: evolutionary comparisons of Cμ, Cδ and Cγ genes and associated switch sequences. *Nucl. Acids Res.* **9**, 4509–4524.

Sakano, H., Huppi, K., Heinrich, G. and Tonegawa, S. (1979). Sequences at the somatic recomination sites of immunoglobulin light chain genes. *Nature (London)* **280**, 288–294.

Sakano, H., Maki, R., Kurosawa, Y., Roeder, W. and Tonegawa, S. (1980). The two types of somatic recombination necessary for generation of complete immunoglobulin heavy chain genes. *Nature (London)* **286**, 676–680.

Watanabe, S., Bainikol, H.U., Hour, J., Bertram, J. and Hilschmann, N. (1973). The primary structure of a monoclonal IgM immunoglobulin (macroglobulin Gal), II: the amino acid sequence of the H-chain μ-type, subgroup HIII. Architecture of the complete IgM molecule. *Hoppe-Seyler's Z. Physiol. Chem.* **354**, 1505–1509.

Gene Expression in Phylogenetically Polyploid Organisms

M. LEIPOLDT and J. SCHMIDTKE

Abteilung Humangenetik, Zentrum Hygiene und Humangenetik
der Universität, D-3400 Goettingen, FRG

EUKARYOTES ARE OF POLYPLOID ORIGIN

Genome evolution is marked by mutational events that lead to changes of the base sequence of the DNA, structural rearrangements of parts of the genetic material and quantitative alterations of the genome composition. The most drastic change of genome size is generated by polyploidy, which was apparently of widespread occurrence during prokaryotic and eukaryotic evolution (Sparrow and Nauman, 1976). Polyploidy creates an abundance of raw genetic material, which can be exploited by subsequent mutation and selection (Ohno, 1970). It seems that after polyploidization there is a strong tendency to evolve into a diploid state by chromosomal rearrangement and by sequence diversification. Since, clearly, most eukaryotic organisms are diploid, but because circumstantial evidence suggests that all organisms have experienced one or more rounds of polyploidization in their phylogenetic past, it can be said that probably all eukaryotes are diploidized polyploids.

CYPRINID AND SALMONOID FISH ARE MODEL GROUPS FOR THE STUDY OF DIPLOIDIZATION

The teleostean family Cyprinidae (carp-like fishes) and the order Isospondyli (which includes herrings and trouts) are excellent model groups to study the diploidization process after tetraploidization. In both these groups several species stand in a diploid-tetraploid relationship to each other, as can be inferred from the number of chromosomes, fundamental number, DNA content per nucleus and the number of expressed structural genes (Tables I to III; literature reviewed by Engel and Schmidtke, 1975; Engel *et al.*, 1975).

TABLE I

DNA Content, Chromosome Number and Fundamental Number in Cyprinid Species (after Ohno, 1974)

	DNA CONTENT (% OF HUMAN LEUKOCYTES)	CHROMOSOME NUMBER	FUNDA-MENTAL NUMBER
2n:			
Barbus tetrazona	20	50	84
Barbus fasciatus	22	50	82
Rutilus rutilus	28	50	78
Tinca tinca	30	48	80
Abramis brama	36	50	80
Leuciscus cephalus	38	50	88
4n:			
Barbus barbus	49	100(44M+56A)*	144
Cyprinus carpio	50	104(46M+58A)*	150
Carassius auratus	53	104(62M+42A)*	168

*A, acrocentric chromosomes; M, metacentric chromosomes.

TABLE II

DNA Content, Chromosome Number and Fundamental Number in Species of the Order Isospondyli (after Engel und Schmidtke, 1975)

	DNA CONTENT (% OF HUMAN LEUKOCYTES)	CHROMOSOME NUMBER	FUNDAMENTAL NUMBER
2n:			
Osmerus esperlanus	19	54	70
Hypomesus pretiosus	20	± 50	± 60
Clupea pallasii	28	52	60
Clupea harengus	28	52	60
Engraulis mordax	43	46–48	48
4n:			
Thymallus thymallus	60	± 94	± 146
Salmo irideus	80	56–68	104
Salmo trutta	80	77–82	100
Coregonus lavaretus	90	± 80	± 100
Salvelinus fontinalis	100	84	100
Salmo salar	103	54–60	72

TABLE III

Number of Gene Loci Coding for Various Isozyme Systems in Species of the Order Isospondyli and the Family Cyprinidae (order Ostariophysi) (after Engel et al., 1975)

ENZYME	GENOME:	NUMBER OF GENE LOCI			
		ISOSPONDYLI		OSTARIOPHYSI	
		$2n$	$4n$	$2n$	$4n$
S-form NADP-IDH		1	2	2	2
SDH		1	2	1	1
S-form AAT		1,2	2	1	1,2
M-form NADP-IDH		1	2	1	2
LDH		2	4	2	2,3
6-PGD		1	1	1	2
a-GPDH		3	3	1	2
PGI		1,2	2,3	2	3,4
S-form NAD-MDH		1	2	1,2	2,3
M-form NAD-MDH		1	2	1,2	2

With regard to all of the parameters listed in Tables I to III, the diploid-tetraploid relationship is not clearcut: Firstly, there is considerable variation of the nuclear DNA content in the diploid and the tetraploid groups. This is due largely to variation in the amount of repetitive sequences (Schmidtke *et al.*, 1979a,b). Secondly, the chromosome count is not always precisely 1 : 2. Apparently, chromosomal rearrangements occurred after the polyploidization events in both the diploids and the tetraploids. Thirdly, the number of gene loci coding for various enzymes in the tetraploids is not always twice that of the number coding for the same proteins in the diploids. This could mean either that sufficient genetic diversification to be detectable by the techniques used has not yet taken place, or that gene duplicates have become functionally silenced by acquiring nonsense mutations.

All these observations suggest that the Cyprinid and Salmonoid tetraploids are engaged in the process of diploidization. The presence of quadrivalents in meiotic metaphase configurations of the Salmonoids (e.g. see Ohno *et al.*, 1965; Ohno, 1974) and the finding that the non-repetitive DNA fractions of several Salmonoid genomes appear to consist of "two-copy" DNA (Schmidtke *et al.*, 1979b) suggest that many Salmonoid tetraploids are less advanced in the process of diploidization than the Cyprinid tetraploids.

DIPLOIDIZATION OF GENE EXPRESSION

A striking observation is that diploidization of the tetraploid occurs also on the level of the quantitation of gene
expression. Although this phenomenon has been known for a
long time (Wettstein, 1937), it was first studied in detail
by Beçak and co-workers (Beçak and Pueyo, 1970; Beçak and
Goissis, 1971) in the frog genus *Odontophrynus*. As in these
frogs, gene expression in the Cyprinid tetraploids is reduced
to the level of the related diploids, with respect to cell
size, protein content, RNA content (predominantly ribosomal
RNA) and enzyme activities (Schmidtke and Engel, 1975;
Schmidtke *et al.*, 1976a; and Fig. 1). Such a reduction of
gene expression was not observed in the less diploidized
tetraploids of the order Isospondyli; here the nucleo-cyto-
plasmic relationships expected from a comparison with the
related diploids are maintained (Schmidtke *et al.*, 1975b,
1976a; Fig. 1). The question then arises as to which mechanism affects genetic activity in the tetraploid Cyprinid
species.

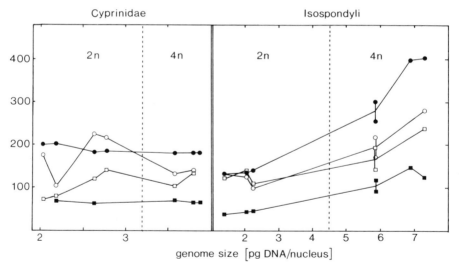

FIG. 1 *Cellular parameters in species of the family Cyprinidae and of the order Isospondyli: erythrocyte volume (●; μm³,
erythrocyte protein content (■; pg/cell), liver cell protein
content (□; pg/cell), liver cell RNA content (○; pg/cell × 10)
(data from Schmidtke and Engel, 1975; Schmidtke et al., 1975a,
1976a).*

AMOUNT AND EXPRESSION OF RIBOSOMAL GENES

It has been suggested that cell size is determined by the number of (28 S + 18 S) ribosomal RNA genes (Pedersen, 1971).
It seemed likely that rRNA genes were lost selectively in the
tetraploid Cyprinids during the diploidization process. This
turned out not to be the case. As can be inferred from Fig.
2 (Schmidtke and Engel, 1976; Schmidtke *et al.*, 1976b), the
diploids and the tetraploids of both Cyprinidae and Isosopon-
dyli, are about 1 : 2 with respect to the number of these
genes. It should be mentioned that a similar observation was
made in diploid and tetraploid *Odontophrynus* frogs (Schmidtke
et al., 1976c).

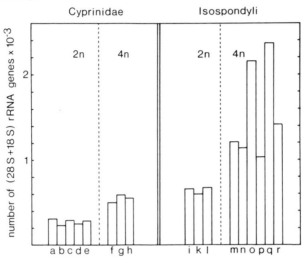

FIG. 2 *Mean number of (28 S + 18 S) ribosomal RNA genes in
species of the family Cyprinidae and of the order Isospondyli.*
(*a*) Rutilus rutilus; (*b*) Tinca tinca; (*c*) Rutilus rutilus ×
Abramis brama; (*d*) Abramis brama; (*e*) Leuciscus cephalus; (*f*)
Barbus barbus; (*g*) Cyprinus carpio; (*h*) Carassius auratus;
(*i*) Osmerus esperlanus; (*k*) Clupea harengus; (*l*) Sprattus
sprattus; (*m*) Thymallus thymallus; (*n*) Salmo irideus; (*o*)
Salmo trutta; (*p*) Coregonus fera; (*q*) Salvelinus fontinalis;
(*r*) Salmo salar. *Data from Schmidtke and Engel (1976);
Schmidtke* et al. *(1976b).*

This finding seemed to suggest that rRNA genes are transcribed less effectively or less frequently (Schmidtke *et al.*,
1975a), and that a regulatory mechanism acts at the level of
transcription of rRNA genes. Transcription of rRNA genes,
for example, could be scaled down by an alteration of the
kinetic properties of RNA polymerase I. Nucleolar RNA

TABLE IV

Kinetic Properties of Nucleolar RNA Polymerase in Diploid and Tetraploid Cyprinid and Salmonoid Fish

ORDER	GENOME	K_M (M–UTP $\times 10^{-5}$)	V_{max} (pmol) UTP/mg PROTEIN)
Ostariophysi (Cyprinidae)	Diploid	1.0 – 5.0	67 – 78
	Tetraploid	1.7 – 2.5	110 – 300
	Diploid	3.4 – 3.5	29 – 140
Isospondyli			
	Tetraploid	2.7 – 4.6	170 – 180

polymerase was therefore assayed for K_M and V_{max} in various species of Cyprinid and Salmonoid fish by introducing whole, purified nucleoli from liver tissue into an *in vitro* transcription test. Tabel IV summarizes the results showing that the values for RNA polymerase from tetraploid Cyprinids lie among the range of values obtained for the diploid Cyprinids and the diploid and tetraploid Salmonoid species. Hence, no change in the kinetic behaviour of the nucleolar RNA polymerase of tetraploid Cyprinids that could be responsible for the diminished activity of ribosomal genes could be found under these conditions. On the contrary, RNA polymerase from tetraploid species of both orders seems to have a considerably higher V_{max} value than is found in diploid species.

STRUCTURAL MODIFICATION OF RIBOSOMAL RNA MOLECULES

As well as by reduced transcriptive activity, the reduced amount of cellular RNA could be arrived at by some regulating mechanism that operates during the processing of pre-rRNA or by accelerated turnover of mature rRNA. The analysis of total cellular RNA by acrylamide gel electrophoresis under non-denaturing and under fully denaturing conditions (Leipoldt and Engel, unpublished results) revealed that the rRNA of some Cyprinid fish carries structural modifications that are similar to the hidden breaks found in the rRNA of insects and other protostomian species (Ishikawa, 1977; Eckert *et al.*, 1978). Whereas 28 S and 18 S rRNA appear to be intact, i.e. are present as uninterrupted molecules upon electrophoresis under aqueous conditions, the treatment and electrophoretic separation of RNA in the presence of formamide (Staynov *et al.*, 1972) showed that both ribosomal RNA components do not exist as a contiguous stretch of covalently bound nucleotides but possess nicks or gaps that lead to the appearance

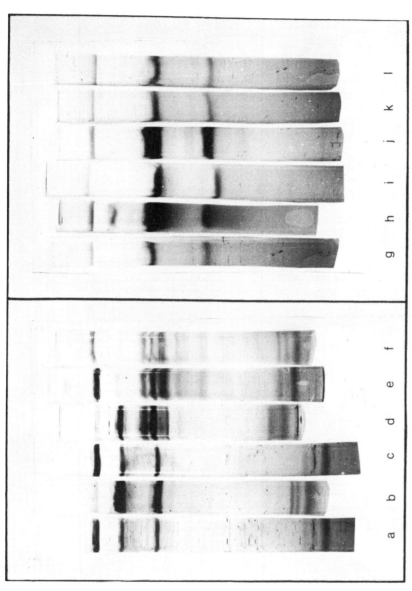

FIG. 3 *Electrophoresis of total, DNAase-treated RNA from liver cells on 3% (w/v) polyacrylamide gels under fully denaturing (lanes a to f) and non-denaturing (lanes g to l) conditions. Lanes a and g, Osmerus esperlanus; lanes b and h, Salmo irideus; lanes c and i, Tinca tinca; lanes d and j, Abramis brama; lanes e and k, Cyprinus carpio; lanes f and l, Carassius auratus.*

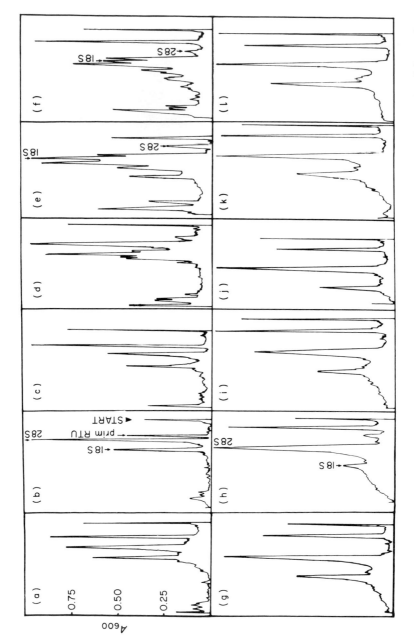

FIG. 4 Densitometer scannings of electrophoretic separation of RNA on polyacrylamide gels stained with methylene blue. RNA probes and arrangement are the same as in Fig. 3. Gel top to bottom is from right to left, respectively.

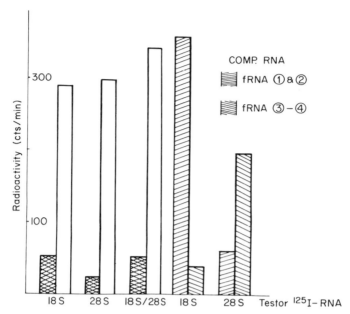

FIG. 5 *Competition-hybridization of RNA fragments (fRNA) and*
^{125}I-*labelled 18 S and 28 S rRNA on total DNA isolated from*
liver cells of Cyprinus carpio. *fRNA was hybridized to DNA*
fixed on nitrocellulose filters and challenged with labelled
18 S and 28 S rRNA as indicated. For classification of fRNA,
compare with Fig. 6.

of distinct RNA fragments if secondary structure hydrogen-
bonding is destroyed (Figs 3 and 4). Competition hybridization experiments of unlabelled RNA fragments and ^{125}I-label-
led 28 S and 18 S rRNA to total DNA showed that the minor RNA
components and ribosomal RNA have a common DNA origin, implying that the RNA fragments are indeed derived from "native"
mature rRNA by eliminating secondary and tertiary structures
(Fig. 5; and Leipoldt, unpublished results). The two main
RNA fragments (f1 and f2; Fig. 6), which are derived exclu-
sively from 28 S rRNA, show molecular weights of about 8.7×10^5 and 5.0×10^5, which is in accordance with the findings of
Cammarano *et al.* (1975), for rRNA from a variety of proto-
stomes and protozoa, that the large rRNA molecules carry one
central hidden break that gives rise to RNA fragments of M_r
6.0×10^5 to 9.6×10^5 and 6.5×10^5 to 7.2×10^5.

Different Cyprinid species seem to differ with respect to
the number of hidden breaks in their rRNA molecules. Quantitation of densitograms of stained gels (Table V) showed that
hidden breaks are not detectable in the rRNA of the diploid

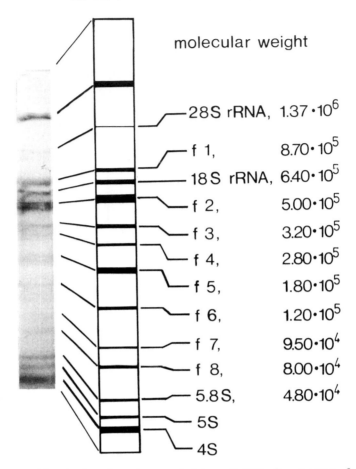

FIG. 6 *Estimated molecular weights of RNA fragments from liver cells of* Cyprinus carpio *on denaturing polyacrylamide gels. Marker RNA (mouse and* Eschericia coli *rRNA, compare with Fig. 7, lane a, was run in parallel. Molecular weights of mouse and* E. coli *rRNA were taken from Ishikawa (1977).*

green tench (*Tinca tinca*), occur in 10% of the 28 S rRNA of the diploid dream (*Abramis brama*(, and in as much as 90% of the 28 S of the tetraploid species carp (*Cyprinus carpio*) and goldfish (*Carassius auratus*). Apart from the fact that Cyprinid fish are the first members of a vertebrate family in which hidden breaks could be shown to exist in a considerable fraction of ribosomal RNA, the striking observation is that in the tetraploid Cyprinid species nearly all the 28 S rRNA molecules and about 50% of the 18 S rRNA molecules are broken.

TABLE V

Densitometric Quantitation of the Fraction of
Ribosomal RNA in Total RNA in Formamide-containing Gels

Order	Genome	Species	28 S	18 S
	Diploid	*Tinca tinca*	0.36(0.36)*	0.24(0.25)
		Abramis brama	0.30(0.36)	0.24(0.24)
Ostariophysi (Cyprinidae)	Tetraploid	*Cyprinus carpio*	0.04(0.34)	0.15(0.27)
		Carassius auratus	0.03(0.32)	0.11(0.22)
Isospondyli	Diploid	*Osmerus esperlanus*	0.42(0.44)	0.25(0.24)
	Tetraploid	*Salmo irideus*	0.44(0.45)	0.29(0.27)

*Numbers in parentheses give values obtained under aqueous conditions.

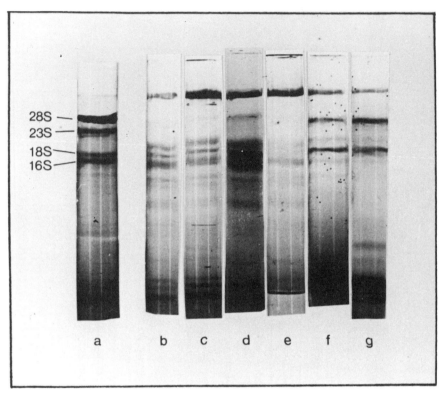

FIG. **7** *Electrophoresis of total, DNAase-treated RNA under de-*
naturing conditions on polyacrylamide gels from different
organs of Cyprinus carpio: *lane b, liver; lane c, kidney; lane*
d, spleen; lane e, heart muscle; lane f, sperm; lane g, oocy-
tes. Lane a, Mouse and E. coli *rRNA served as markers run in*
parallel.

In order to rule out the possibility that nicked rRNA may
be a unique feature of liver cells from tetraploid Cyprinid
species, the electrophoretic rRNA pattern was analyzed in
various tissues of the carp. Ribosomal RNA containing hidden
breaks occurs in all somatic tissues tested to date, as revealed by electrophoresis under denaturing conditions
(Fig. 7, lanes b to e). Surprisingly, rRNA with latent breaks
could be detected in oocytes and sperm to only a negligible
extent (Fig. 7, lanes f and g). The question of tissue specificity of latently nicked rRNA was subsequently studied in
a variety of Salmonoid and Cyprinid fish, including *Salveli-*
nus fontinalis as a representative of tetraploid Salmonoid
species, *Leuciscus cephalus* and *Barbus tetrazona* as

representatives of diploid Cyprinids, and *Barbus barbus* and *Carassius auratus* as representatives of tetraploid Cyprinid species (Leipoldt and Kellner, unpublished results). Ribosomal RNA containing hidden breaks could be detected in the somatic tissues of the barbel (*B. barbus*) and the goldfish (*C. auratus*). Like the situation in the carp, nearly all rRNA molecules exhibit occult intramolecular nicks in these two tetraploid Cyprinid species. The RNA fragments produced by treatment with formamide show exactly the same size and electrophoretic pattern, indicating that the generation of hidden breaks is not a random process with respect to their location on the RNA molecule. Nicked RNA was found in only very small amounts in somatic tissues of *L. cephalus* and *B. tetrazona*, not at all in somatic tissues of *S. fontinalis* nor in sperm or oocytes of either species.

It has been suggested that the generation of breakpoints in 28 S rRNA could be the initial degradation step in the turnover of 28 S rRNA and could thus be related to the ageing process of rRNA and ribosomes, as latent breaks are predominantly located in "old" rRNA species of rat liver cells (Kokileva *et al.*, 1971; Awata and Natori, 1977). According to this view, the existence of latent breakpoints in almost all 28 S rRNA molecules of tetraploid Cyprinids can be interpreted as the result of a very rapid turnover of mature ribosomal RNA. This could explain the reduced amount of rRNA per cell in tetraploid Cyprinids as a consequence of excessive and rapid degradation *via* RNA fragments to oligo- and mononucleotides, rather than as a result of a diminished rate of synthesis. It can be speculated that the rapid turnover of rRNA leads to a reduced amount of ribosomes, which consequently influences the overall rate of protein biosynthesis in the cell.

RIBOSOME FUNCTION IN TETRAPLOID CYPRINIDS

The function of ribosomes during translation could be influenced directly by latently nicked RNA, if (1) the nicks are located within regions of the molecule that are essential for the reaction steps in translation, or if (2) the nicks influence the assembly of rRNA and proteins during ribosome formation, leading to a modified conformation or spatial orientation of ribosomal proteins at distinct sites of functional importance. To substantiate this speculation, ribosomes were isolated from liver tissue of a diploid (green tench) and a tetraploid (carp) Cyprinid fish, and assayed for the capability *in vitro* to incorporate phenylalanine into acid-insoluble material when directed by poly(U). Following the time-course of incorporation of ^{3}H-labelled phenylalanine (Fig. 8), it is

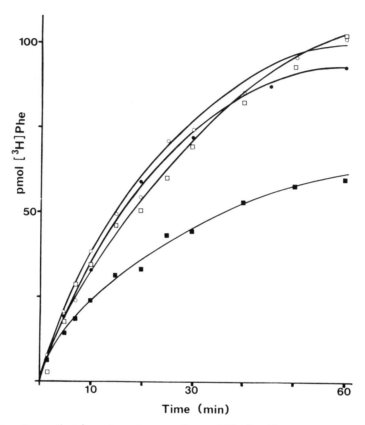

FIG. 8 *Translation* in vitro *of purified ribosomes on poly(U) template from rat liver (●), green tench liver (O), carp liver (■) and carp oocytes (□). Isolation of ribosomes and the translation assay were performed according to Staehelin and Falvey (1971).*

obvious that a given amount of ribosomes from carp liver promotes about half as much poly(Phe) synthesis compared to ribosomes from the liver of the green tench (Leipoldt and Engel, unpublished results).

This result suggests that a correlation could exist between ribosomes that have nicked RNA molecules as an integral part and a reduction in the efficiency of protein synthesis *in vitro*.

If the presence of rRNA molecules with hidden breaks leads to the observed reduced translational performance of ribosomes from liver cells, then ribosomes isolated from oocytes should not show such a restricted rate of translation. Indeed, the rate of incorporation of phenylalanine promoted by ribosomes from carp oocytes resembles very much that of ribosomes from liver cells of the green tench, and exceeds that of ribosomes from carp liver by a factor of about two (Fig. 8).

CONCLUSIONS AND SPECULATIONS

There is ample experimental evidence to tentatively propose a regulatory mechanism that leads to a reduction of protein content per cell, to a reduced cell size and, consequently, to the diploidization of gene expression in tetraploid Cyprinid fish.

Hidden breaks have been interpreted to represent the result of initial degradation steps in the turnover of rRNA. According to that, it seems that a prevailing fraction of rRNA in somatic cells of tetraploid Cyprinid fish (90%) is degraded very rapidly; i.e. at least has run through the first turnover steps. Despite the fact that the degradative fragmentation of RNA molecules does not alter its native higher structure as deduced from its behaviour in non-denaturing electrophoresis, there seem to be consequences for the function of the ribosome. The efficiency *in vitro* of ribosomes whose ribonucleic acid portion carries hidden nicks, to translate poly(U) into poly (Phe) is reduced about twofold. However, it can not be decided in what way nicked RNA molecules might cause this functional reduction. The supposed role of nicked rRNA in translation is supported by the fact that ribosomes isolated from oocytes of a tetraploid Cyprinid, in which only a small amount of nicked RNA is detectable, promote translation *in vitro* at a rate similar to that found in diploid species, and exceeds the rate found in liver ribosomes by a factor of two.

Hidden breaks have not been found exclusively in the rRNA of tetraploid Cyprinids. A minor fraction of rRNA (10%) was found to contain latent nicks in some diploid species, indicating that the occurrence of hidden breaks is a common characteristic of Cyprinid rRNA. It appears that the fraction of nicked RNA is small in diploid species, whereas the rRNA in tetraploid Cyprinids is predominantly present as nicked molecules.

Germ cell rRNA seems to be essentially free from hidden breaks in all species studied. This could be due to the following reasons: (1) a specific ribonuclease is expressed only in somatic tissues and not in germ cells; (2) two populations of genes for rRNA (28 S + 18 S) are expressed in somatic and germ cells, similar to the situation found for 5 S RNA genes of somatic cells and oocytes in amphibia and fish (Ford and Southern, 1973; Denis and Wegnez, 1977). Each of these two sets of rRNA genes is transcribed almost exclusively in either somatic or germ cells. The base sequences of rRNA genes might differ at a few specific sites, thus predetermining the generation of breaks at distinct sites in the rDNA transcript. It could be speculated that a subset of rRNA genes, which occur in a small amount in somatic cells and codes for intact rRNA, is amplified selectively during oocyte development.

Although the data of course do not prove that the structural modification of rRNA molecules cause the diploidization process of gene expression in tetraploid Cyprinids, there are experimental clues that suggest a rapid ageing of ribosomes due to nicking of ribosomal rRNA molecules, leading to a restricted capacity for protein synthesis. As a consequence, the overall output of gene products in tetraploid Cyprinids could be scaled down to a level similar to that found in closely related diploid species. This mechanism would enable the tetraploid organism to utilize the benefits of polyploidization by retaining the increased nuclear DNA amount, in that it counteracts the disadvantageous and often deleterious effects of an increased cell size. It has been discussed (Schmidtke and Engel, 1975; Engel and Schmidtke, 1975) that, in larger cells with greater amounts of DNA, the cell metabolism seems to be relatively reduced and that, as the increase of cell size in polyploid animals tends to be concomitant with a reduction in cell number, larger cells could have severe consequences for early embryogenesis.

REFERENCES

Awata, S. and Natori, Y. (1977). Turnover of rat liver 28 S ribosomal RNA. Nicking as the initial step of degradation. *Biochim. Biophys. Acta* **478**. 486-494.

Beçak, W. and Goissis, G. (1971). DNA and RNA content in diploid and tetraploid amphibians. *Experientia* **27**, 345-346.

Beçak, W. and Pueyo, M.T. (1970). Gene regulation in the polyploid amphibian *Odontophrynus americanus*. *Exp. Cell Res*. **63**, 448-451.

Cammarano, P., Pons, S. and Londei, P. (1975). Discontinuity of the large ribosomal subunit RNA and rRNA molecular weights in eukaryote evolution. *Acta biol. med. germ.* **34**, 1123-1135.

Denis, H. and Wegnez, M. (1977). Oocytes and liver cells of the Teleost fish *Tinca tinca* contain different kinds of 5 S RNA. *Develop. Biol.* **59**, 228-236.

Eckert, W.A., Kaffenberger, W., Krohne, G. and Franke, W.W. (1978). Introduction of hidden breaks during rRNA maturation and ageing in *Tetrahymena pyriformis*. *Eur. J. Biochem.* **87**, 607-616.

Engel, W. and Schmidtke, J. (1975). Die Bedeutung von Gendup-
   likationen für die Evolution der Wirbeltiere. *In* "Humangenetik. Ein kurzes Handbuch in fünf Bänden" (Becker, P.
   E., ed.), vol. III/2, pp. 618-654. Thieme, Stuttgart.
Engel, W., Schmidtke, J. and Wolf, U. (1975). Diploid-tetraploid relationships in teleostean fishes. *In* "Isozymes IV.
   Genetics and Evolution" (Markert, C.E., ed.), pp. 449-462.
   Academic Press, New York, San Francisco, London.
Ford, P.J. and Southern, E.M. (1973). Different sequences for
   5S RNA in kidney cells and ovaries of *Xenopus laevis*. *Nat-
   ure New Biol*. **241**, 7-12.
Ishikawa, H. (1977). Evolution of ribosomal RNA. *Comp. Bio-
   chem. Physiol*. **58B**, 1-7.
Kokileva, L., Mladenova, I. and Tsanev, R. (1971). Differen-
   tial thermal stability of old and new ribosomal RNA of rat
   liver. *FEBS Letters* **12**, 313-316.
Ohno, S. (1970). Evolution by gene duplication. Springer,
   Berlin, Heidelberg, New York.
Ohno, S. (1974). Cytogenetics of chordates, protochordates,
   cyclostomes and fishes. Borntraeger, Stuttgart.
Ohno, S., Stenius, C., Faisst, E. and Zenzes, M.T. (1965),
   Post-zygotic chromosomal rearrangements in rainbow trout.
   (*Salmo irideus* Gibbons). *Cytogenetics* **4**, 117-129.
Pedersen, R.D. (1971). DNA content, ribosomal gene multipli-
   city and cell size in fish. *J. Exp. Zool*. **177**, 65-78.
Schmidtke, J. and Engel, W. (1975). Gene action in fish of
   tetraploid origin. I. Cellular and biochemical parameters
   in Cyprinid fish. *Biochem. Genet*. **13**, 45-51.
Schmidtke, J. and Engel, W. (1976). Gene action in fish of
   tetraploid origin. III. Ribosomal DNA amount in Cyprinid
   fish. *Biochem. Genet*. **14**, 19-26.
Schmidtke, J., Zenzes, M.T., Dittes, H. and Engel, W. (1975a).
   Regulation of cell size in fish of tetraploid origin. *Nat-
   ure (London)* **254**, 426-427.
Schmidtke, J., Atkin, N.B. and Engel, W. (1975b). Gene action
   in fish of tetraploid origin. II. Cellular and biochemi-
   cal parameters in Clupeoid and Salmonoid fish. *Biochem.
   Genet*. **13**, 301-309.
Schmidtke, J., Schulte, B., Kuhl, P. and Engel, W. (1976a).
   Gene action in fish of tetraploid origin. V. Cellular RNA
   and protein content and enzyme activities in Cyprinid,
   Clupeoid and Salmonoid species. *Biochem. Genet*. **14**, 975-
   980.
Schmidtke, J., Zenzes, M.T., Weiler, C., Bross, K. and Engel,
   W. (1976b). Gene action in fish of tetraploid origin.
   IV. Ribosomal DNA amount in clupeoid and Salmonoid fish.
   Biochem. Genet. **14**, 293-297.
Schmidtke, J., Beçak, W. and Engel, W. (1976c). The reduction

of genic activity in the tetraploid *Odontophrynus americanus* is not due to a loss of ribosomal DNA. *Experientia* **32**, 27-28.

Schmidtke, J., Schmitt, E., Leipoldt, M. and Engel, W. (1979a). Amount of repeated and non-repeated DNA in the genomes of closely related fish species with varying genome sizes. *Comp. Biochem. Physiol.* **64B**, 117-120.

Schmidtke, J., Schmitt, E., Matzke, E. and Engel, W. (1979b). Non-repetitive DNA sequence divergence in phylogenetically diploid and tetraploid teleostean species of the family Cyprinidae and the order Isospondyli. *Chromosoma (Berlin)* **75**, 185-198.

Sparrow, A.H. and Nauman, A.F. (1976). Evolution of genome size by DNA doublings. *Science* **192**, 524-529.

Staehelin, T. and Falvey, A.K. (1971). Isolation of mammalian ribosomal subunits active in polypeptide synthesis. *In* "Methods in Enzymology" (Colowick, S.P. and Kaplan, N.O., eds) vol. 20, pp. 433-446. Academic Press, New York.

Staynov, D.Z., Pinder, J.C. and Gratzer, W.B. (1972). Molecular weight determination of nucleic acids by gel electrophoresis in non-aqueous solutions. *Nature New Biol.* **235**, 108-110.

Wettstein, E. (1937). Experimentelle Untersuchungen zum Artbildungsproblem. I. Zellgrössenregulation und Fertilwerden einer polyploiden Bryum-Sippe. *Z. Ind. Abst. Vererb.* **74**, 34-53.

PART III
NUCLEAR ORGANIZATION AND DNA CONTENT

Nucleotypic Basis of the Spatial Ordering of Chromosomes in Eukaryotes and the Implications of the Order for Genome Evolution and Phenotypic Variation

MICHAEL D. BENNETT

Plant Breeding Institute, Cambridge, England

INTRODUCTION

Like many other biologists, I have often questioned what is the meaning (if any) of first, the large interspecific variation in C value, and, second, the amazing quantitative and qualitative variation in the copy numbers and arrangements of DNA sequences. Because I work at the Plant Breeding Institute my approach to these matters has been to ask, "What are the present consequences for the plant of major differences in DNA C value and sequence copy number?", rather than to ask how such differences arose in evolution? Putting the question in this form means that I am trying to discover whether DNA C value and sequence copy number are characters amenable to selection or manipulation in crop plant breeding programmes to produce predictable improvements in useful agronomic phenotypic characters such as yield and quality.

My research has led me to conclude that large absolute variation in DNA C value, and large variation in DNA sequence organization, seen as variation in the size of C bands, is often of great significance in affecting the biology of organisms. In particular, I have shown that the DNA C value plays an important role in determining the maximum rate of development in angiosperms, and hence where they can survive, and the range of life-cycle types they can display in a particular environment. It seems worth mentioning here four sets of the results that lead me to these conclusions.

First, after Van't Hof and Sparrow (1963) found a positive linear correlation between minimum cell cycle time and DNA C value, I found a similar relationship between the duration of meiosis and DNA C value (Bennett, 1977a). Thus, in

diploid angiosperms the duration of meiosis at 20°C ranges
from less than one day in species whose C DNA value is about
1 to 2 pg, to more than two weeks in species whose C DNA
value exceeds about 50 pg.

Second, a positive relationship between DNA C value and the
rate of somatic and meiotic cell development would lead inevitably to the existence of a positive relationship between DNA
C value and minimum generation time, and hence between DNA C
value and life-style. For example, as meiosis at 20°C in
plants with C values of 1 pg and 80 pg contributes about 1
and 18 days, respectively, to their minimum generation times,
it is obvious that ephemeral species capable of the shortest
minimum generation times (about 6 weeks) must have low C
values; while species with high C values must be incapable of
displaying an ephemeral life-style. The available data fully
support these expectations, and show a relationship between
DNA C value and minimum generation time (Bennett, 1972; Smith
and Bennett, 1975). Thus, species with medium or high C values cannot be ephemerals, and species with very high C values
are obligate perennials. In other words, the DNA C value
sets a limit on the present range of phenotypes that can be
achieved by genic control. These results show that nuclear
DNA can affect the plant phenotype by its mass or size, and I
called such effects "nucleotypic" (Bennett, 1971). Thus, the
nucleotype can be defined as those non-genic characters of the
nuclear DNA that affect or control the phenotype, independent
of its encoded informational content.

Third, the relationship between DNA C value and rate of
development can determine which species survive in particular
environments. Recent work (Bennett and Lewis-Smith, unpublished results) indicates a complete absence of angiosperms
with high C values per diploid genome from the continuously
cool environments of South Georgia and the Antarctic penin-
sula. The probable cause is that all species with high DNA
C values, and hence slow development, cannot establish from
seed in the short growing season in these environments. If
so, this result will further illustrate an important point;
namely, that the DNA C value of an organism is probably often
limited by its consequences at a single stage of the life-
cycle.

Further evidence that DNA C values affect where plants
grow, or rather grow best, is provided by a study on DNA
amount and the latitude of cultivation of the world's major
agricultural crop plant species, which shows a striking
correlation between the DNA amount per diploid genome and
the chosen latitude for cultivation (Bennett, 1976). The
cause of the cline is unknown, but insofar as one object of
agriculture is to maximize yield per unit and per unit time,

then clearly even this is somehow related with DNA C value, and perhaps its component sequences.

Fourth, seed development is adversely affected by the presence of late-replicating blocks of telomeric heterochromatin on the rye chromosomes in the intergeneric wheat-rye hybrid crop plant *Triticale*. The 33% difference in C value between the diploid genomes of wheat and rye, which determines different rates of development in the parent species, is directly responsible for nuclear instability and/or the abortion during endosperm development in *Triticale* seeds (Bennett, 1977b). Selecting for a progressive reduction in the size of the troublesome rye C bands in *Triticale* (equivalent to a 7% reduction in the rye C value) has resulted in highly significant improvements both in nuclear stability in young endosperm and in the yield and quality of mature seeds (Gustafson and Bennett, 1982). This seems to be a good example of selection for a lower C value acting specifically against sequences located in telomeric segments of particular chromosomes.

SPATIAL ORDERING OF CHROMOSOMES

The purpose of this paper is to describe some results from my unpublished work, which demonstrate an important role of the nucleotype in determining the architecture of karyotypes and the spatial disposition of the chromosomes. I want first, to establish that chromosome disposition is normally highly ordered; second, to show that simple rules determine chromosome disposition; and third, to discuss some implications of these facts for genome evolution and phenotypic variation.

There has been continued interest in trying to answer the question as to whether the dispositions of chromosomes in interphase nuclei and mitotic cells of higher organisms is normally ordered. The relevant literature is very extensive and often controversial (useful reviews are given by Comings, 1968; Avivi and Feldman, 1980), and it contains numerous examples of various particular non-random chromosome arrangements (e.g. see Miller *et al.*, 1963; Kempanna and Riley, 1964; Horn and Walden, 1978, Ashley, 1979). Nevertheless, it has remained unclear as to whether chromosome disposition is normally ordered or not. Moreover, those examples where chromosome disposition is non-random have together yielded no widely accepted fundamental understanding of any general features of chromosome order, beyond the assertion that heterochromatin is sticky (Schmid *et al.*, 1975). (N.B. It is generally agreed that chromosomes usually retain the relic-telophase arrangement during interphase with centromeres and telomeres roughly associated at opposite poles, and also that telomeres are often attached to the nuclear membrane. However, in the present work the term "chromosome order" is used to mean much more than these limited generalizations. Here chromosome order means specific spatial associations between each individual chromosome and one or more other particular chromosome(s).) If the disposition of chromosomes is normally ordered according to some widely applicable biological law, then there must have been critical deficiencies in either the techniques used to prepare the experimental material, or in questions asked when analysing the data.

Most studies of chromosome disposition use mitotic cells that have been heavily pretreated (often with colchicine) and then heavily squashed. Both treatments introduce large amounts of random chromosome redisposition (Horn and Walden, 1978; Rodman *et al.*, 1980). Thus, the fear has been expressed that looking for order in such metaphase squashes is like trying to reconstruct an egg that has been thrown at a door, while for colchicine-pretreated squashes, the simile is scrambled egg! Clearly, it would be better to study chromosome dispositions in unsquashed cells with undisturbed chromosomes. Serial section electron microscopy allows such studies. However, its successful application to questions of chromosome disposition depends on an ability to identify unequivocally each individual chromosome in single cells seen in serial electron micrographs, and hence to establish their real spatial relationships.

Pachytene chromosomes seen in serial electron micrographs can be individually identified on morphological grounds in single cells of several organisms, including man (Holm and Rasmussen, 1977). I have carefully developed three techniques, which use morphological characters to identify individual chromosomes, or chromosome segments, in serial electron micrographs of single somatic cells at mitosis or at interphase.

First, in suitable materials (e.g. *Hordeum vulgare* cv. Tuleen 346), all the chromosomes are identified in most single mitotic metaphase cells using relative chromosome volume, the ratio of the volumes of their two arms, and the presence of a nucleolar organizer (unpublished results), all characters routinely used to identify chromosomes in light microscope metaphase squashes.

Second, major C bands can be shown up in electron micrographs. Thus, in suitable materials (e.g. *Secale africanum*), chromosome identification is also rigorously established by the presence and relative size of lighter staining segments corresponding exactly to the major C bands in Giemsa-stained light microscope metaphase squashes (unpublished results). While such regions are occasionally seen in material grown at 20°C, their differentiation is regularly enhanced in material

grown for 24 hours in water at about 1°C. While this treatment prevents or delays anaphase separation, surprisingly it
does not inhibit congression of chromosomes on to the metaphase plate in our material (Heslop-Harrison and Bennett, unpublished results).

Third, the volume of the centromere is often directly proportional to the volume of the chromosome to which it belongs
(Jenkins and Bennett, 1981; Bennett, unpublished results).
The correlation is close enough to allow the unequivocal
identification of some or even all chromosomes in single
cells of suitable materials. Moreover, this method has the
advantage that it can be applied to interphase nuclei.

These techniques have been applied in studies of chromosome disposition in cereals.

GENOME SEPARATION IN INTERGENERIC HYBRIDS

About two years ago, Dr Finch and I noticed that in *Hordeum* ×
Secale hybrids, the two haploidal parental genomes often
appeared separate in light microscope mitotic metaphase
squashes. The DNA *C* values of the two diploid parents differ
by about 33 to 49%, and the seven *Secale* chromosomes are all
readily distinguished by their larger size from the seven
smaller *Hordeum* chromosomes. At first, we questioned whether
this genome separation was real and, if so, whether it was an
artefact of squashing. However, we showed that it was real,
and significantly so, in most cells of the hybrid, barley ×
S. africanum, either cold pre-treated or un-pretreated, both
in squashes viewed by light microscopy and in serial section
electron micrographs of unsquashed cells (Finch *et al.*, 1981;
and unpublished results). Figure 1 shows polar views of the
positions of the 14 centromeres on congressed metaphase plates
of nine unsquashed, cold pre-treated, hybrid metaphase cells
These views, obtained by computer rotation of the co-ordinates
of centromeres measured, using axes in or at right-angles to
the plane of serial sectioning, show obvious genome separation. Comparing the mean distance of the *Hordeum* and *Secale*
centromeres from the centre of the metaphase place (the mid-
centromere point) in 14 reconstructed root-tip cells of two
Hordeum × *Secale* hybrids gave equally striking results. *Hordeum* centromeres were closer to the mid-point in every cell,
and significantly so in 12 out of 14!

These observations pose the question as to the cause of
genome separation. It was not due to chromosome size, since
comparisons of *Secale* chromosomes along and of *Hordeum* chromosomes alone both showed negative correlations between chromosome size and distance from the mid-centromere point in most
cells. Perhaps the separation of haploid genomes in *Hordeum*

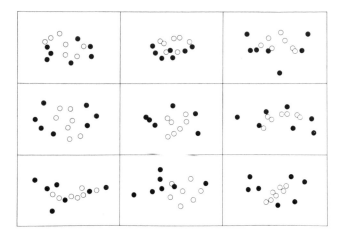

FIG. 1 *Polar views of the distribution of* Hordeum *(open circles) and* Secale *(closed circles) centromeres on the congressed plates of 9 reconstructed root tip metaphase cells of the hybrid* H. vulgare × S. africanum.

TABLE I

The mean distance (μm) of the Hordeum *and* Secale *centromeres from the mid-centromere point in serial section reconstructions of single root tip cells of two intergenomic hybrids*

HYBRID	REPLICATE	*SECALE*		*HORDEUM*	$t_{(13)}$	P[†]
	1	4.49	>	2.53	5.04	***
	2	4.58	>	2.60	3.83	***
H. vulgare ×	3	4.97	>	2.44	4.81	***
S. africanum	4	5.41	>	2.87	2.88	***
(from Dr Finch,	5	5.01	>	2.93	2.41	***
PBI, Cambridge)	6	4.37	>	3.12	1.31	N.S.
$2n = 2x = 14$	7	6.30	>	2.62	6.07	***
	8	3.25	>	2.60	1.26	N.S.
	9	4.26	>	3.01	2.31	*
H. vulgare ×						
S. africanum	1	3.76	>	2.47	2.67	**
(cv. 20 × R)	2	4.38	>	2.64	2.43	*
(from Dept of Botany,	3	5.52	>	2.81	4.73	***
University of Edin-	4	4.60	>	2.82	3.07	**
Edinburgh)	5	5.11	>	2.85	2.95	**
$2n+1 = 2x+1 = 15$						

[†]$P > 0.001$ (***); > 0.01 (**); > 0.02 (*); > 0.02 (N.S.)

× *Secale* hybrids is due to the greater similarity of DNA
sequences between chromosomes of the same parental species?

   If chromosome associations normally reflect the hierarchy
of gross sequence similarity, then in a diploid species,
somatic association of homologues would be expected, since
these share virtually identical sequences. On the other hand,
side-by-side separation of haploid genomes would be unexpected, since this arrangement would separate homologues. We
tested these expectations against the dispositions of the 14
centromeres in a diploid barley, *H. vulgare*, Cultivar Tuleen
346, a triple interchange homozygote stock with the decided
advantage over normal barley that all seven of its chromosomes can be identified on morphological grounds in the great
majority of cells, both in the light and electron microscope.
In serial electron micrographs of nine root tip cells, the
mean separation of homologues (4.66 μm) was greater, though
not significantly so, than the mean separation of non-homologues (4.45 μm).

   Clearly, diploid Tuleen 346 did not show somatic association of homologues. Does it show haploid genome separation?
The answer is probably yes. Modelling shows that in a large
population of cells, each with 14 centromeres arranged at random on a flat metaphase plate, two haploid genomes are separated by chance by a single straight line drawn on the polar
view of the plate in about 21% of cells. However, it was possible to separate two haploid genomes in this way in five out-
of 11 root tip cells of Tuleen 346 (i.e. about 45%), and preliminary studies using other materials have yielded similar
results. Although these preliminary results are together significant ($P < 0.01$), they must be viewed with some caution
until it is confirmed that haploid genomes are separated more
frequently than would occur by chance using larger samples of
nuclei in each material. However, if the early trends are
maintained, confirmation should be forthcoming. Examination
of the polar views of metaphase plates of reconstructed Tuleen 346 root tip cells suggests that there may be a strong
tendency for a side-by-side separation of the two haploid
genomes (see Fig. 2).

   As genome separation occurred in both an established diploid species and new intergenomic hybrids, it is probably the
norm. Recent analysis of the spatial dispositions of human
chromosomes may support this view (Coll *et al.* 1980). The
side-by-side arrangement of two haploid genomes (Fig. 2) in
diploid barley and the large separation of homologues clearly
shows that the separation of haploid sets is independent of
gross DNA sequence similarity. Both haploid sets share virtually identical sequences, yet they are separated; while
homologues, which have the greatest sequence similarity are,

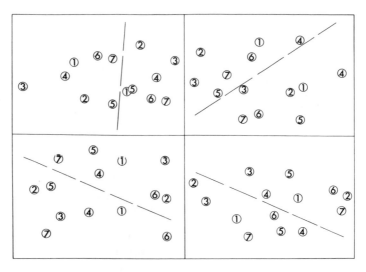

FIG. 2 *Polar views of the distribution of centromeres of the 7 pairs of homologues on the congressed plates of 4 cold-treated root tip metaphase cells of* H. vulgare *cv. Tuleen 346. N.B. Numbers are the genetic linkage groups of barley.*

on average, further apart that non-homologues with lower sequence similarity.

SPECIFIC ORDERING OF CHROMOSOMES IN THE HAPLOID GENOME

The next question is "What is the basis of haploid genome separation?" (or to put it another way "How does an haploid set assemble itself?"). I think I can begin to answer this question, but to do so I must first describe a model for predicting the position of each chromosome in a haploid genome and also the analysis that was used to test for non-random chromosome dispositions in general, and for the unique prediction of the model in particular.

1. A Model for Predicting Chromosome Order

Shchapova (1969) claimed that in many organisms with non-telocentric chromosomes, the chromosome arms are highly ordered with respect to their associations in pairs to form chromosomes. In essence, she asserted that different-sized arms are distributed between chromosomes of the haploid complement so that given telomere-to-telomere links between the longest long arm and the shortest short arm, the second longest long arm and the next to shortest short arm, etc., the sum of the lengths of each pair of adjacent arms on

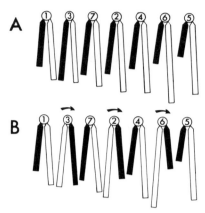

FIG. 3 *The 7 chromosomes of the haploid genome of* S. cereale *cv. UC 90 ordered according to (A) Shchapova's model and (B) Bennett's model.*

different chromosomes is virtually constant. Assembling a haploid chromosome set in a closed chain according to Shcha-pova's model, linking only associations of long and short arms, gives a unique prediction of the disposition (affini-ties) of the chromosomes of the haploid genome. (Fig. 3A)

In my view, the arrangement of chromosomes in Shchapova's model, while perhaps of functional significance at some stage of nuclear development, would not function mechanically in normal somatic cells. However, it occurred to me that if alternate chromosomes were excised from Shchapova's model, rotated by 180° and reinserted, then this would bring together the most similarly sized pairs of long arms, and pairs of short arms, throughout the haploid complement, except at one point of discontinuity (Fig. 3B). This arrangement which should work mechanically in real nuclei, gives the same unique prediction for chromosome order in the haploid genome as Shchapova's model, but attains it by specific associations of pairs of long arms, or pairs of short arms. Strong sup-porting evidence for my model is provided by several observa-tions (notably Wagenaar (1969); Ashley (1979)) of chromosomes showing regular, specific, telomere-to-telomere links between pairs of long, and pairs of short arms. Is this arrangement the basis of genome separations, and how can this possibility be tested?

2. *Analysis Testing for a Predicted order*

Seven chromosomes, the basic number in *Hordeum* and *Secale*, can be arranged in a closed chain in 360 (factorial 6 × ½) different ways, ignoring the 360 mirror images. Each

chromosome was considered as being located at its centromere.
The three-dimensional co-ordinates and the identity of each
centromere in a haploid genome were entered into our Kontron
Videoplan micro computer. The computer then calculated for
each of the 360 possible orders, the distance travelled in
passing once through each centromere, while starting and
finishing at the same centromere. It then ranked the 360
orders in order of increasing length and stored the list.
This was repeated for several replicate cells. Finally, the
computer summed the rank of each order from each replicate
cell, reranked the 360 sums, and printed them out in increasing size order.

   If the chromosomes are arranged according to the unique
prediction of my model, then the predicted order of centromeres should occur high on the list of 360 sums, being achieved by a significantly shorter route than other orders low on
the list of 360 ranked sums. Furthermore, order at or close
to position 360 should be rich in "anti-orders" with respect
to the prediction.

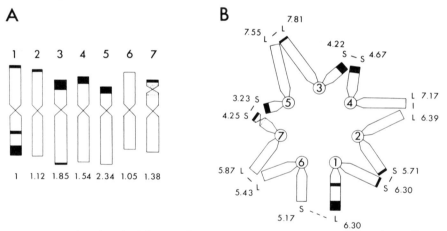

FIG. 4 *The haploid complement of* S. africanum *showing the
major C bands (A) arranged in order of decreasing total chromosome size (from left to right; arm ratios are given below
each chromosome); (B) ordered according to Bennett's model
(numbers by arms are their mean volumes in* μm³*).*

3. *Secale africanum*

Figure 4(A) shows the haploid complement of diploid *S. africanum*. All seven chromosomes are readily identified in single
cells using relative size, arm ratio and the presence and size
of telomeric heteromatic regions. Using my model, placing the

TABLE II

A comparison of the order of chromosomes in the haploid genome of S. africanum predicted using Bennett's model (1243576 – see Fig. 4), with a part of the list of ranked sums of ranks for the 360 possible orders in 11 replicate haploid sets from root tip cells

Rank	Order	Rank in replicate cells											Ranked sum of ranks
		1	2	3	4	5	6	7	8	9	10	11	
1	1243576	80	320	9	303	36	240	47	45	16	38	49	1183
2	1245376	114	323	23	75	14	203	160	132	13	15	130	1202
3	1267534	185	115	38	306	34	346	60	72	25	28	4	1213
4	1426735	246	164	87	62	66	164	117	73	3	153	52	1222
5	1624357	76	133	2	122	299	217	35	116	84	35	117	1236
356	1723654	251	20	322	236	304	223	339	187	327	313	201	2723
357	1723654	262	102	251	165	339	258	293	174	324	244	356	2768
358	1725463	163	197	289	39	296	305	272	228	321	237	346	2803
359	1564723	283	243	337	281	353	7	225	189	303	327	298	2846
360	1527463	181	199	336	211	267	73	271	309	328	340	338	2853

most similarly sized pairs of long arms and pairs of short
arms together, the predicted order of chromosomes (and hence
of their centromeres) in the haploid genome is 1-2-4-3-5-7-6-
(Fig. 4(B)).

The computer ranked the 360 orders by length for 11 haploid
genomes of *S. africanum*, summed the ranks for each order, and
reranked, as described above. Table II shows part of the
resulting list. Summed ranks for individual orders differed
widely from a minimum of 1183 to a maximum of 2853, and t-
tests showed highly significant differences between lines of
11 replicate rank numbers for orders high and low on the list.
Clearly, centromeres in this haploid genome were not randomly
distributed on the metaphase plate. The predicted order,
1-2-4-3-5-7-6-, ranked first out of 360. Also, as predicted,
the five orders with the longest routes (ranked 356 to 360)
were all anti-orders containing no pair of adjacent chromosomes, in common with the prediction. For example, -1-5-2-7-
4-6-3-, (order 360), puts 1 by 5, 5 by 2, 2 by 7, 7 by 4, 4 by
6 and 6 by 3; none of which occurs in major prediction 1-2-4-
3-5-6-7-.

This result was gratifying, but "Would the model work for
another situation, for example a diploid species rather than
a hybrid?" What about Tuleen 346 barley, which apparently
showed haploid genome separation?

4. *Tuleen 346 Barley*

In a diploid with two homologues of each chromosome, there
are not only 360 possible orders of the seven chromosomes
within a closed haploid set, but there are 64 ways of assembling each of the 360 orders, depending on which homologue is
chosen each time. Our computer constructed all 64 possibilities, ranked them in order of increasing distance travelled
in assembling two haploid sets as closed chains, and then dis-
carded all but the one whose assignment of centromeres to two
haploid genomes used the shortest routes. These "best fit"
values were obtained for all 360 possible orders, and there-
after processed as already described for haploid genomes in
hybrid cells.

The predicted order for normal barley (Fig. 5(A))
according to my model is -4-1-2-3-6-5-7-. However, if chromosomes are ordered in the haploid genome purely according to
arm length, then Tuleen 346 has a different expectation, because the translocations have greatly rearranged the complement.

Four adjacent arms in the prediction for normal barley
(Fig. 5(C)) are unaffected by translocations in Tuleen 346,
namely; 7-long, both arms of 4, and 1-long, a region

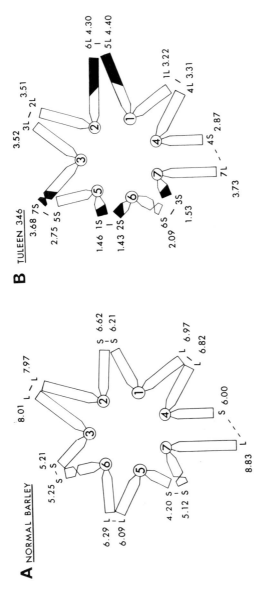

FIG. 5 The haploid complements of (A) normal barley, and (B) Tuleen 346 barley, ordered according to Bennett's model (numbers by arms of normal barley are relative lengths from Noda & Kasha (1978); the length of chromosome 6 minus the satellite = 100 arbitrary units). The numbers by the arms of Tuleen 346 chromosomes are their mean volumes (in μm³) in reconstructed root-tip metaphase cells (Bennett, unpublished results). Note that all 7 pairs of associated telomeric segments are identical in both predictions.

TABLE III

A comparison of the order of chromosomes in the haploid genome of Tuleen 346 barley predicted using Bennett's model (412567 – see Fig. 5), with a part of the list of ranked sums of ranks for the 360 possible orders in replicate root tip metaphase cells

Rank	Order	Rank in replicate cells									Ranked sum of ranks
		1	2	3	4	5	6	7	8	9	
1	4123567	158	63	301	37	142	27	36	34	55	853
2	4123576	134	30	332	14	150	3	134	41	22	860
3	4123765	255	61	212	15	138	7	122	67	53	930
4	4123675	32	149	178	2	134	16	272	116	36	935
5	4127635	35	146	30	63	250	19	201	114	99	957
356	4736152	18	245	184	324	241	203	208	360	351	2234
357	4517362	59	331	54	298	397	254	314	289	137	2243
358	4735162	50	287	279	326	316	174	108	269	262	2271
359	4715263	48	353	231	339	149	358	75	341	191	2285
360	4371526	56	262	311	358	154	296	73	319	186	2315

containing the important discontinuity between 7 long and 4
short. If this region is left intact, then the new predicted
order for Tuleen is -4-1-2-3-5-6-7. This order (Fig. 5(B)) is
also uniquely desirable, because all seven pairs of telomeres
are associated exactly as in the prediction for normal barley,
even though six of the seven chromosomes have been involved in
translocations. The summed ranks for the 360 orders in nine
cells (Table III) differed widely, ranging from 853 to 2315,
and t-tests showed highly significant differences between
lines of nine replicate rank numbers for orders near the top
and bottom of the list. Again, centromeres were not randomly
ordered on the metaphase plate. Most significantly, the favoured prediction -4-1-2-3-5-6-7- was ranked first out of 360.
Also as expected, the 360th order (4-3-7-1-5-2-6-) is an anti-
order containing no adjacent pair of numbers in common with
the prediction (-4-1-2-3-5-6-7-).

5. *Secale cereale*

The spatial disposition of chromosomes in diploid rye (*Secale
cereale*) will probably also fit the prediction of my model
when sufficient replicate cells are analysed. Figure 6 shows
the order of chromosomes in the haploid genome of rye predicted by my model; namely, -1-3-7-2-4-6-5-. Computer analysis

TABLE IV

*A comparison of the order of chromosomes in the haploid gen-
ome of rye (S. cereale cv. UC 90) predicted using Bennett's
model (1372465 - see Fig. 6) with parts of the lists of ranked
sums of ranks for the 360 orders in haploid genomes from
(a) 5 reconstructed root tip metaphase cells and (b) 4 diploid
rye cv. King II root tip metaphase squashes*

MATERIAL	RANK	ORDER	RANK IN REPLICATE CELLS					RANKED SUM OF RANKS
			1	2	3	4	5	
	1	1372654	35	73	73	38	98	317
20 R	79	1372465	202	153	77	258	49	739
	360	1643527	237	346	293	309	321	1506
	1	1537642	45	67	14	167	–	293
	2	1537246	59	154	83	43	–	339
King II	25	1372465	98	137	142	99	–	476
	360	1432657	348	346	343	199	–	1236

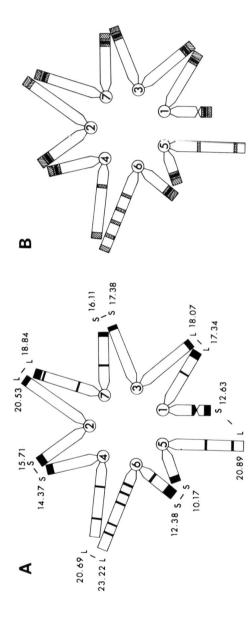

FIG. 6 The haploid complement of S. cereale including major C bands (A) arranged according to Bennett's model and (B) with the distribution of major sites for 4 families of repeated DNA sequences (Hutchinson et al., 1981) superimposed (numbers by arms of chromosomes in A are their relative lengths in arbitrary units (J.P. Gustafson, personal communication)).

of the three-dimensional co-ordinates for the rye centromeres
in five root-tip metaphase cells of an aneuploid intergeneric
hybrid called 20 × R (barley × rye), known to show separation
of haploid genomes (Table I), placed the prediction 79th out
of 360 (Table IV), but comprising ranking numbers significantly lower from those for the order ranked 360.   However,
the order whose sum ranked first, -1-3-7-2-6-5-4-, differed
from the prediction only by having centromere 4 in a different
position.   Analysis of just four randomly chosen light microscope C banded metaphase squashes of the rye variety King II
ranked the prediction 25th out of 360.

CONCLUSIONS

In view of the above results, I feel justified in asserting
that the disposition of chromosomes is normally highly ordered.
Moreover, the basic unit in diploids seems to be the haploid
genome within which the individual chromosomes show a very
strong tendency to be arranged in a highly ordered unique
sequence of near neighbours.   Already, the model has successfully predicted orders with significantly closer arrangements
than random in three out of three materials, and in two of
these (S. africanum and H. vulgare cv. Tuleen 346) the unique
prediction was ranked first out of 360 possibilities.   Clearly
the model works, and therefore it should be a powerful tool
for studying genome size and architecture.   It seems likely
that the model can predict what that unique sequence is for
many materials given (1) the number of chromosomes in a unit,
and (2) accurate measurements of their relative sizes and arm
ratios.
    The results just described allow certain other conclusions
of central relevance to the theme of this volume.   First,
there seems little doubt that the relative disposition of each
chromosome in the haploid genome is determined primarily by
the size of its arms.   Now, it has been carefully shown in
several species, including man, that relative chromosome size
and relative chromosome DNA content are directly proportional.
Thus, in view of its prime biological role, it is arguable
that chromosomes are spatially ordered in the haploid genome
according to the DNA contents of their arms.   If so, the
determination of ordered chromosome dispositions is, at least
partly, a nucleotypic phenomenon, insofar as the mass of DNA
in the various arms is independent of the encoded information
it contains.   Arm size determines the chromosome position,
but what determines arm size in the short and long terms?
Any nucleotypic control of chromosome order will doubtless
interact with genic control of chromosome behaviour, and it
seems probable that there are genes that control the relative

size of the component segments of the genome (Bennett *et al.*, 1981).

Second, insofar as the chromosomes have ordered dispositions, so do all the genes that they carry. It has long been recognised that genes on the same chromosome are spatially associated (we say "linked"). Ordered dispositions of chromosomes means that genes on heterologues also tend to be associated. It is reasonable to assume that the spatial associations of genes, not only along a chromosome, but also on the super-domains of associated arms not all sharing the same centromere, reflect strong selection for their economic and efficient action. If so, major structural changes that disrupt the selected pattern of domains will normally reduced the efficiency of gene action and hence be selected against. However, even when occasional structural rearrangements improve the efficiency of gene action, there will often still be a strong selection to retain or regain the original shapes of the chromosomes, because the rules governing the special ordering of chromosomes remain unchanged. Thus, the appearance of a karyotype will often tend to be conserved, even if homologous chromosomes in related diploids come to have different positions in the haploid genome.

Third, the rules governing a chromosome's position in the ordered haploid genome according to its arm lengths will often place narrow constraints on its ability to undergo unilateral changes in size without also changing its position. If the original order of chromosomes and disposition of gene domains within the haploid genome is to be retained as the C value changes, so must the original ranking of arm lengths. Thus, as C value increases, there will often be selection for a balanced increase in the DNA content of each chromosome and chromosome arm. Presumably, this selection would not be directed at particular sequences but, equally presumably, there may well be particular sequences able to react more quickly to such pressures and retain the form of an efficient karyotype design. Moreover, adjustments in C value are expected usually to involve sites that disrupt gene dispositions within or between chromosomes least. Such sites are located primarily at or near telomeres and/or centromeres, which is just where most of the largest C bands known in higher plants occur. (It is also important to recognise that the ordering of chromosomes in the haploid set according to my model does not exclude polymorphisms. Rather, it should help to explain them, since the model predicts which arms are free to vary in size without upsetting the order, and which are not. For example, the longest long and short arms can both become longer, and the shortest long and short arms can both become shorter, without upsetting the order.

Moreover, the two arms at the discontinuity can often also vary in length without upsetting the order. Thus, up to six arms can vary greatly in size (within the rules governing order) without changing the order. It follows, therefore, that karyotypes with the same order can have markedly different appearances, especially as the number of chromosomes in the haploid set is small (i.e. less than 7). These facts, alluded to here, will be set out fully in a later publication.) Consequently, the distribution and size of C bands can probably reveal a lot more about the spatial ordering of chromosomes than has been realised hitherto.

If chromosomes are ordered as my model suggests, then adjacent pairs of long arms, or pairs of short arms should often display similar changes in their DNA content, and hence similarly sized C bands. The evidence is that they do. For example, superimposing the C bands in *S. africanum* and rye on their predicted order of chromosome arms (Figs 4 and 6) shows that the presence or absence and the relative size of C bands matches on adjacent pairs of long or short arms (other than, significantly, at the discontinuity).

The ordered arrangement of chromosome arms in the haploid complement also has mechanical consequences for chromosome behaviour. For example, Fig. 7(A) superimposes data for the frequency of transmission as a trisome of each barley chromosome on the predicted order of chromosomes in this species. Transmission frequency is clearly related to chromosome position in the haploid genome order. Lack of space prevents my multiplying similar examples here. However, I believe that in order to function correctly, DNA sequences directly concerned with controlling mechanical aspects of chromosome behaviour *must* be sited in the mechanically right place. The *minute* gene sited near the end of the short arm of the chromosome 4, which greatly affects genome separation at anaphase in barley (Takahashi, 1975), may be a good example (Fig. 7(B)).

The mechanism whereby associated chromosome arms recognise each other, and the means whereby order in the haploid genome is maintained, are both unknown. Once established, a particular spatial arrangement of chromosomes might be maintained by spindle microtubules at anaphase, and at interphase, by protein elements of the nuclear skeleton, and perhaps by the intranuclear bundles of fibrillar material found to link pairs of telomeric segments in rye and wheat (Bennett *et al.*, 1979). Riley and Flavell (1977) suggested that recognition leading to meiotic pairing may involve particular DNA sequences, and the same may be true of somatic ordering within the haploid genome, except that the process would involve sequences on heterologous chromosomes in somatic cells, but on homologues at meiosis. If, as seems likely, a similar order

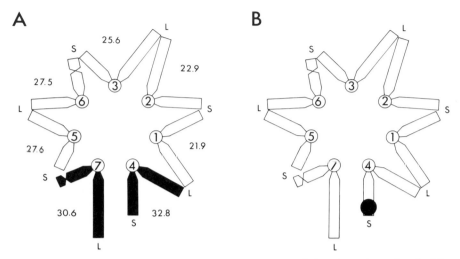

FIG. 7   *(A) The frequency of occurrence (%) of trisomic individuals after selfing each of the 7 trisomics of* H. vulgare
and (B) the site of the min *gene, superimposed on the haploid
complement of* H. vulgare *ordered to Bennett's model.*

of chromosomes in two haploid genomes is an essential prerequisite for meiotic pairing in a diploid, then DNA controlling the first steps of recognition in the pairing process
will probably be located at sites involved in the mechanics
of switching from the normal separation of homologues seen in
somatic cells to the different meiotic configuration with
associated homologues. Chromosome ends (that is telomeres),
and end arms at the discontinuity in the ordered haploid genome, are likely sites for first recognition events (on mechanical grounds). Thus, it is interesting to note that in all
three materials where the order of chromosomes is experimentally verified, the C banding of a chromosome at the discontinuity has unique features. First, in normal barley, chromosome 4 (located at the discontinuity) has a much greater
length of C bands than the other six chromosomes (Linde-
Laursen, 1978). Second, in *S. africanum* (Fig. 4), the long
arm of chromosome 1 (which is at the discontinuity) has the
largest C band in the complement, and the only large C band
on a long arm in the complement. It also has a large intercalary band, which stains uniquely with Wright's stain (Seal
and Bennett, unpublished results). Third, in rye (Fig. 6),
the long arm of chromosome 5 (which is at the discontinuity)
has the only telomere lacking an *in situ* binding site for any
of the four highly repeated DNA sequences studied by Dick
Flavell and his colleagues (Hutchinson *et al.*, 1981).

In conclusion, I believe that the existence of spatial
ordering of chromosomes has important implications for many
fields of biology, especially cytogenetics. For some of us
it may increase our faith, and hopefully our understanding
too, of the roles of law-abiding DNA in ordering genome size
and architecture.

REFERENCES

Ashley, T. (1979). Specific end-to-end attachment of chromo-
    somes in *Ornithogalum virens*. *J. Cell Sci.* **38**, 357-367.
Avivi, L. and Feldman, M. (1980). Arrangement of chromosomes
    in the interphase nucleus of plants. *Human Genet.* **55**,
    281-295.
Bennett, M.D. (1971). The duration of meiosis. *Phil. Trans.*
    Roy. Soc. ser. B **178**, 277-299.
Bennett, M.D. (1972). Nuclear DNA content and minimum genera-
    tion time in herbaceous plants. *Proc. Roy. Soc. ser. B*
    181, 109-135.
Bennett, M.D. (1976). DNA amount, latitude and crop plant
    distribution. *Enviro. Exp. Bot.* **16**, 93-108.
Bennett, M.D. (1977a). The time and duration of meiosis.
    In "A discussion on the meiotic process" (Riley, R.,
    Bennett, M.D. and Flavell, R.B., eds), *Phil. Trans. Roy.*
    Soc. Ser. B **277**, 201-226.
Bennett, M.D. (1977b), Heterochromatin, aberrant endosperm
    nuclei and grain shrivelling in wheat-rye genotypes.
    Heredity, **39**, 411-419.
Bennett, M.D., Smith, J.B., Simpson, S. and Wells, B. (1979).
    Intranuclear fibrillar material. *Chromosoma (Berlin)* **71**,
    289-332.
Bennett, M.D., Smith, J.B., Ward, J. and Jenkins, G. (1981).
    The relationship between nuclear DNA content and centro-
    mere volume in higher plants. *J. Cell Sci.* **47**, 91-115.
Coll, M.D., Caudras, C.M. and Egozcue, J. (1980). Distribu-
    tion of human chromosomes on the metaphase plate. Symmet-
    rical arrangement in human male cells. *Genet. Res. Camb.*
    36, 219-234.
Comings, D.E. (1968). The rationale for an ordered arrange-
    ment of chromatin in the interphase nucleus. *Amer. J.*
    Human Genet. **20**, 440-460.
Finch, R.A., Smith, J.B. and Bennett, M.D. (1981). *Hordeum*
    and *Secale* mitotic genomes lie apart in a hybrid. *J. Cell*
    Sci. In the press.
Gustafson, J.P. and Bennett, M.D. (1982). The effect of telo-
    meric heterochromatin from *Secale cereale* L. on triticale
    (× *Tritiosecale* Wittmack) I. The influence of the loss of
    several blocks of telomeric heterochromatin on early endo-

sperm development and kernel characteristics at maturity. *Canad. J. Genet. Cytol.* In the press.

Holm, P.B. and Rasmussen, S.W. (1977). Human meiosis I. The human pachytene karyotype analysed by three dimensional reconstruction of the synaptonemal complex. *Carlsberg, Res. Commun.* 42. 283-323.

Horn, J.D. and Walden, D.B. (1978). Affinity distance values among somatic metaphase chromosomes in maize. *Genetics* 88, 181-200.

Hutchinson, J., Jones, J. and Flavell, R.B. (1981). Physical mapping of plant chromosomes by *in situ* hybridisation. *In* "Genetic Engineering" (Setlow, J.K. and Hollaender, A., eds), vol. 3, pp. 207-222, Plenum Press, New York and London.

Jenkins, G. and Bennett, M.D. (1981). The intranuclear relationship between centromere volume and chromosome size in *Festuca scariosa* × *drymeja*. *J. Cell Sci.* 47, 117-125.

Kempanna, C. and Riley, R. (1964). Secondary association between genetically equivalent bivalents. *Heredity* 19, 289-299.

Linde-Laursen, I.B. (1978). Giemsa C-banding of barley chromosomes. *Hereditas* 88, 55-64.

Miller, O.J., Mukherjee, B.B., Breg, W.R. and Gamble, A. Van N. (1963). Non-random distribution of chromosome in metaphase figures from cultured human leucocytes I. The peripheral location of the Y chromosome. *Cytogenetics* 2, 1-4.

Noda, K. and Kasha, K.J. (1978). A proposed barley karyotype based on C-band chromosone identification. *Crop Science* 18, 925-930.

Riley, R. and Flavell, R.B. (1977). A first view of the meiotic process. *In* "A discussion on the meiotic process" (Riley, R., Bennett, M.D. and Flavell, R.B., eds), *Phil. Trans. Roy. Soc. ser. B* 277, 191-199.

Rodman, T.C., Flehinger, B.J. and Rohlf, F.J. (1980). Metaphase chromosome associations: Colcemid distorts the pattern. *Cytogenet. Cell Genet.* 27, 98-110.

Schmid, M., Vogel, W. and Krone W. (1975). Attraction between centric heterochromatin of human chromosomes. *Cytogenet. Cell Genet.* 15, 66-80.

Shchapova, A.I. (1969). On the karyotype pattern and the chromosome arrangement in the interphase nucleus. (Russian with English summary). *Tsitologia* 13, 1157-1164.

Smith, J.B. and Bennett, M.D. (1975). DNA variation in *Ranunculus. Heredity* 35. 231-239.

Takahashi, R. (1975). Minute Barley Genetics Newsletter 2, 185.

Van'T Hof, J. and Sparrow, A.H. (1963). A relationship be-

tween DNA content, nuclear volume, and minimum mitotic
cycle time. *Proc. Nat. Acad. Sci., USA.* **49**, 897–902.
Wagenaar, E.B. (1969). End-to-end chromosome attachments in
mitotic interphase and their possible significance to meiotic chromosome pairing. *Chromosoma (Berlin)* **26**, 410–426.

Repeated DNA Sequences and Nuclear Structure

LAURA MANUELIDIS

Section of Neuropathology, Yale University School of Medicine, U.S.A.

The genome of higher eukaryotes contains excess DNA that
neither codes for specific proteins, nor has any demonstrable
control function for the transcription of adjacent gene
regions. In considering the roles of this "other" DNA, the
fact that eukaryotic chromosomes are highly organized architectural units, and that the interphase nucleus is an ordered
three-dimensional matrix, is often overlooked. It is likely
that at least some of this additional DNA is necessary for
the structure of chromosomes and their positional order in the
nucleus. A simple phenotypic function of selective advantage
in evolution would thus not be carried by this DNA. However,
more general capabilities of nuclei, such as an orderly
heterochromatinization of selected chromosome segments in different cell types, may depend in part on the presence and
organization of this DNA.

Repeated DNAs constitute a rather large proportion of this
excess DNA. Some repeated DNAs appear to define specific
chromosomal domains, such as centromeres or telomeres.
Furthermore, the assignment of specific repeated DNA sequence
subsets to different chromosomes or chromosome domains within
a species may occur (e.g. see Manuelidis, 1978b; Steffenson *et
al*., 1981). Specific repeated DNA subsets, such as "satellite
DNAs", in different species can differ in exact nucleotide
sequence, and can vary enormously with respect to the proportion of the total genome that they occupy. However, we do not
know the *minimum* copy number, repeat length or configuration
that is necessary for the "structural" roles that we postulate
these DNAs perform. Such functions could include: (1) the
definition of chromosome substructure or segments; (2) the
registration of chromosomes in specific three-dimensional
locations in interphase nuclei; or (3) a contribution to the
variety of chromosome forms that occur in different species
(e.g. plant and animal). In these roles, groups of DNA
sequences may participate in concert with specified proteins
(see Brutlag, 1980).

DEFINITION OF REPEATED DNA SUBSETS

From the preceding considerations, the position and distribution of repeated DNA subsets on specific chromosomes becomes
an important feature for their classification. Although
originally some repeated DNAs were defined by their unique
buoyant density and extreme simplicity of sequence, i.e. centromeric satellite DNAs, examples of centromeric sequences
that are more complex in sequence and with little base bias
have been found (e.g. see Wu and Manuelidis, 1980). Thus far,
such sequences have the common features of (1) tandem sequence
repetition with arrangement in long uninterrupted arrays and
(2) centromeric preponderance. In evolution, and in the
delineation of chromosome domains, it is not known if such
tandem repeats are related to the considerably shorter tandem
DNA repeats adjacent to gene coding regions, for example as
those seen between ribosomal genes in Xenopus (Boseley *et al*.,
1979) or as those seen near the immunoglobulin genes, which
may be involved in homologous recombination (T. Rabbitts, this
volume). Another way of examining the possibility that such
shorter and longer repeated DNAs are related, is to consider
if classical centromeric sequences (e.g. mouse satellite DNA)
can reside in smaller copy numbers, and possibly different
configurations, on chromosome arms. This example is considered in more detail below.

A second group of repeated DNAs has been designated interspersed. These sequences may have higher copy numbers than
those of some satellite DNAs. For example, in man, the Alu
interspersed sequences (Houck *et al*., 1979) make up approximately 3% of the genome, as compared to the Y chromosome
sequence (Cooke, 1976) or the family of A+T-rich satellites
(Manuelidis, 1978a), which each constitute less than 1% of
the genome. The Alu sequences have a complexity (190 to 300
base-pairs: Rubin *et al*., 1980; Pan *et al*., 1981) that is
similar to some satellite centromeric sequences (Hsieh and
Brutlag, 1979; Wu and Manuelidis, 1980). Precisely what
mechanism operates to initially distinguish each of these
two sequence groups is not known.

It seems useful to subdivide the territory of interspersed
repeated sequences into at least three subgroups: (1) dispersed short (<1 kb*) sequence elements such as the Alu family,
which are found associated with virtually all large restriction
fragments and that may also be transcribed (Jelinek *et al*.,
1980; Weiner, 1980; Pan *et al*., 1981); (2) "en bloc" or segmental larger (1-20 kb) DNA repeats that may occupy specific dom-

---

*Abbreviations used: kb, 10^3 base-pairs; bp, pase-pair; NOR,
nucleolus organizer region.

ains on chromosome arms; and (3) mobile large elements (>3 kb) that can vary in copy number and location in cell lines derived from the same species (Levis *et al.*, 1980; Finnegan, this volume). These transposable elements in *Drosophila* have flanking direct and short inverted repeats.

A given specific DNA sequence of the interspersed family may not be confined exclusively to one of the categories outlined. For example, some shorter Alu family sequences with terminal direct repeats may also be capable of transposition (Van Arsdell *et al.*, unpublished results). Configuration and location of a given repeated DNA sequence may also modify its behavior. In this context, the observation of transcription of some centromeric satellite DNAs is pertinent (Varley *et al.*, 1980, Diaz *et al.*, 1981).

In evolution, some repeated DNA families, such as centromeric arrays, are apparently amplified in copy number, and subsequently they may be relegated to obscurity. Thus residual rare copies of such sequences are also likely to be "interspersed" in the chromosomes in a different configuration and copy number than their sequence homologues found in long arrays at centromeres in different species. These may be considered as relic sequences that form part of a common library for different genomes; alternatively, one may assume they are constantly being formed *de novo* in different species.

The chromosome arms are large and require elements for their reproducible registration. We therefore supposed there should be "interspersed" DNA repeats with structural characteristics, that could be associated with a Giemsa-like banding pattern, or with a distinctive location on chromosome arms. Such sequences are likely to be maintained in proportion to the rest of the DNA if they participate in the processes such as sister chromatid exchange, translocation, or the delineation of specific regions on the chromosome arms, such as secondary constrictions. These non-centromeric sequences may also be more conserved in evolution than either centromeric sequences or transposable elements. We describe one candidate for this interspersed subset in detail below.

In summary, although knowledge of the exact nucleotide sequence and relative abundance of repeated DNA subsets allows their definition on one level, the configuration and organization of these elements in higher order or three-dimensional architecture may be an important parameter for their final definition. Some repeated DNA subsets may significantly contribute to three-dimensional chromosome order. A more dynamic view of their flexibility in evolution is warranted.

NOVEL CLASSES OF REPEATED DNAS

1. *Non-satellite Repeated DNAs in the Mouse*

In light of the above, it became important to identify
repeated DNA subsets that were not in the main confined to
centromeric domains but were present in significant copy number elsewhere. In mouse cells, where a DNA satellite occupies
a large portion of almost all centromeres (Pardue and Gall,
1970), it was possible to identify two distinct non-homologous
repeats in restriction enzyme digests (Manuelidis, 1980a). On
digestion of total nuclear DNA with the restriction enzyme
*Bst*NI, two repeated non-satellite DNA sequences of ~1.7 and
1.5 kb in length were seen, and these showed no obvious
simple internal (tandem) repetition of smaller sequence units
by further restriction analysis. The 1.7 kb *Bst*NI fragment
was homologous to a shorter 1.3 kb fragment visualized after
restriction with endonuclease *Eco*RI. When these bands were
isolated from gels and used as a probe for hybridization *in
situ*, it was found that both sequences were widely dispersed
on many chromosome arms, although without cloning the 1.7 and
1.5 kb DNAs, other contaminating fragments such as those of
the Alu family co-migrating with these sequences could add to
the impression of interspersion.

2. *A Novel Class of Human Repeated DNA*

We have also identified a novel class of human DNA that is
resolved from most other DNA after digestion with *Hind*III,
and has not been previously described. The DNA in this band
is ~1.9 kb long, and constitutes ~0.25% of the total DNA as
determined by densitometry of ethidium-stained gels. When
this band was eluted and digested with a second enzyme (e.g.
with *Hae*III, *Xba*I, *Bgl*II, *Hin*f, *Hinc*II and *Alu*I), as in the
above studies with the mouse *Bst*NI bands, there was no clear
evidence for an internal simple repeat (data not shown).
   In order to study the chromosomal position of this DNA
(*vide infra*), the *Hind*III gel band DNA was cloned in bacteriophage lambda and subcloned in plasmid pBR322 (L. Manuelidis
and P.A. Biro, unpublished results). A clone was identified
corresponding to the major (real) repeat of this band and was
designated c-*Hind* R. When used to probe whole human DNA
digested with *Hind*III, this cloned DNA highlighted the 1.9 kb
fragment with considerably less background hybridization than
the DNA from the eluted *Hind* band (Fig. 1(A), lanes 1 and 3).
On hybridization of c-*Hind* R to an *Eco*RI digest of total DNA,
a larger restriction fragment (~3kb) was highlighted as well
as a smear (rather than a ladder) of other higher molecular
weight material (Fig. 1(A), lane 2). This pattern indicated

(1) that this sequence was associated with many other DNA fragments, and (2) that the basic repeating unit could be more than 1.9 kb. *Xba*I digests also showed an ~3 kb hybridized band (Fig. 1(A), lane 4).

The c-*Hin*d R sequence was different from previously identified major human repeated DNAs. The 340 bp repeated human multimers gave a different hybridization pattern to *Hin*dIII and *Xba*I-digested whole nuclear DNA than c-*Hin*d R (Fig. 1(A), lanes 3 to 6). Blots of *Hae*III-digested human DNA probed with c-*Hin*d R also hybridized at different positions than the 340 bp multimers (Fig. 1(C)), and the c-*Hin*d R probe also failed in *Hae*III digests to highlight the human Y satellite DNA or satellite I or II sequences, all of which are relatively undigested by *Hae*III (Cooke, 1976; Manuelidis, 1978a) and consequently run at higher positions than the bands highlighted by c-*Hin*d R. The c-*Hin*d R DNA also did not have the *Hin*f restriction pattern of satellite II (Manuelidis, 1978a).

No obvious difference in the amount of the R band was seen in DNA from male and female cells. More than 1,800 bp of the c-*Hin*d R insert have been sequenced (unpublished results) and are different from the conserved portion of the Alu sequence (Pan *et al.*, 1981). Additionally, in preliminary studies of transcription (with A. Weiner), we could not demonstrate any detectable band of RNA complementary to the c-*Hin*d R sequence, and this contrasts with similar studies on the human Alu family of repeats (Weiner, 1980).

3. Novel Repeated Subsets of Mouse and Man Share Sequence Homologies

Since we postulated above that "en bloc" interspersed sequences should be relatively conserved, we tested the sequence homologies of these novel repeated DNA subsets in these two species. First, each of the two non-homologous mouse *Bst*NI bands (1.7 and 1.5 kb) were labelled and used to probe *Hin*dIII-digested human DNA. The 1.5 kb fragment highlighted the *Hin*d R 1.9 kb band as well as a smear of other material in *Hin*dIII digests (Fig. 1(A), lane 10). It also highlighted the identical series of bands in *Xba*I digests as c-*Hin*d R (hybridization not shown). The 1.7 kb *Bst*NI mouse fragment did not label the human *Hin*d R band (Fig. 1(A), lane 7) but it did highlight another discrete and distinct set of bands in *Xba*I digests (Fig. 1(A), lanes 4 and 8).

In order to prove cross-species sequence homologies more conclusively, clones of human DNA were used to probe whole mouse DNA digests. This reverse experiment ruled out the possibility that other minor contaminating DNAs in the mouse *Bst*NI bands were responsible for the hybridization patterns

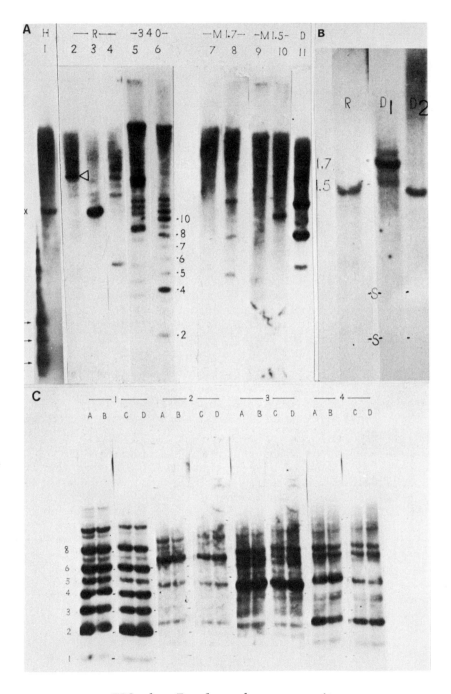

FIG. 1.   For legend see opposite.

FIG. 1 (A) *Human nuclear DNA immobilized on filters after cleavage with restriction enzymes and hybridized to nick-translated, [32P]-labelled DNA probes in 4×SSC (SSC is 0.15 M-NaCl,
0.015 M-sodium citrate, pH 7) at 61°C for 16 h as described
(Manuelidis, 1978a). In order to verify exact positions of
hybridized bands, after autoradiography, [32P]DNA was eluted
from filters, which were then hybridized to a second probe;
hybridization to virgin blots gave comparable results. Lane
1, [32P]Hind R band from a gel hybridized to whole human placental DNA digested with HindIII. A band at X is highlighted, which represents the position of the eluted band in
ethidium bromide-stained gels (not shown). Lanes 2, 3 and 4
represent, respectively, EcoRI, HindIII and XbaI digests
hybridized to [32P]c-Hind R. A band is seen in the EcoRI
digest (open triangle) as well as a smear of other material.
Note, there is considerably less background hybridization
using the cloned c-Hind R probe (lane 1 versus lane 3).
c-Hind R highlights a distinctive set of bands in XbaI digests;
EcoRI and XbaI digests both show bands of ~3kb. Lanes 5 and 6
show HindIII and XbaI digests hybridized with [32P]-labelled
340 bp multimers, and show a different banding pattern than
that obtained with the c-Hind R clone. Bands 2 to 10 are
molecular weight multiples of 170 bp highlighted by this tandem repeat. Lanes 7 and 8 are HindIII and XbaI digests of
human DNA probed with the [32P]-labelled mouse 1.7 kb BstNI
fragments. Lanes 9 and 10 are EcoRI and HindIII digests of
human DNA probed with the mouse [32P]-labelled 1.5 kb Bs+N1 fragments, the Hind R human band hybridizes to this mouse isolate.
Lane 11 (from a different gel profile) is a XbaI digest of
human DNA probed with [32P]-labelled c-globin D DNA. This
probe hybridized to 3 bands that were equivalent to those
highlighted by the mouse 1.7 kb fragment (lane 8 versus lane
11). The c-Hind D clone gave a comparable XbaI digestion
pattern as seen in lane 11. The impure [32P]HindR fragment
additionally highlighted mouse satellite DNA bands on an
adjacent filter strip containing mouse nuclear DNA (arrows).
Reverse hybridizations, of purified [32P]-labelled mouse satellite DNA to human DNA (not shown) failed to reveal any detectable hybridization on long radiographic exposures.*
(B) *Hybridization of human DNA probes to digests of mouse
nuclear DNA immobilized on filters (conditions as above).
The [32P]c-Hind R probe highlighted the mouse 1.5 kb BstNI
band, seen in lane R. The c-globin D probe highlighted the
1.7 kb BstNI fragment (lane D1) as well as its 1.3 kb
EcoRI homologue (lane D2). Neither c-Hind R nor c-globin D
in BstNI digestions hybridized to mouse satellite DNA (positions, from ethidium bromide-stained gels, marked S).*
(C) *Lanes A, B, C and D represent, in each case, DNA blots of
the following human isolates: placental DNA, sperm DNA, cloned*

Legend to Fig. 1 continued

glioblastoma tissue culture line 526-C2b and cloned neuro-
blastoma line TC 691c, all digested with HaeIII.   *DNA probes*
were hybridized and eluted as above.   Lanes 1, A to D show
hybridization to the 340 bp tandem repeats; 1 to 8 are molecu-
lar weight multiples of 170 bp.   Hybridization of [^{32}P] c-Hind
R (same filters) is seen in lanes 2, A to D.   Lanes 3, A to D
are the same filters hybridized to [^{32}P]c-Hind *D.   Lanes 4,*
A to D are the same filters hybridized to [^{32}P]c-globin D.
There are minor band differences that distinguish the c-Hind
D and c-globin D hybrids.   Bands representing the Y sequence
as well as satellites I and II are not highlighted by any of
*these probes.   Xba*I *digests (not shown) hybridized to the*
cloned Hind *D and* Hind *R sequences also showed no obvious dif-*
ferences in band pattern or relative abundance of each of
these sequences in any of the DNA isolates tested.

---

obtained.   The c-*Hin*d R DNA as expected, hybridized to the
mouse 1.5 kb fragment (Fig. 1(B), lane R).
    Two additional human clones were identified, which gave
the same *Xba*I pattern as that obtained when the mouse 1.7 kb
fragment was used to probe human DNA.   One clone, a fragment
D from the globin cluster (a gift from S. Weissman and P.A.
Biro) is designated c-globin D (see Fig. 1(A), lane D).   The
second, obtained from a DNA in the *Hin*dIII 1.9 kb gel band,
is designated c-*Hin*d D.   Further analysis showed that these
two clones were similar, but not identical, as determined by
their hybridization patterns to *Hae*III (Fig. 1(C) and *Mbo*I
digests of human DNA.   Both of these clones, as in the above
cases with c-*Hin*d R, were also clearly distinguishable from
human centromeric DNA satellites and the Alu family of dispersed repeated DNAs in filter hybridization studies.   When
the human c-globin D clone was used to probe mouse DNA, there
was positive hybridization demonstrated to the mouse 1.7 kb
*Bst*NI band and to its 1.3 kb *Eco*RI sequence counterpart (Fig.
1(B), lanes D1 and D2).   Thus each of the two novel repeated
interspersed sequence subsets, originally identified in the
mouse, are represented in appreciable copy numbers in the
human genome, and are relatively conserved in evolution.
    In human cells, both the R and D sequence subsets are also
maintained proportionally in aneuploid glioma tumor cells,
which display high chromosome numbers and abnormal or
rearranged chromosomes.   No obvious difference in copy number
could be detected by Southern hybridization of these probes
to *Hae*III digests of cloned human neuroectodermal tumor cell
lines as compared to DNA from sperm and placenta (Fig. 1(C)).
In *Xba*I digests, where a band of ~3 kb is seen using the c-
*Hin*d R probe, again no evidence for rearrangement or copy

number could be demonstrated. Thus, there is as yet no basis
for identifying these repeated DNA subsets as independent
transposable or mobile elements (*vide supra*).

One possibility was that these two DNAs were involved in
the delineation of larger, integral architectural units. Such
units could be exchanged or transposed en bloc with other
sequences. An example of such a chromosomal compartment would
be a Giemsa band, or a bromodeoxyuridine-induced sister chromatid exchange unit. The proportional maintenance of these
two subsets in aneuploid tumor cells would be consistent with
this view. Other large chromosomal blocks, such as centromeric arrays, have also been shown to be maintained proportionally in glioma tumor lines (e.g. see Manuelidis and Manuelidis, 1979). Additionally, in a study of *Mus musculus* and
Mus spretus, where there is a marked difference in the amount
of centromeric satellite DNA, a striking conservation of the
amount of the 1.7 kb *Bst*NI sequence was observed (Brown and
Dover, 1981). The above finding of sequence homology between
species as divergent as man and mouse may indicate that with
these two sequences, unlike the centromeric repeats, exact
nucleotide sequence is a more important feature in determining their role in the nucleus.

One intriguing observation was that the impure human *Hin*d
R band before cloning hybridized to mouse satellite DNA (Fig.
1(A), arrows). However, no hybridization was detected when
purified mouse satellite was hybridized to human DNA. One
interpretation of these results is that the human genome contains one or very few copies of the mouse satellite sequence,
possibly representing a relic sequence in the library that
has been relegated to obscurity in the human genome. In this
context, the finding of different amounts and patterns of
satellite DNA in *M. musculus* and *M. spretus* (Brown and Dover,
1980) could represent an example intermediate in this process
of evolutionary recruitment or obsolescence of a repeated DNA.

ARE REPEATED CENTROMERIC AND INTERCALARY
SEQUENCES MUTUALLY EXCLUSIVE?

The above hybridization experiments indicated that the c-*Hin*d
R and c-*Hin*d D sequences were likely to be interspersed. The
definition of bands in blots distinguished at least a component of these sequences from Alu interspersed sequences that
do not highlight discrete bands upon hybridization. They
were also unlikely to be predominantly centromeric, as indicated by comparison with previously identified centromeric
arrays. Since we postulated these sequences could be non-
randomly arranged on chromosome arms, we studied them by
chromosomal hybridization *in situ*. Using cloned DNA, we were

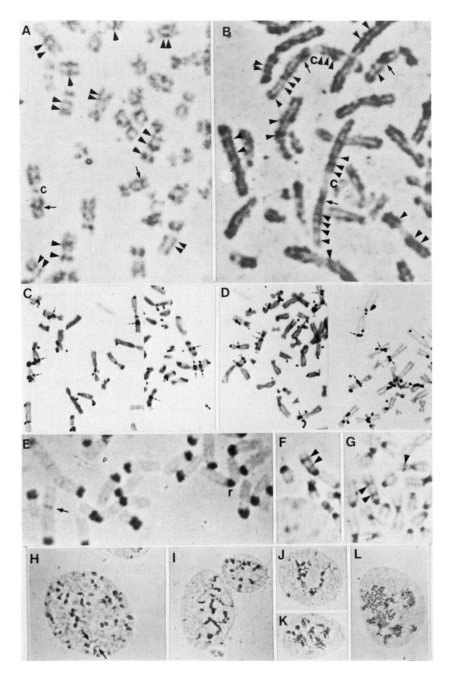

FIG. 2. For legend see opposite.

FIG. 2 (A) *Chromosomal hybridization* in situ *of biotin-labelled c-Hind R DNA to human chromosomes, using immunological detection, indicated a banding pattern. Arrowheads point to regions of bands that were brown (peroxidase stained). Two secondary constrictions were also identified in the spread (arrows). Although most centromeres did not show appreciable hybridization, one centromere (at c) shows a small amount of hybridization to this DNA.*
(B) *Hybridization of biotin-labelled c-Hind R DNA to human chromosomes using intensification with PAP. The brown-staining regions (arrowheads) appear to delineate bands in the chromosome arms. Some telomeres are also labelled. Secondary constrictions are labelled in some instances (arrows). Two long chromosomes with absence of hybridization to centromeres (c) show a comparable banding pattern, suggesting that these c-Hind R sequences can be discretely and reproducibly organized on the chromosome arms.*
(C) *and* **(D)** *Conventional* in situ *hybrids of ³H-labelled-c-Hind R sequences detected by autoradiography as described (Manuelidis, 1978b); (C) using c-Hind R in λ, (D) with c-Hind R in pBR322. In both instances regions on the chromosome arms, sometimes appearing as bands, were labelled (arrows). Occasional telomeres (open triangles) and centromeres (c) also appeared labelled. With longer exposures, there were many grains over the chromosome arms that could not be resolved from each other.*
(E) *to* **(G)** *Hybridizations* in situ *of biotin-labelled purified mouse satellite DNA to mouse chromosomes, detected using peroxidase-labelled antibodies without PAP. In addition to the intense peroxidase staining of the centromeres, a region that appears to be a secondary construction in E showed brown reaction product (arrow); a chromosome in this tumor cell line (TC 509) appears to have undergone Robertsonian fusion (r). (F) and (G) Intercalary bands on 2 similar chromosomes in different spreads (double arrowheads); an additional band on one chromosome arm is also seen (single arrowhead).*
(H) *to* **(L)** *Biotin-labelled purified mouse satellite DNA hybridized to squash preparations of mouse nuclei, detected with immunological techniques. In (H) one can see balls and clusters of peroxidase-staining satellite regions in the nucleus and from these, chromosome arms appear to extend (arrows). In some nuclei, satellite stained regions appear to form strings or associated features around the nucleolus or center of the nucleus (I) and (J). In a few cases, the satellite staining regions were somewhat more diffuse but associated as a matrix of one or more groups (K and L). These latter forms were seen less frequently.*

able to exclude the possibility that the observed hybridization *in situ* results were due to other groups of interspersed
sequences.

Both the c-*Hin*d R and c-*Hin*d D DNAs were shown by hybridization *in situ* and autoradiography to be dispersed on many
chromosome arms. Exposure times *in situ* for these sequences
required five to more than ten times the exposure times of
comparable hybridizations with the human cloned Alu sequences
(a gift from C. Duncan and S. Weissman) to achieve similar
grain densities. The c-*Hin*d R sequences were studied in more
detail using biotin-labelled DNA probes with detection by
rabbit anti-biotin antibody (Langer and Ward, 1981), and peroxidase-labelled goat anti-rabbit antibodies to enhance the
detection of small quantities of sequences (L. Manuelidis,
P.R. Langer and D.C. Ward, unpublished results). With these
methods it was possible to localize these sequences more precisely than with autoradiography (Fig. 2). The *Hin*d R sequences yielded a highly ordered banding pattern, largely on chromosome arms, and at or near some telomeres. A few centromeres
of some chromosomes also revealed a small amount of label,
although many centromeres showed no detectable hybridization
to this sequence (Fig. 2). The banding pattern on one large
chromosome was checked and shown to be reproducible in many
spreads. The hybridization signal at these sites was enhanced
by applying the peroxidase-anti-peroxidase immune complex
(PAP) technique of Sternberger (1979). An example using this
enhancement procedure is shown in Fig. 2. As a control, hybridizations of cloned 18 S ribosomal DNA were processed iden-
ticaly with the same chromosome spreads at the same time. No
similar banding pattern to that observed with the c-*Hin*d R
probe was detected. With the 18 S ribosomal DNA probe, the
expected labelling of the acrocentric nucleolus organizing
chromosomes was seen (data not shown). It therefore appears
that at least some of the *Hin*d R interspersed repeated DNA may
be defined as a subset with a segmental or en bloc pattern.
It is this type of repeated DNA sequence that we postulate
could play a role in the alignment of chromosome arms, and
processes such as sister chromatid exchange between defined
chromosome segments.

Mouse satellite DNA was also studied using the more sensitive and higher resolution immunological method of hybridization *in situ*. These classically centromeric sequences could
be found as minor bands on mouse chromosome arms of one or
two specific chromosomes even without the PAP enhancement procedure (Fig. 2). One of the intercalary mouse satellite
bands appeared at a region compatible with a secondary con-
striction. Preparations greater than 98% pure as judged by
analytical centrifugation, and used for unambiguous

sequencing of mouse satellite DNA (Manuelidis, 1981) were
employed in these studies. It is unlikely that a contaminating main band repeated DNA could account for these results,
since these DNAs give a widely dispersed chromosome hybridization pattern (Manuelidis, 1980a); however, cloned satellite
DNA could be used to confirm these results. In this context,
it is of interest that with gene amplification in the mouse,
both repeated DNAs and satellite DNA may be concomitantly
increased (C. Bostock, this volume), suggesting some role for
repeated DNAs in higher organization of chromosome segments.
Studies using independent methods of cloning (Bostock and
Clark, 1980) also confirm the presence of mouse satellite at
some intercalary positions.

From the above experiments, it appears that centromeric
and some intercalary repeated sequences are not mutually
exclusive.   Both subsets appear to organize highly specified
chromosome segments. It is possible that the configuration
and amount of a given repeated sequence at different chromosomal sites is different, and thus the same sequence may participate in different modes of higher organization, possibly in
conjunction with specific proteins.

THE INTERPHASE POSITION OF SPECIFIC CHROMOSOME REGIONS

Even in squash preparations, where nuclear structure is not
well-preserved, centromeric satellite DNA appeared in several
morphological arrays suggesting the centromeres bearing these
sequences were ordered or associated in interphase. Adjacent
~2000 Å balls of satellite DNA were often seen, from which
parallel chromosome fibers appeared to extend (Fig. 2(H)).
In other nuclei, satellite hybridizing (centromeric) regions
were sometimes associated as continuous features (Fig. 2(I)
and (J)) suggesting they were arrayed in a definite position
relative to each other, and in some cases the satellite
regions appeared as an aggregate or unified, more extended
nuclear region or matrix (Fig. 2(K) and (L)). The matrix
type of configuration of centromeres in nuclei may possibly
vary as a function of cell cycle (e.g. late replication of
satellite DNA could be reflected in the rarer matrix configuration). Hybridization of mouse satellite DNA to three-
dimensionally preserved nuclei by thin sectioning and electron
microscopy also indicated, in preliminary studies, that
satellite-stained chromosome regions were generally oriented
toward the nucleolus and away from the nuclear membrane. In
thin sections, centromeric regions also appeared to be associated. The alignment or association of these centromeric
regions in interphase suggests that there are discrete positions delineated by these sequences in the nucleus.

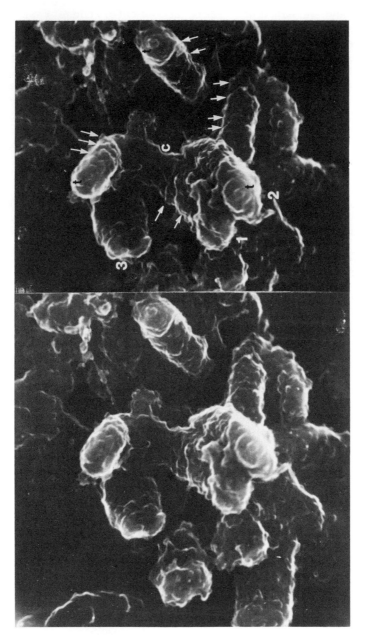

FIG. 3. For legend see opposite.

Association of mouse centromeric satellite could be of help in explaining the occurrence of the Robertsonian centromeric fusions that are frequently observed in metaphase in mouse tumor cell lines.

2. *Scanning Electron Microscopy of Prophase and Anaphase Cells*

Studies of prophase, metaphase and anaphase nuclei have indicated that there are discrete positions for each chromosome, and that these ultimately play a significant role in defining the shape of the interphase nucleus and the exact position of the nucleolus (Laane *et al.*, 1977; Manuelidis *et al.*, 1980). Alignment of centromeric and Giemsa-like bands on chromosome arms could be facilitated by some repeated DNA subsets, such as those described here, and indeed such sequences could participate in the orderly alignment and association of individual chromosomes, as well as three-dimensional recruitment of different adjacent chromosome segments for orderly transcription or heterochromatinization.

During metaphase, chromosomes of a given arm length appear to be aligned (Fig. 3) with concomitant alignment of substructured regions of adjacent metaphase chromosomes. In glioma tumor cells, long marker chromosomes appear at discrete consistent positions in prophase and anaphase. We have proposed previously that metaphase chromosomes represent a helical wind of the 2000 Å interphase chromosome fibers (Sedat and Manuelidis, 1978), with a simple transition from metaphase to interphase. In this model, metaphase order is a reflection of the functional interphase organization, with alignment of adjacent chromosome segments (Sedat, Agard, Manuelidis, unpublished results). In keeping with this, specific identified metaphase chromosomes in plants appear to have tightly ordered spatial relationships (M. Bennett, this volume).

The association or three-dimensional ordering of specific chromosomes in interphase has several ramifications or

FIG. 3 (opposite) *Scanning electron microscopy of chromosomes from one cell isolated and prepared as described (Manuelidis et al., 1980). Chromosomes labelled 1 and 2, appear to be aligned with each other. Chromosome substructure is seen in stereo pairs (arrows), consistent with the model of helical winding. In chromosomes labelled 1 and 2, these substructured regions also appear aligned. Note that the other arm of the chromosome labelled 1, whose centromere is at c is parallel to a chromosome of similar arm length (labelled 3). Regions labelled t are likely to represent telomeres. Metaphase chromosome width measured ~6000Å.*

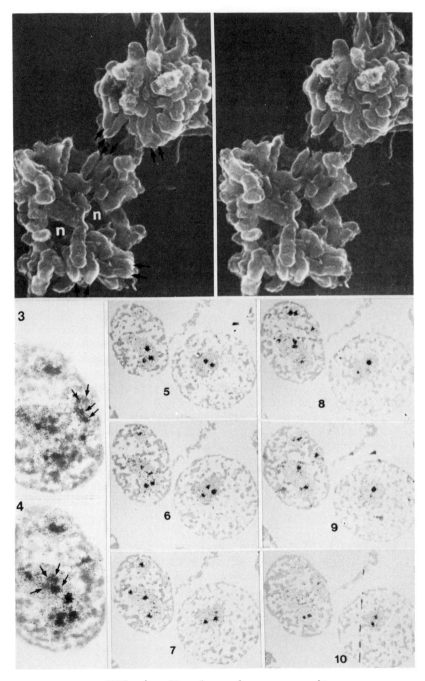

FIG. 4. For legend see opposite.

predictions. One of these is that functional units (in terms
of genetic coding regions) will have a meaningful three-
dimensional position relative to each other; e.g. human globin
genes on two different chromosomes (6 and 11) might be expected to have related interphase positions, or be neighbors, so
that the "puffing" of one transcribing region can influence
the configuration and transcription of its adjacent related
chromosome. Similarly, contraction of adjacent chromosome
fibers, and meaningful heterochromatinization or inactivation
of larger gene units (chromosome segments) may occur. This
latter feature will also modify the final nuclear shape and
size in different cell types, where the relative position of
each chromosome may be fairly constant.

*3. Study of Nucleolus Organizing Regions in Metaphase Chromo-
some and Interphase Nuclei*

In order to further test the specific location and association
of specific gene regions in interphase we studied the nucleolus organizer regions in detail. These gene segments were
chosen because (1) they occupy known metaphase chromosomal
sites that are adjacent to the five human acrocentric chromosome centromeres (Evans *et al.*, 1966), (2) at least two
repeated sequences are closely associated with these sites
(Manuelidis, 1978b), and (3) we have found that these sites
can be mapped precisely with high resolution and little dis-
tortion of chromosome structure; such sites represent staining
of a protein-DNA complex (L. Manuelidis 1980b; and unpublished
results). The studies covered briefly here are based on modifications of the silver staining of nucleolus organizer regions previously identified on metaphase chromosomes at the

---

FIG. 4 (opposite) *Top, stereo pair; chromosomes in anaphase
curve to form the shape of an interphase nucleus. Note that
chromosomes of a given length appear to be aligned in each of
these 2 anaphase groups (arrows). In the lower anaphase group,
the formation of the nucleolus (n) is also defined by the
position of the chromosomes. 3 and 4 represent superimposed
serial thin electron microscope sections (shown in part in
adjacent serial sections 5 to 10) stained with silver to delineate the nucleolus organizing regions. These regions measured ~2000 Å in diameter and appear as balls or continuous
2000 Å features. In front and back views of superimposed
serial sections, arrows show clusters of silver-staining
chromosome regions that appear to make up closely associated
or coherent units, that were better appreciated in stereo
(not shown).*

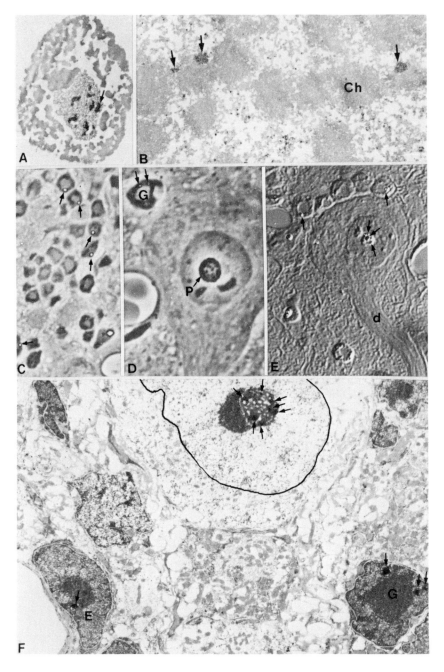

FIG. 5. For legend see opposite.

light microscopic level (Goodpasture and Bloom, 1976); these
are thought to represent transcriptionally active chromosomal
regions (Miller *et al.*, 1978).

Study of isolated metaphase chromosomes, as those used for
scanning electron microscopy, showed localization of NORs at
clustered positions compatible with a defined position of the
nucleolus or nucleoli as seen by scanning electron microscopy
(Fig. 4). In whole metaphase cells, 2000 Å NOR features were
observed on individual chromosomes by transmission electron
microscopy (Fig. 5(B)). In interphase nuclei, chromosome
fibers of 2000 Å were defined in and around the nucleolus
(Fig. 5(A)). Isolated metaphase chromosome arrays, as those
seen in scanning electron microscopy above, also showed
defined and coherent positions for the NORs. Serial sections
of human interphase nuclei (Fig. 4) showed that each of these
2000 Å fibers or balls were closely associated or formed

FIG. 5 (opposite)   (A) *Isolated nuclei stained for NORs show*
~2000 Å thick strands coursing through the nucleolus (as at
arrow). The rest of the interphase chromosomes are not
stained with silver.
(B) *NORs in metaphase chromosomes in whole cells by electron*
microscopy. Two silver-stained regions ~2000 Å in diameter
and the edge of a third NOR are seen (arrows). Ch shows meta-
phase chromosome region not binding silver.
(C) *Silver-staining NORs are at the edge of the dense central*
heterochromatin in internal granule cell neurons (arrows).
The position of these acrocentric chromosome regions in these
cells was discrete and consistent.
(D) *Extensive silver-stained NORs in cerebellar Purkinje cell*
nucleus (P, arrow) appeared at the periphery and center of
this cell's large nucleolus. A granule cell neuron with its
heterochromatic central region labelled G, also shows 2 smal-
ler NORs. Note, the "active" NOR region is more extensive in
the Purkinje cell compared to the granule cell.
(E) *Differential interference microscopy also delineated the*
silver-staining NORs within the nucleolus of a Purkinje cell
(3 arrows) whose dendrite is labelled d. Two additional NORs
are noted in 2 adjacent granule cell neurons (arrows).
(F) *Transmission electron microscopy shows in more detail the*
distribution and delineation of these NORs (arrows). E is an
endothelial cell adjacent to a vessel. G is a cerebellar
granule cell neuron with its heterochromatin labelled G. The
nucleus of a Purkinje cell is outlined in black. In all these
cells the NOR formed coherent or associated 2000 Å features,
but in each cell the position and extent of the NOR was dis-
tinct. A defined location for the NOR in each cell type was
consistently seen.

continuous elements in and around the nucleolus, as judged by three-dimensional reconstruction (Fig. 4) and stereoanalysis (data not shown). Through focus and television monitoring of unsectioned nuclei also confirmed the coherent and continuous nature of these 2000 Å chromosome features in interphase. The position of nucleolus organizing centromeric regions were always facing the center of the nucleus and were never observed directly on the nuclear membrane. In other species, such as mouse, hamster and guinea pig, where ribosomal genes are on different chromosomes, a similar orientation and association of NORs were observed. The classical fibrillar elements of the nucleolus in each case represented the chromosome elements bearing the ribosomal genes.

4. Nucleolus Organizer Regions and Nuclear Structure in Differentiated Cells

For analysis of the distribution of nucleolus organizer chromosome regions in different cell types, we chose to study these defined chromosome segments in highly differentiated cells of the nervous system. In each cell type, a characteristic location of the NORs was observed. For example, cerebellar Purkinje cell neurons displayed an extensive major central nucleolus organizing region (Fig. 5(D), (E) and (F) with occasionally a second smaller NOR closer to the nuclear membrane. In contrast, granule cell neurons showed several smaller NORs, each closer to the edge of the nucleus, and surrounding the characteristic central heterochromatin of these neurons (Fig. 5(C) through (F)). Astrocytes (glial cells), endothelial cells and developing sperm cells also each had their characteristic three-dimensional organization and localization of nucleolus organizing chromosomes in interphase. Brain tumor cells with high chromosome numbers displayed considerably more NORs than normal diploid cells, with multiple centers and variable positions, consistent with their aneuploid rearrangements. The recruitment of different nucleolus organizers in different cell types, and their three-dimensional rearrangements could occur initially during mitosis. We were able to exclude the possibility that the non-chromosomal nucleolar elements were transposed independently in position from the active NOR chromosome regions in these different cell types.

Thus, in interphase, the nucleolus organizer regions on specific chromosomes appear at distinct positions, and there is three-dimensional association of groups of chromosomes bearing these related gene regions. It is likely that the relative position of the NORs in different diploid cell types is modified by the relative contraction (heterochromatinization) of specific arrays of three-dimensionally aligned chromosomes (e.g. granule cell neurons) or by extension of these same chromosomes (e.g. Purkinje cells). Chromosome rearrangements were also reflected in interphase nuclei, as observed in the arrangement of NORs in aneuploid tumor cells; these three-dimensional rearrangements are likely to contribute to the unusual biological properties of tumor cells.

The specific arrangement of NORs is in keeping with the defined interphase organization of mouse satellite DNA. Further study of the alignment of specific repeated DNA subsets in metaphase and interphase will help to define their role in nuclear structure more clearly.

SUMMARY

We have suggested here that some repeated DNA subsets may participate in the precise organization of interphase chromosomes. On a linear level, exact nucleotide sequence, length, and relative abundance are important parameters for repeated DNA subset definition. However, chromosomal and three-dimensional interphase assignments may indicate the role of each subset more precisely. It is suggested that repeated sequences should not be defined too rigidly, since for a given sequence subset, chromosomal position and configuration may vary, and with this there may be a concomitant alteration of the role for this sequence in the nucleus.

Interspersed repeated sequences may be subdivided into several classes; we have defined and purified at least one novel subset that is relatively conserved, that appears to be non-randomly distributed on chromosome arms with a specific banding pattern, and that may participate in alignment or association of specific chromosome segments or chromosome arms. Three-dimensional chromosome structural alignments have complex biological consequences, rather than a simple or single (linear) phenotypic function. Such consequences can include recruitment, as well as heterochromatinization, of large related gene clusters, and the demarcation of chromosome segments. In delineating blocks or regions on chromosomes, repeated sequences may participate in processes such as transposition, Robertsonian fusion, sister chromatid exchange, and the chromosome reorganizations that occur in tumor cells.

ACKNOWLEDGMENTS

The author thanks David Ward and P.A. Biro for their careful and helpful advice on the manuscript, and is grateful to A. Coritz for help with the photography. This work was supported by National Institutes of Health grant CA15044.

REFERENCES

Boseley, P., Moss, T., Machler, M., Portmann, R. and Birnstiel, M. (1979). Sequence organization of the spacer DNA
in a ribosomal gene unit of *Xenopus laevis*. *Cell*. 17. 19–
31.

Bostock, C.J. and Clark, E.M. (1980). Satellite DNA in large
marker chromosomes of methotrexate-resistant mouse cells.
Cell 19, 709–715.

Brown, S.D.M. and Dover, G. (1980). Conservation of segmental
variants of satellite DNA of *Mus musculus* in a related species: *Mus spretus*. *Nature (London)* 285, 47–49.

Brown, S.D.M. and Dover, G. (1981). The organization and evolutionary progress of a dispersed repetitive family in widely
separated rodent genomes. *J. Mol. Biol.* 150, 441–465.

Brutlag, D.L. (9180). Molecular arrangement and evolution of
heterochromatic DNA. *Annu. Rev. Genet.* 14, 121–144.

Cooke, H. (1976). Repeated sequence specific to human males.
Nature (London) 262, 182–186.

Diaz, M.O., Barsacci-Pilone, G., Mahon, K. and Gall, J.G.
(1981). Transcripts from both strands of a satellite DNA
occur on lampbrush chromosome loops of the Newt *Notophtal-
mus*. *Cell* 24, 649–659.

Evans, H., Buckland, R. and Pardue, M.L. (1966). Location of
the genes coding for 18S and 28S ribosomal RNA in the human
genome. *Chromosoma (Berlin)* 48, 405–426.

Goodpasture, C. and Bloom, S.E. (1975). Visualization of
nucleolus organizer regions in mammalian chromosomes using
silver stain. *Chromosoma* 53, 37–50.

Houck, C.M., Rhinehart, F.P. and Schmid, C.W. (1979). A ubiquitous family of repeated DNA sequences in the human
genome. *J. Mol. Biol.* 132, 289–306.

Hsieh, T. and Brutlag, D.L. (1979). Sequence and sequence
variation within the 1.688 g cm^{-3} satellite DNA of *Drosophila melanogaster*. *J. Mol. Biol.* 135, 465–481.

Jelinek, W.R., Toomey, T.P., Leinwand, L., Duncan, C.H., Biro,
P.A., Choudary, P.V., Weissman, S.M., Rubin, C.M., Houck,
C.M., Deininger, P.L. and Schmid, C.W. (1980). Ubiquitous
interspersed repeated sequences in mammalian genomes.
Proc. Nat. Acad. Sci., U.S.A. 77, 1398–1402.

Laane, M.M., Wahlstrom, R. and Mellem, T.R. (1977). Scanning
electron microscopy of nuclear division stages in *Vicia
faba* and *Haemanthus cinnabarinus*. *Heridatas* 86, 171–178.

Langer, P.R. and Ward, D.C. (1981). A rapid and sensitive immunological method for *in situ* gene mapping. *ICN-Symposium*.
In the press.

Levis, R., Dunsmuir, P. and Rubin, G.M. (1980). Terminal
repeats of the *Drosophila* transposable element *copia*:

nucleotide sequence and genomic organization. *Cell* **21**, 581–588.

Manuelidis, L. (1978a). Complex and simple sequences in human repeated DNAs. *Chromosoma (Berlin)* **66**, 1–21.

Manuelidis, L. (1978b). Chromosomal localization of complex and simple repeated human DNAs. *Chromsoma (Berlin)* **66**, 23–32.

Manuelidis, L. (1980a). Novel classes of mouse repeated DNAs. *Nucl. Acids Res.* **8**, 3247–3258.

Manuelidis, L. (1980b). Ultrastructure of nucleolus organizers in normal and neoplastic neuroectodermal cells. *J. Neuropath. Exp. Neurol.* **39**, 373.

Manuelidis, L. (1981). Consensus sequence of mouse satellite DNA indicates it is derived from tandem 116 b.p. repeats. *FEBS Letters.* 129, 25–28.

Manuelidis, L., Sedat, J. and Feder, R. (1980). Soft X-ray lithographic studies of interphase chromosomes. *Ann. N. Y. Acad. Sci.* **342**, 304–325.

Miller, D.A., Breg, W.R., Warburton, D., Dev, V.G. and Miller, O.J. (1978). Regulation of rRNA gene expression in a human familial 14pt marker chromosome. *Human Genet.* **43**, 289–297.

Pan, J., Elder, J.T., Duncan, C.H. and Weissman, S. (1981). Structural analysis of interspersed repetitive polymerase III transcription units in human DNA. *Nucl. Acids Res.* **9**, 1151–1170.

Pardue, M. and Gall, J.G. (1970). Chromosomal localization of mouse satellite DNA. *Science* **168**, 1356–1358.

Rubin, C.M., Houck, C.M., Deininger, P.L., Friedmann, T. and Schmid, C.W. (1980). Partial nucleotide sequence of the 300 nucleotide interspersed repeated human DNA sequences. *Nature (London)* **284**, 372–374.

Sedat, J. and Manuelidis, L. (1978). A direct approach to the structure of eukaryotic chromosomes. *Cold Spring Harbor Symp. Quant. Biol.* **42**, 331–350.

Steffensen, D.M., Appels, R. and Peacock, W.J. (1981). The distribution of two highly repeated DNA sequences within *Drosophila melanogaster* chromosomes. *Chromosoma* **82**, 525–543.

Sternberger, L.A. (1979). *Immunocytochemistry*, 2nd edit., John Wiley and Sons, New York.

Varley, K.M., Macgregor, H.C. and Erba, H.P. (1980). Satellite DNA is transcribed on lampbrush chromosomes. *Nature (London)* **283**, 686–688.

Weiner, A.M. (1980). An abundant cytoplasmic 7S RNA is complementary to the dominant interspersed middle repetitive DNA sequence family in the human genome. *Cell* **22**, 209–218.

Wu, J.C. and Manuelidis, L. (1980). Sequence definition and organization of a human repeated DNA. *J. Mol. Biol.* 142, 363–386.

Assays of the Phenotypic Effects of Changes in DNA Amounts

H. REES, G. JENKINS

Department of Agricultural Botany, U.C.W., Aberystwyth

A.G. SEAL and J. HUTCHINSON

Plant Breeding Institute, Cambridge

INTRODUCTION

Most problems in genetics start from observations on variation in the phenotype. There follows a search for the genetic material, the particular DNA sequences that are responsible for the variation. We are now confronted with a problem of a very different kind. The chromosomes of eukaryotes carry astonishing amounts of DNA. Much of it consists of highly repetitive base sequences that are not translated into protein. We know also that the divergence and evolution of eukaryote species, particularly among flowering plants, are often accompanied by massive quantitative changes in this supplementary, non-coding, DNA fraction. The search, now, is for variation in the phenotype that is dependent on these massive changes in DNA amount.

It may be, of course, that much of the DNA in eukaryote chromosomes is genetically inert and "parasitic" (Doolittle and Sapienza, 1980; Orgel and Crick, 1980). In the light of present evidence, certainly, it is a proposition that cannot be discounted. What evidence we have on this score comes from two kinds of assay. The first is based on comparisons of the phenotypes of species with different amounts of nuclear DNA; the second on measuring the consequences of the recombination and segregation of quantitative DNA differences in hybrids and hybrid derivatives within and between species.

INTERSPECIFIC VARIATION IN CELL CYCLE AND CELL SIZE

It is well-known that the duration of the mitotic cycles

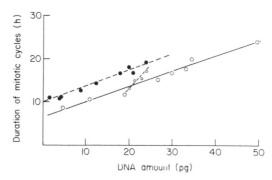

FIG. 1 *The duration of mitotic cycles at 20°C in root meristems of dicotyledonous (●) and monocotyledonous (○) species, plotted against the 2C nuclear DNA amounts. Small circles are rye plants with 1 to 4 B chromosomes. Data are from Evans* et al. *(1972).*

increases with increasing nuclear DNA amount (Evans *et al.*, 1972; Bennett, 1972). Figure 1 gives the duration of mitotic cycles in plant root meristems of monocotyledonous and dicotyledonous species. In both groups, the duration of the cell cycle increases linearly with increase in nuclear DNA, even though the composition of the DNA, its quality, is vastly different for different species. It would appear, therefore, that the duration of the cell cycle is, at least mainly, dependent on the DNA amount *per se*. Bennett (1972) has dubbed such effects nucleotypic, to distinguish them from the effects of structural genes, which implicate DNA sequences of particular composition, of particular quality.

It will be observed from Fig. 1, however, that while the rate of increase in the duration of the mitotic cycle (0.38 h/pg) is the same for monocotyledons and dicotyledons, the duration of the cycle in monocotyledons with the same amount of DNA is about four hours less than in dicotyledons. It will be observed also that extra DNA due to the addition of B chromosomes in rye increases the duration of the mitotic cycle at a significantly higher rate (0.41 h/pg) than the addition of DNA within normal, A chromosomes. While the element of predictability about the duration of mitotic cycles is impressive, there is, nevertheless, a strong indication that, to a limited degree at least, the DNA quality as well as the quantity is a significant ingredient of the controlling mechanism.

One final point is worth making about these surveys. It will be seen that the linear regression lines for both monocotyledons and dicotyledons are a remarkably good fit. It means that infinitely fine adjustments to the duration of cycles are possible by the addition or deletion of nuclear

DNA. In this respect, the delicacy of control is reminiscent
of polygenic systems.

A survey among flowering plants has shown that cell size in
root meristems is also strongly dependent on nuclear DNA amount
(Martin, 1966).

HYBRID AND HYBRID DERIVATIVES

In the Graminae, as in many families of flowering plants,
there is a substantial variation in DNA amount among species.
Figure 2 shows the distribution of DNA amounts within two genera, *Lolium* and *Festuca*. It will be observed that much of the
DNA variation is independent of polyploidy.

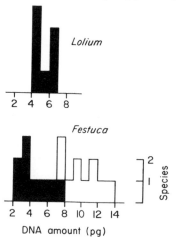

FIG. 2. *The 2C nuclear DNA amounts of species of* Lolium *and*
Festuca. *Open blocks represent polyploids.*

1. Lolium

(a) DNA segregation. Inbreeding *Lolium* species such as *L.
temulentum* have about 40% more DNA (6.23 pg/2C nucleus) than
outbreeders such as *L. perenne* (4.16 pg/2C nucleus). Most of
the "extra" DNA in *L. temulentum* is comprised of repetitive
sequences (Hutchinson *et al.*, 1980). All are diploids with 14
chromosomes. Despite the large DNA difference, the cross *L.
temulentum* × *L. perenne* yields viable hybrids and fertile hybrids at that. Figure 3 shows the distribution of DNA amounts
among parents, F_1 and backcross progenies. It will be seen
that the DNA distribution among backcross individuals follows
a normal curve, and ranges from the "low" DNA parent to the
mid-parent value. We conclude that the recombination and
segregation of the supplementary DNA fraction that

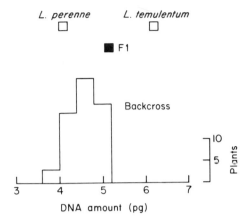

FIG. 3 *The distribution of nuclear DNA amounts in* L. *temulen-tum, the F1 and in the progenies of the backcross F1 × L. per-enne.*

distinguishes *L. temulentum* from *L. perenne* has, surprisingly, no detectable effect on the viability either of gametes or of zygotes (Hutchinson *et al.*, 1979). There is one qualification to make. The mean nuclear DNA amount of backcross progenies (4.64 pg) is slightly lower than expectation (4.68 pg). Even so, the effect of the segregating DNA fraction is, at most, marginal.

(b) Chiasma frequency in pollen mother cells. An assay of phenotypic characters in backcrosses and F_2s derived from *Lolium* hybrids has been described by Hutchinson, Rees and Seal (1979). We wish to concentrate on one of these charac-ters, the chiasma fequency in pollen mother cells. The ques-tion is; what effect has the extra, mainly repetitive DNA from *L. temulentum* upon chromosome pairing and chiasma formation in the hybrid and backcross derivatives? Two kinds of effect may be envisaged, indeed expected. First, the extra DNA in *L. temulentum* chromosomes may be expected to affect their structural homology, their ability to pair effectively with the smaller *L. perenne* homoeologues. Second, the extra DNA may carry determinants that affect chromosome pairing and chiasma formation.

    Figure 4 shows the distribution of chiasma frequencies in the parents, the F_1 and backcross progenies plotted against the nuclear DNA amount. It will be observed that the chiasma frequency of the F_1 hybrids is not significantly different from the *L. perenne* parent, despite the gross structural dis-similarity between the parental chromosomes. There is

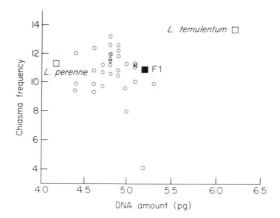

FIG. 4 *The mean chiasma frequency in pollen mother cells of the parents, F_1 and backcross progenies (O) plotted against the nuclear DNA amount.*

a suggestion that the massive DNA difference between the *L. temulentum* and *L. perenne* chromosomes has no effect upon the efficiency of pairing or upon chiasma formation, at least when measured as the chiasma frequency. This is confirmed by reference to the backcross results. It will be observed that there is a wide range of chiasma frequencies among the back-cross individuals. This is to be expected as a consequence of the segregation of structural genes affecting chiasma frequency. There is, however, no correlation between the chiasma frequencies and the nuclear DNA amounts in the backcross. There is, in fact, no evidence that the extra DNA deriving from *L. temulentum* affects the chiasma frequency. We can draw another conclusion from these assays. The fact that the F_1 hybrids produce viable gametes in the first place, tells us that chiasma formation and crossing-over at meiosis must, to a large degree at least, be between homologous regions of the chromosomes. The consequences, otherwise, would be deficiencies and duplications of structural genes and other essential determinants. If this is indeed the case, by what mechanism is the effective pairing achieved? To answer this question, we have switched attention from *Lolium* to a *Festuca* hybrid. The switch was made because the pachytene chromosomes in the *Festuca* hybrid are more amenable to analysis than those of the *Lolium* hybrid.

2. Festuca

Festuca drymeja ($2n$ = 14) has approximately 50% more nuclear

DNA than *F. scariosa* (also $2n = 14$). The 2 CDNA amounts are
7.17 and 4.73 pg, respectively. At metaphase of meiosis in
their F_1 hybrid, the chiasma frequency is low, about 3.3 per
pollen mother cell. Not surprisingly, the hybrid is sterile.
The fact that there is widespread failure of chiasma formation
in this hybrid does not, paradoxically, detract from its usefulness in establishing how homoeologous chromosomes of vastly
different lengths and DNA contents achieve effective pairing
at pachytene. On the contrary, we can find out directly to
what extent the amount and distribution of DNA differences
within different pairs of homoeologues may contribute to the
failure of pairing such as we find in this particular hybrid.

(A) Light microscopy. Figure 5 shows that the bivalents and
homoeologous univalents at first metaphase in the hybrid are
asymmetrical, as expected. The large and small chromosomes
of *F. drymeja* and *F. scariosa*, respectively, are readily identified. The pachytene in Fig. 6 is atypical in one sense, in

FIG. 5 *First metaphase of meiosis in the* F. scariosa × F. drymeja *hybrid. The smaller the chromosomes, the greater is the
asymmetry of bivalents and univalent pairs. The bar represents
10 µm.*

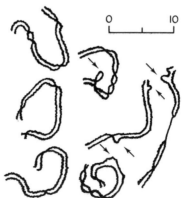

FIG. 6 *Late pachytene in the* F. scariosa × F. drymeja *hybrid.
Arrows point to unpaired loops. The bar at top right represents 10 µm*

that all chromosomes are paired. In most pachytenes there are four or more univalents. The cell, however, is typical in that complete, parallel associations within the three largest pairs of chromosomes may be accomplished without prominent unpaired segments, by way of interstitial loops or loose ends. This is despite the fact that there are, between members of each pair, DNA differences ranging from 25 to 42%. The remaining four, smaller and most asymmetrical pairs show characteristic loops in one arm. For this reason, it is perhaps not surprising that failure of chiasma formation is greater among the smaller pairs with loops than large pairs without. This is to suggest, of course, that extreme differences in DNA amount between homoeologues may indeed impede effective pairing.

Even so, the pachytene configurations in general, particularly of the large chromosomes, are surprisingly complete and normal in appearance when we take account of the structural differences between homoeologues. In order to establish whether the pairing is indeed complete and effective, we need more detailed information. A prerequisite for effective pachytene pairing is the synthesis and organisation of synaptonemal complexes. We have investigated the formation and distribution of synaptonemal complexes in this hybrid by an analysis of "reconstructed" chromosome configurations, as revealed in serial sections under the electron microscope.

(b) Electron microscopy. Figure 7 is a section through a late pachytene nucleus. It shows segments of synaptonemal complexes and, as well, the chromatin mass on either side of the lateral elements of the complexes. It is possible to trace the synaptonemal complexes, and, equally important, to trace and estimate the volume of chromatin associated with the complexes throughout the length of each chromosome. For the present purpose, we shall confine our observations to the longest chromosomes. These, it will be recalled, are more often paired at pachytene than the short chromosomes. The large chromosomes are readily identified at pachytene because they have larger centromeres (Jenkins and Bennett, 1981).

Figure 8 shows one of the three largest chromosome pairs from each of six nuclei at late pachytene. It will be observed that:

(1) The synaptonemal complex is normal in structure and uninterrupted throughout the length of the associations represented in (a) to (d). The pairing is effective, despite the massive DNA and size differences between homoeologous chromosomes and chromosome arms.
A good, albeit approximate, estimate of the DNA differences between chromosomes and between chromosome arms is provided

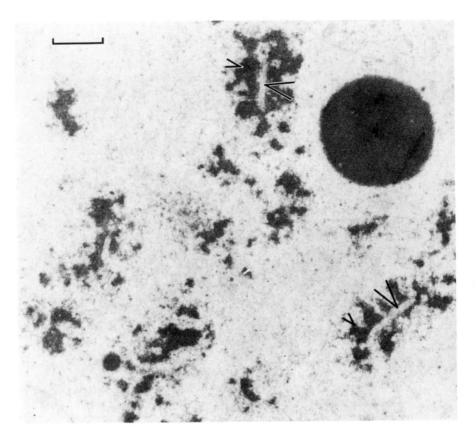

FIG. 7 *An electron microscopic section through a pollen mother cell nucleus of* F. scariosa × F. drymeja *at pachytene of meiosis. Synaptonemal complexes are indicated by large arrows, the chromatin mass by smaller arrows. The bar at top left represents 1 μm.*

by the chromatin volumes. Differences in chromatin volumes between homoeologous chromosomes range from 18 to 57% and between homoeologous arms from 14 to 73%. Yet, as we see, the pairing appears normal and complete.

(2) The synaptonemal complexes in (e) and (f) are restricted to the long arm.

The failure of complete synaptonemal complex formation cannot be explained by a greater difference in size between the members of the pairs (e) and (f). The difference in chromatin volume is less overall, for example, than in (d). There is, however, a massive percentage difference in chromatin volume between the unpaired arms in (e) and (f). It would appear

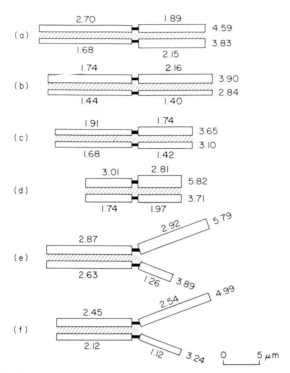

FIG. 8 *Pairing configurations and synaptonemal complexes in a long chromosome at late pachytene in each of 6 pollen mother cells. Solid lines are centromeres. Hatching represents the synaptonemal complexes. Volumes (μm³) of chromatin are indicated for each chromosome part and for each chromosome. The* F. drymeja *and* F. scariosa *chromosomes are the upper and lower in each pair, respectively. The horizontal scale is represented by the bar at bottom right. The vertical scale is arbitrary.*

that the distribution of the size and DNA differences within the pairs is the decisive factor.

There is confirmation for our earlier suggestion that a localised, extreme disparity in DNA content *does* prevent effective pairing at pachytene. However, what is perhaps more deserving of emphasis overall is the completeness of the synaptonemal complexes, the effectiveness of pairing, particularly in Fig. 8 (a) to (d), despite, in each case, a massive disparity in length and DNA content between the homoeologous chromosomes.

SUMMARY AND CONCLUSIONS

(1). There is, surprisingly, effective pachytene pairing and chiasma formation at meiosis between *L. temulentum* chromosomes and their *L. perenne* homoeologues, despite an average difference of 40% in DNA amount.

(2) Observations at pachytene in the *Festuca* hybrid confirm the formation of parallel configurations, complete with uninterrupted synaptonemal complexes, between homoeologous chromosomes with comparable DNA differences.

(3) On the basis of the viability of gametes and zygotes deriving from the *Lolium* species hybrid, we inferred that the formation of chiasmata between homoeologous chromosomes must, at least frequently, involve homologous segments. Figure 9 suggests a likely model, whereby this is achieved.

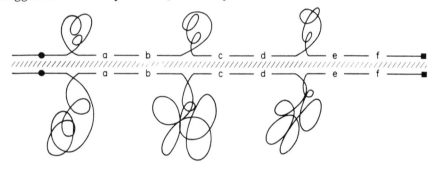

FIG. 9 *A model showing how pairing between homologous segments in the vicinity of the synaptonemal complex (hatching) may be achieved between homoeologous chromosomes of different lengths.*

(4) Where differences in DNA amount between homoeologous chromosomes or chromosome segments are disproportionately large, it is clear from the *Festuca* observation that pairing at pachytene becomes ineffective.

(5) These observations allow us to conclude with some confidence that supplementary, non-coding DNA affects chromosome pairing and chiasma formation at meiosis. The most important conclusion in our view, however, must be that such supplementary DNA has, in general, such a small effect on pairing; which is another way of saying that its effects are achieved only by massive, localised changes in amount. While such localised changes may in certain circumstances be of functional and even adaptive significance in respect of crossing-over and chiasma formation (cf. John and King, 1980), it is difficult to attribute such functional significance to the supplementary DNA in general. To date, therefore, our assays confirm only that a large-scale variation in DNA amount has

astonishingly little effect on many aspects of growth and development of the phenotype.

REFERENCES

Bennett, M.D. (1972). Nuclear DNA amounts and minimum generation time in herbaceous plants. *Proc. Roy. Soc. ser. B* **191**, 109-135.

Doolittle, W.F. and Sapienza, C. (1980). Selfish genes, the phenotype paradigm and genome evolution. *Nature (London)* **284**, 601-603.

Evans, G.M., Rees, H., Snell, C.L. and Sun, S. (1972). The relationship between nuclear DNA amount and the duration of the mitotic cycle. *In* "Chromosomes Today", vol. 3, pp. 24-31, Longman, London.

Hutchinson, J., Rees, H. and Seal, A.G. (1979). An assay of the activity of supplementary DNA in *Lolium*. *Heredity* **43**, 411-421.

Hutchinson, J., Narayan, R.K.J. and Rees, H. (1980). Constraints upon the composition of supplementary DNA. *Chromosoma* **78**, 137-145.

Jenkins, G. and Bennett, M.D. (1981). The intranuclear relationship between centromere volume and chromosome size in *Festuca scariosa* × *drymeja*. *J. Cell Sci.* **47**, 117-125.

John, B. and King, M. (1980). Heterochromatin variation in *Cryptobothrus chrysophorus* III Synthetic hybrids. *Chromosoma* **78**, 165-186.

Martin, P.G. (1966). Variation in the amounts of nucleic acids in the cells of different species. *Expt. Cell. Res.* **44**, 84-90.

Orgel, L.E. and Crick, F.H.C. (1980). Selfish DNA: the ultimate parasite. *Nature (London)* **284**, 604-607.

PART IV
GENOME EVOLUTION AND SPECIES SEPARATION

Sequence Amplification, Deletion and Rearrangement: Major Sources of Variation During Species Divergence

RICHARD FLAVELL

Plant Breeding Institute, Cambridge, England

The genomes of eukaryotes are bewilderingly complex. The variation in DNA content between species is immense (Hinegardner, 1976). Even within certain genera, the DNA content of different species can vary up to tenfold (Rees and Hazarika, 1969; Jones and Brown, 1976). In spite of the great complexity and diversity of DNA sequences within and between species, a general understanding of the origins of the complexity and diversity is emerging. Much can be explained by recognising that small pieces of DNA become amplified, mutated, deleted, rearranged and translocated to new sites in chromosomes, and that these "macromutations" spread through populations over time-scales that can cause certain segments of a genome to change rapidly during evolution.

In this contribution I wish to illustrate (1) how these "macromutations" have played a major role in determining the structure and sequence composition of complex eukaryotic genomes, and (2) how fixation of different "macromutations" creates substantial differences between the genomes of closely related species: differences that probably contribute to the failure of chromosome pairing and recombination in interspecies hybrids. I will use as illustrations results from studies carried out on some higher plant species from the *Aegilops, Triticum, Secale, Hordeum* and *Avena* genera of the Gramineae. My references to related studies on other organisms that lead to similar conclusions are inevitably incomplete but other contributions in this volume remedy most of these deficiencies.

The haploid sizes of the Gramineae species range between 3.6 and 8.8 pg and are considerably larger than those, for example, of *Drosophila melanogaster* (0.14 pg) and man (3 pg). The larger genome implies that very little (1%) of the DNA consists of coding and other sequences carrying out sequence-specific functions (Flavell, 1980), and that most of the DNA is "secondary" DNA (Hinegardner, 1976). Much of the secondary DNA appears not to be maintained under strong, sequence-dependent selection. It is therefore particularly useful for learning about the kinds of changes that DNA undergoes during evolution.

THE CONTRIBUTION OF SEQUENCE AMPLIFICATION AND TRANSLOCATION TO GENOME STRUCTURE

The dominant role of sequence amplification in the determination of cereal genome structure is obvious from the fact that over 75% of the total DNA consists of repeated sequences (Rimpau *et al.*, 1978,1980). These repeated sequences can be classified into hybridisation families, where sequences belonging to the same family have sufficient base sequence homology to form stable duplexes under defined conditions. There are probably hundreds (or even thousands) of such families in a cereal genome, a figure that provides a guide to the *minimum* number of amplification events whose products have survived. As we shall see later, this minimum value is a great underestimate. The number of repeats in a hybridisation family varies from very few to over 10^6 (Flavell and Smith, 1976; Smith and Flavell, 1977), which illustrates that the extent of amplification of a sequence can vary enormously.

Repeats of some families are clustered together in long tandem arrays, and arrays of the same family are often present on several or all chromosomes of the complement, proof of sequence translocation to non-homologous chromosomes (Bedbrook *et al.*, 1980a,b; Gerlach and Peacock, 1980; Dennis *et al.*, 1980a,b; John and Miklos, 1979). Interestingly, the positions of the arrays are often but not always similar in many of the chromosomes, e.g. at centromeres or telomeres. The sites of arrays of a 120 base-pair repeating unit on the chromosomes of *Aegilops speltoides* are shown in Fig. 1 as an example. Notice that the sequences are on *all* the chromosomes near or at the telomeres or at a distal interstitial site.

Other arrays including the repeating unit G-A-A-G-A-G and its variants, are clustered at interstitial sites on the chromosomes of this species (Gerlach and Peacock, 1980; Dennis *et al.*, 1980a,b; Hutchinson and Lonsdale, 1982). Most of these sites stain preferentially with Giemsa after certain treatments and coincide with regions recognised as heterochromatin (Gerlach, 1977; Gerlach and Peacock, 1980). There is evidence from many species to show that chromosomes assume the more folded structure of heterochromatin where long

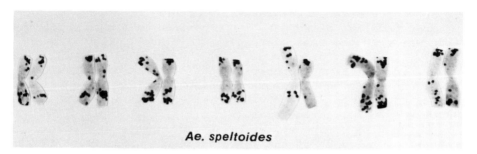

Ae. speltoides

FIG. 1 *Localisation of the major arrays of repeats of a single family in* Aegilops speltoides. *The repeats were localised by hybridising a ³H-labelled repeat unit of the family to metaphase chromosomes followed by autoradiography. The 120 base-pair repeat unit family has been described by Bedbrook et al. (1980a). The photograph was supplied by J. Jones (Jones, 1981).*

tandem arrays of repeated sequences are clustered (John and Miklos, 1979; Peacock *et al.*, 1981; Bedbrook *et al.*, 1980a).

Why are tandem arrays of sequences frequently clustered at similar positions on chromosomes? It seems reasonable to suggest that arrays might be preferred or tolerated only at specific locations. In addition, their similar location might reflect the mechanisms by which they have moved between chromosomes. Chromosomes often appear to be arranged such that centromeres lie at one pole of a nucleus, while telomeres are clustered at the other (see e.g. Appels *et al.*, 1978). If the transfer of sequences between non-homologous chromosomes involves direct chromosome contact, transfer and recurrent exchange might be expected to occur at similar chromosomal regions. Furthermore, where chromosomes lie in a specific order such that each chromosome is much closer to one or two of the complement than to all the others, exchange between these chromosomes would be expected to be much more common and their arrays of repeats to be more similar. Evidence supporting this line of argument for some cereal species is presented in the contributions to this volume by Bennett (1982). The spatial disposition of rye chromosomes in the somatic nucleus has been determined. Telomeres of some "associated" chromosomes appear more alike in their arrays of repeated sequences than those of "non-associated" chromosomes (see also Jones and Flavell, 1982).

Telomeres also sometimes aggregate in leptotene bouquets (Thomas and Kaltsikes, 1976), and may interact with one another to complete their DNA replication (Holmquist and Dancis, 1979). These are other occasions when the exchange of sequences between telomeres may occur.

Most families of repeats are organised in a much more complex way than the tandem arrangement (Flavell, 1980; Flavell *et al*., 1981; Davidson *et al*., 1973; Wensink *et al*., 1979; Young, 1979). Some of their members are interspersed with short non-repeated DNA segments and/or unrelated repeated sequences. Members of many such families hybridise to sequences on all chromosomes (Flavell *et al*., 1981). The major fraction of the genome organised in this way (50%) emphasises how much of the linear sequence arrangement in the genome is due to the rearrangement of short pieces of DNA (Flavell *et al*., 1981).

Members of a hybridisation family of repeats are not identical. Small mutations such as base changes, small deletions and insertions accumulate in the family, creating sequence variation. Sometimes a variant is propagated when individual repeats or subsets of repeats are reamplified (Bedbrook *et al*., 1980b; Flavell *et al*., 1977; Anderson *et al*., 1981; Dennis *et al*., 1980a,b; Pech *et al*., 1979) or when members of the family are "replaced" by a variant sequence (see Dover, 1981, Dover *et al*., 1982). Propagated variants are recognised as subsets of a hybridisation family by their different length or complement of restriction endonuclease sites. The reamplification of DNA sequences gives rise to a cyclical model of repeated sequence evolution as shown in Fig. 2.

When a member of a dispersed family of repeats is reamplified, it may be amplified together with a neighbouring single copy or unrelated repeated sequence. This creates "compound" repeating units (Flavell, 1980; Bedbrook *et al*., 1980b,1981). (See Fig. 2.)

The presence of many reamplified variants within a hybridisation family means, as stated earlier, that the number of hybridisation families is a gross underestimate of the total number of amplification events that have occurred in the evolution of a genome. Furthermore, it is likely that many amplification events are never detected because they are deleted and not fixed in the genome.

The amplification of a sequence, however rarely it occurs in evolution, implies that the sequence departs from strict Mendelian inheritance over evolutionary time periods. Translocation coupled with amplification enables a sequence to spread to other chromosomes and eventually to all chromosomes in a sexually reproducing population (Dover, 1981, Dover *et al*., 1982). This departure from Mendelian inheritance may be very important in spreading repeats through an interbreeding population relatively rapidly, without selection

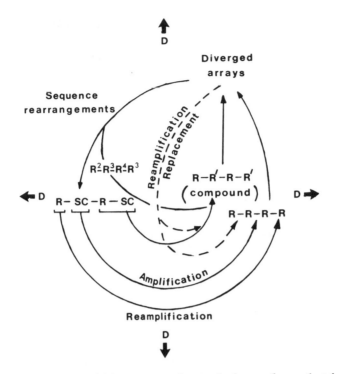

FIG. 2 *A scheme to illustrate the origin and evolution of repeated sequence organisation (Flavell, 1980). New arrays of repeats arise by the amplification of single copy (SC) DNA or the reamplification of a repeat (R). Compound repeating units arise when DNA segments containing repeat and single copy (R-SC) or different repeats (R^3-R^4) are amplified. Reamplification sometimes may be achieved by mechanisms that result in replacement or conversion of old sequences to new sequences (Dover et al, 1982). Sequence rearrangements may occur within the fraction of the genome where repeats and non-repeats are interspersed, in segments containing only arrays of repeats or between both fractions. Rearrangement commonly involves the transposition of short segments of DNA into new sites. The 4 "deletion" (D) arrows are meant to signify that deletion can occur from any of the DNA organisation patterns.*

for the sequence through the phenotype. Amplification and translocation mechanisms, therefore, may be responsible for the origin of most sequences in the cereal genomes and possibly for the rapid fixation of these sequences in the species, when they do not accumulate as a result of selection, drift or other drive mechanisms.

HOW ARE SEQUENCES AMPLIFIED AND TRANSLOCATED?

So far I have used the words amplification and translocation
but have made little reference to the nature of the mechanisms
involved. We are in need of much more information here. The
modulation of copy number of sequences in tandem arrays can
and probably certainly does occur by unequal crossing over
(Smith, 1973,1976). This process is a direct consequence of
the arrangement of the repeats and, in principle, is likely
to be independent of the sequence of the repeating unit.
Rapid amplification may occur by the excision of DNA followed
by replication on a rolling circle-type model and reintegration (Hourcade *et al.*, 1973). Alternatively, aberrant replication may occur *in situ* to produce localised tandem arrays.

I have already made some comments on the translocation of
arrays, or sequences within them, to non-homologous chromosomes based on mechanisms that require direct physical contact between chromosomes for a reciprocal (or non-reciprocal)
exchange event. But how do dispersed sequences change rapidly in copy number? What is the dispersal or translocational
mechanism(s)? Here, the amplification mechanism may be coupled with the translocation process, as implied by the studies
on the transposable elements (repeated sequences) of *Drosophila* (Finnegan *et al.*, 1978) and yeast (Cameron *et al.*, 1979)
described by Finnegan *et al.* (1982) in this volume.

The amplification-transposition mechanism of these elements seems to depend on features of the structure of the
element. This makes these pieces of repeated DNA examples of
selfish DNA (see Doolittle, 1982, this volume); i.e. DNA
that accumulates in the genome because of structural features
that promote its amplification. The details of how these DNA
elements move is not established (see Finnegan, 1982, this
volume). However, their similarity to the proviruses of
retroviruses raises the tantalising possibility that they
amplify by "reverse transcription" of an RNA transcript, followed by reinsertions of the DNA copy into a new chromosomal
site.

Are all the dispersed repeats in cereal genomes "selfish"
transposable elements or descended from them? Probably not.
However, further research should obviously be directed at
whether members of dispersed repeat families have structural
features of transposable elements, are transcribed in germ
line cells and if a reverse transcriptase exists. Other
plausible dispersal mechanisms can be hypothesised. For
example, circles of DNA could excise by recombination, move
to a new site and reinsert by the reverse of the excision
event.

SEQUENCE DELETION AND REPLACEMENT

Sequence amplification results in an increase in the DNA content of chromosomes. This clearly cannot go on indefinitely without the genome becoming too large. As genomes increase in size, then the maximum rate of cell development through mitosis and meiosis is reduced and cell size increases (Rees *et al.*, 1982; Macgregor, 1982; Bennett, 1972; Cavalier-Smith, 1978). These reasons alone are sufficient to conclude that genome growth will be selected against in many circumstances. This is achieved by deletion, either of the most recently amplified DNA or of other DNA previously accumulated. Deletion of "older" DNA, while the "newer" DNA is retained implies a "turnover" of DNA during evolution (Flavell *et al.*, 1980, 1981; Flavell, 1980; Thompson and Murray, 1980; Dover, 1981, Dover *et al.*, 1982).

DNA turnover or replacement of repeats also may result from unequal crossing over within tandem arrays of these repeats. One variant sequence can be amplified (expanded) at the expense of another to create a new family of variants or a subfamily. Unlinked members of a repeat family, even on non-homologous chromosomes, can be replaced by a new variant to create, when the same variant is propagated, a new homogeneous *dispersed* family of repeats (Scherer and Davis, 1980). This is another kind of sequence turnover postulated to explain the "concerted" evolution of dispersed repeats and is discussed more fully by Dover (1981) and Dover *et al.* (1982).

Therefore, deletion and replacement processes need to be superimposed on the amplification-reamplification cycle discussed earlier to describe the evolutionary turnover of a hybridisation family of repeats. This scheme is depicted in Fig. 2.

GENOME EVOLUTION AND SPECIES DIVERGENCE

Many of the observations summarised above on the structure and organisation of hybridisation families of repeats are illustrated in the evolutionary scheme of Fig. 2. The model emphasises DNA turnover, created by amplification and deletion. It shows that families of repeats arise by amplification of non-repeated, repeated or combinations of non-repeated and repeated DNAs and undergo recurrent rounds of amplification. This reamplification may result in sequence "replacement" when all or most members of the family evolve together (concerted evolution). Different families become intertwined in the evolutionary cycles when, after sequence rearrangement, segments of two or more families are amplified as part of a single new (compound) repeat unit.

The scheme implies that at any one time related sequences, i.e. in the same hybridisation family, may be organised in many ways throughout the genome and their copy number, position and sequence heterogeneity reflects the various stable molecular changes that have occurred to the sequences.

There are probably hundreds or thousands of hybridisation families evolving as depicted in Fig. 2 in large complex genomes like those of cereal plants. The rates of evolution of the different families and also for different members of the same family are likely to be very variable. The parameters determining the rates are very complex and one can do little except list some of them: the structures of the DNA sequences and their position; the frequencies of amplification, translocation, mutation and deletion; the rates of fixation in the population by stochastic processes; the effects of the changes on fitness and the components of selection.

Some appreciation of the rates at which the products of amplification, translocation, mutation and deletion are fixed in populations can be gained by comparing the genomes of related species. If repeats turn over and reorganise rapidly compared with speciation, then many differences would be expected between the repeated DNA fractions of related species.

We have made quantitative assessments of the differences between the cereal species that have evolved from a common

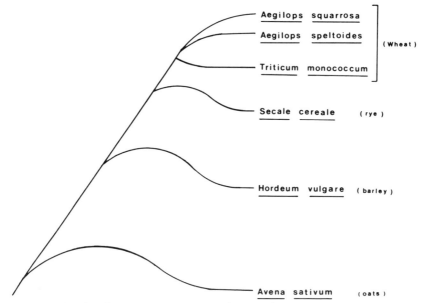

FIG. 3 *Evolution of some cereal species from a common ancestor.*

ancestor as shown in Fig. 3.   All the species have the same
numbers of chromosomes.

The repeated DNA relationships between the two *Aegilops*
species and *Triticum monococcum*, which form the most closely
related group, are considered first.   Afterwards, the genomes
of more distantly related *Secale*, *Hordeum* and *Avena* species
are compared.

Essentially all the highly repeated DNA families of *Aegilops squarrosa* are present in *Ae. speltoides* and *T. monococcum*
DNAs (Flavell *et al.*, 1979).   A detectable but small proportion of the highly repeated DNA in *T. monococcum* (0.5% of the
total DNA) is not present in the highly repetitive DNA fractions of *Ae. speltoides* and *Ae. squarrosa* but a significantly
larger proportion of the *Ae. speltoides* genome (2 to 3% of the
total DNA) consists of repeated DNA families not found in the
highly repeated fractions of *Ae. squarrosa* and *T. monococcum*
(Flavell *et al.*, 1979).   The chromosomal organisation and
location of most of these *Ae. speltoides* sequences has been
determined by Gerlach and Peacock (1980), Dennis *et al.*
(1980a,b), Hutchinson and Lonsdale (1982) and Jones (1981).
They reside in tandem arrays on all of the *Ae. speltoides*
chromosomes.   Thus each of the *Ae. speltoides* chromosomes
can be distinguished from its homologue in *Ae. squarrosa* and
T. monococcum by major localised, repeated sequence DNA differences.   The sequences in high copy number in *T. monococcum*
distinguishing the DNA of this species from the *Aegilops* DNAs
have not been studied in detail.

Thus, although there are families that can effectively distinguish the three genomes qualitatively, most of the repeated
DNA in these three diploids belong to families that are common to all three species, and which were probably in the common ancestral species.   The subfamilies and organisation of
sequences within a few of these families have been studied
using repeats purified from the wheat genome at random by
molecular cloning.   In Fig. 4A, the DNAs from the three diploid species have been restricted with endonuclease *Hin*dIII,
transferred to nitrocellulose and hybridised to two different
cloned sequences.   With the repeat fragment on the plasmid
pTa87, *Ae. squarrosa* shows only very low hybridisation compared with the other two species.   *Ae. speltoides* and *T.
monococcum* both have a major subfamily represented by the
band of 1100 base-pairs.   Each also shows species-specific
bands or subfamilies, which have presumably evolved since the
species diverged.   Thus, although the three species have the
pTa87 family, sequences within the family are in very different copy numbers and some of the sequences are organised differently in the three species.

The hybridisation banding pattern is also species-specific

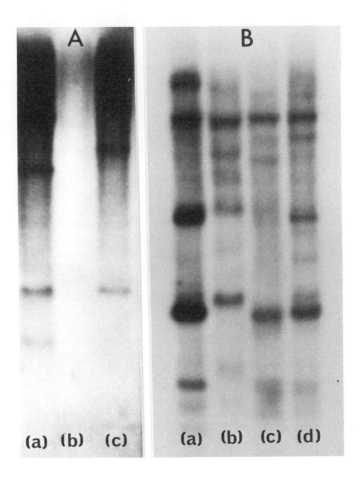

FIG. 4 *Hybridisation of repeats cloned from wheat to restriction digests of (a)* Ae. speltoides, *(b)* Ae. squarrosa, *(c)* T. monococcum *(c) and (d) wheat.   The 3 DNAs were digested with* HindIII, *fractionated by electrophoresis in 1% (w/v) agarose, transferred to nitrocellulose and hybridised with* 2×10^6 *cts/min of* ^{32}P-*labelled pTa87 (A) or pTa82 (B).*

when these three DNAs are probed with the plasmid pTa82, which contains a different repeat from the wheat genome (Fig. 4B).   On the basis of these results and similar hybridisations with other cloned sequences, it can be concluded that some of the sequences in a high proportion of the common families belong to species-specific subsets and are organised differently in the chromosomes, i.e. many molecular events have occurred in the turnover cycle (Fig. 2) for many families of repeats during species divergence.

T. monococcum, *Ae. speltoides* and *Ae. squarrosa* are closely
related to the three diploid progenitors of hexaploid bread-
wheat *Triticum aestivum*. Hybridisation studies between wheat,
rye (*Secale cereale*), barley (*Hordeum vulgare*) and oats (*Avena
sativum*) DNAs have shown that 16%, 22%, 28% and 58% of the
DNAs, respectively, are species-specific repeated sequences
(Rimpau *et al.*, 1978,1980) that have probably arisen by the
amplification of single copy DNA since species divergence
(Flavell *et al.*, 1977). Their repeated sequence DNA homology
is therefore much less than between the *Aegilops* and *Triticum*
species. Some of the species-specific repeats in wheat, rye
and barley are interspersed between DNA segments belonging to
common families (Rimpau *et al.*, 1978,1980). This must be
due to either the transposition of species-specific or common
repeats into new domains or to the amplification in each species of compound units. When the proportions of interspersed
common DNA are added to the proportions of species-specific
DNA, it can be concluded that the segments carrying interspersed species-specific repeats occupy 27% of the wheat genome, 34% of the rye genome, 44% of the barley genome and 66%
of the oat genome. Because these segments are present on all
chromosomes, there are clearly major differences in sequence
homology between wheat and rye chromosomes and even more between wheat or rye and barley chromosomes. Most of the repeats in the oats genome appear unrelated to those in the
other three species. The lower amount of species-specific
DNA between wheat and rye than between wheat or rye and barley
is consistent with wheat and rye being closer phylogenetically
than wheat or rye and barley (Fig. 3; see also Flavell *et al.*,
1977).

These estimates of the species-specific regions are by no
means the full assessment of sequence divergence because, as
for the *Aegilops*/*Triticum* comparisons, there are major differences between the families present in all three species.
Figure 5 shows the hybridisation patterns of pTa87 and pTa82
to wheat and rye DNAs after digestion of the DNAs with
*Hin*dIII. There are copy number and band differences between
the two species, implying many sequence structure and organi-
sational differences in the segments of the chromosomes
possessing sequences in these families.

Different subfamilies between species arise from the
amplification of different variants. When variant sequences
reanneal together, the thermal stability of the duplexes is
less than that of duplexes between perfectly base-paired
molecules (Ullman and McCarthy, 1973). The thermal stabilities of duplexes between wheat repeats, between wheat and
rye, wheat and barley and wheat and oats repeats are shown in

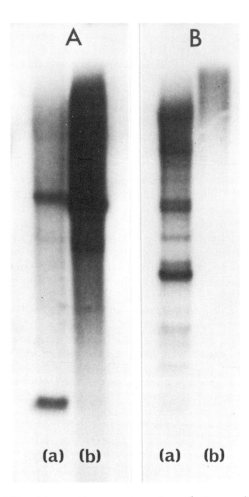

FIG. 5 *Hybridisation of repeats cloned from wheat to wheat and rye. (a) Wheat DNA or (b) rye DNA were digested with* HindIII, *fractionated by electrophoresis in 1% (w/v) agarose, transferred to nitrocellulose and hybridised with 2 × 10⁶ cts/ min of* ³²P-labelled pTa87 (A) *or* pTa82 (B).

Table I. The significant reduction in thermal stability of duplexes formed within a genome is consistent with the exis- tence of many subfamilies. The even greater reduction in the interspecies duplexes also implies the presence of dif- ferent subfamilies in diverged species (Flavell *et al.*, 1977; Rimpau *et al.*, 1978,1980). These different subfamilies have undoubtedly arisen by different variants emerging from the cycles of Fig. 2 in the separate species.

TABLE 1

*Thermal stabilities of reannealed repeated sequences
of wheat and of duplexes between repeats from
wheat, rye, barley and oats*

t_m(°C) Repeats reannealed with wheat repeats

WHEAT	RYE	BARLEY	OATS
83	77	74	73

Data taken from Smith and Flavell (1974).

When the variation between species due to the sequence-
specific regions is combined with the variation in common
families, it is obvious that molecular events in the turnover
cycle (Fig. 2) have changed most of the DNA since *Hordeum* and
Secale divergence, and a substantial proportion since *Triticum*
and *Secale* divergence. There is probably very little linear
homology between any substantial proportion of wheat/rye and
barley chromosomes, and probably little between wheat and rye
chromosomes.

There is no doubt, therefore, that sequence amplification,
translocation, mutation and deletion are responsible for the
major differences in genome structure between these species.

These mechanisms are also the most likely sources of variation affecting chromosome size. The size differences between
Secale cereale and *Secale silvestre* chromosomes are due prin-
cipally to arrays of repeated sequences (Bedbrook *el al.*,
1980a). As already stated, the total amount of DNA per cell
affects many important phenotypic characters. Furthermore,
the amount of DNA per chromosome arm may influence many properties of individual chromosomes, including their position
in the cell (Bennett, 1982).

GENOME DIVERGENCE AND REDUCTIONS IN MEIOTIC CHROMOSOME PAIRING

In the previous section, assessments were made of the DNA
sequence homologies between the genomes of related cereal
species. There is a good correlation between the crude estimates of nucleotide sequence differences and the reduction in
chiasma formation in F_1 hybrids between these species. In
hybrids between species that contain little species-specific
DNA when they are compared in DNA/DNA hybridisation experiments, chromosome pairing and some chiasma formation occurs

TABLE II

Chiasmata formation in Ae. speltoides × T. boeticum *hybrids*

	UNIVALENTS	BIVALENTS	OTHER	CHIASMATA
Hybrid (− B chromosome)	0.33–7.7	3.0–5.9	0.1–0.5	3.7–11.2
Hybrid (+ B chromosome*)	12.5–14.7	0.2–1.7	0.0	0.2–1.3

Taken from Riley *et al.* (1973)
*The B supernumerary chromosome was introduced into the F_1
hybrid *via* the *Ae. speltoides* parent.

but it is always below the level found in intraspecies hybrids. For example, in the 14-chromosome F_1 hybrid of *Ae.
speltoides* × *T. boeticum* (similar to *T. monococcum*) the frequency of bivalents, trivalents and quadrivalents was between
three and six per cell, and up to eight univalents were
scored (Table II; and see Riley *et al.*, 1973). Seven bivalents are normally found in intraspecific hybrids.
    In haploid wheat, assessments of pairing at meiosis provide
a measure of the homology between the genomes of the three
constituent genomes. Virtually no bivalents are found (Table
III; and see Riley, 1966; Miller and Chapman, 1976). However,

TABLE III

*Chiasmata formation in haploid wheat and in F_1 hybrids
between wheat and related diploid species*

	CHROMO- SOME NUMBER	UNI- VALENTS	BI- VALENTS	TRI- VALENTS	OTHERS
Haploid wheat	21	20.5	0.2	0.0	0.0
Haploid wheat (−5B)	20	8.4	3.5	1.4	0.07
Wheat(−5B) × *Ae. longissima*	20+7	7.5	7.6	0.7	0.6
Wheat(−5B) × *Ae. caudata*	20+7	6.8	5.8	2.1	0.6
Wheat(−5B) × *S. cereale*	20+7	18.0	3.2	0.7	0.1

Taken from Riley (1966) and Miller and Chapman (1976).

the extent of pairing and chiasma formation between the wheat
and related genomes is regulated by a number of specic loci,
the predominant one being the Ph locus on chromosome 5B
(Sears, 1976). When this chromosome is removed (in a nulli
5B haploid), more bivalents are seen (Table III) but the
number is considerably less than expected if the genomes were
completely homologous (Miller and Chapman, 1976). In F_1 hybrids between nulli 5B haploid wheat and other *Aegilops* species, e.g. *Ae. longissima* and *Ae. caudata* (Table III; and see
Riley, 1966), the number of chiasmata indicates some homology
between the chromosomes of the *Aegilops* parent and the chromosomes of the *Aegilops* species in wheat, but again the number
of chiasmata is below that expected between completely homologous genomes.

This partial homology between *Aegilops* and *Triticum* chromosomes is in marked contrast to the very much lower chiasmata
formation between *Aegilops* or *Triticum* and *Secale* chromosomes.
In F_1 hybrids between nulli 5B wheat and *S. cereale* (Table
III; and see Riley, 1966), the level of bivalent and multi-
valent formation is not higher than in the nulli 5B haploid
stock alone. Similar results are found in wheat × *Hordeum*
hybrids (Islam *et al.*, 1975; Fedak, 1977; Martin and Chapman,
1977). Therefore, genome homology is low, as found in the
DNA sequence comparisons.

An example of a correlation between DNA sequence variation
and chiasma formation for a specific chromosome comes from
studies of chromosome 4A of hexaploid wheat. This chromosome
possesses arrays of repeated sequences not found in the homologous chromosome of *Triticum* species closely related to the
A genome parent of wheat (Gerlach and Peacock, 1980; Dennis
et al., 1980a,b; Hutchinson and Lonsdale, 1982). When telo-
centric derivatives of all the A genome chromosomes of wheat
were assayed for their ability to pair with a chromosome of
Triticum urartu, all except those of chromosome 4A showed high
levels of chiasma formation (Chapman *et al.*, 1976).

This correlation between the level of chromosome pairing
and crossing over and the extent of DNA sequence homology suggests that the DNA turnover and rearrangement depicted in Fig.
2 is responsible for the reduction in chromosome pairing between species. Whether all sequences of a chromosome contribute to the homology necessary for recombination or whether
only a subset is involved is unknown. However, if the subset
also changes by the turnover mechanisms, then the causal
correlation between rapid sequence turnover and reduction in
chromosome pairing/recombination remains an attractive hypothesis.

GENOME EVOLUTION AND SPECIATION

Genetic isolating mechanisms that initiate speciation may
involve many different kinds of mutations or genomic changes.
The mutations may create pre-mating or post-mating barriers
(for a review, see White, 1977). When two species are able
to form an F_1 hybrid, but it is infertile due to meiotic
chromosome pairing or recombination failure, then the genomic
changes causing the meiotic abnormalities may be the genetic
isolating mechanisms.

I have argued above that many of the genomic differences
between certain cereal species are due to rapid sequence
turnover. If some of these differences were contributors to
loss of pairing and recombination in hybrids between individuals *before* speciation, then sequence turnover could have
been a major contributor to speciation.

Mutations are known that prevent the formation of fertile
hybrids between some species combinations, e.g. between wheat
and rye (Riley and Chapman, 1967), but these mutations are
not present in all isolates of the species. They may therefore have arisen after speciation, although equally they could
have caused speciation and then been lost following the accumulation of other pre- or post-mating barriers that have
kept the species distinct.

In all discussions of speciation there is the problem (but
often overlooked) of how the mutation(s) or genomic change(s)
causing speciation spreads through the population when the
individual in which it arises is unable to mate (or produce
sufficient fertile offspring) with other members of the population. The common solution to this problem is to postulate
the emergence of the new species from a small segment of the
population, which "struggles" through the infertility barrier.
This difficulty is much less severe in a model of speciation
suggested by the data described here for some of the cereal
species.

The hypothesis that accumulated differences between genomes reduces fertility in hybrids has been made already. We
have also seen that the resulting reduced recombination in
the hybrids is dependent on the genotype. It is much more
marked in the presence of chromosome 5B of wheat or of a B
supernumerary chromosome (Tables II and III). Therefore, it
is possible to envisage a population of individuals in which
many genomic differences had accumulated, but which were all
interfertile because the genotype "suppressed" the differences. Suppose that in an individual, a dominant mutation
arose that no longer allowed "suppression" of the chromosomal
differences. This individual could still produce offspring
with the large number of individuals that had very similar

genomes, but its hybrid with individuals with less homologous genomes would be relatively sterile. The subsequent spread of the mutation through the population of similar genomes by drift, selection or a drive mechanism would divide the population into separate breeding groups or species. But the new emerging species would not necessarily have started from bottlenecks of very small gene pools. The validity of this sort of speciation model for cereal species is, of course, difficult to test. The special purposes in outlining the model in this contribution are (1) to highlight a possible role of repeated sequence turnover, not necessarily affecting genes, in the separation of species, and (2) to illustrate that the effect of repeated sequence variation may be dependent on the activity of particular genes. Aspects of this model have been discussed by others, including Waines (1976), Riley (1981) and Dvorak and McGuire (1981).

Repeated sequence differences may result in reduced fitness of hybrids between individuals for reasons other than chromosome pairing failure at meiosis. Variation in gene expression could result from changes in families of repeated sequences (Flavell, 1981), although clear cases need to be proven. Variations in chromosome behaviour also result from major repeated sequence differences. For example, in wheat-rye hybrids with both the wheat and rye chromosome complements in the disomic condition to maintain fertility, grain development is initiated by a series of very rapid nuclear cycles of replication and division in a syncytium. The rye chromosomes sometimes fail to divide, and this is the cause of large nuclear aggregates (Bennett, 1973). In lines lacking the telomeric heterochromatin consisting of arrays of repeated sequences not found in the wheat chromosomes, the frequency of these aggregates is much reduced (Bennett, 1977, 1981). Therefore, it appears very likely that the heterochromatin on the rye chromosomes that is late-replicating, at least in root cells (Ayonoadu and Rees, 1973), prevents the rye chromosomes from completing the replication cycle sufficiently rapidly during endosperm development to permit normal development. This problem occurs only in a wheat-rye hybrid. Development obviously occurs normally in rye.

This example therefore illustrates how different chromosome behaviour due to blocks of reiterated sequences can influence the fitness of hybrids. Some wheat-rye hybrids with the rye terminal heterochromatin blocks do not show aberrant grain development. This is presumably due to genetic variation that suppresses the effects of the chromosomal differences; another illustration of where the effects of repeated sequence differences are dependent on the genotype.

CONCLUDING REMARKS

From examination of the structure and organisation of families
of repeats within and between genomes, the evidence is now
very substantial that gross changes, "macromutations", accumulate relatively rapidly in the chromosomes. Whether their
origins are "selfish" or not, it is hard to believe that these
changes are unimportant in evolution. Many may be neutral
with respect to gene activity and spread through populations
by stochastic drive mechanisms, while others survive by selection. However, by changing the structure, organisation,
sequence composition and size of chromosomes, all make a contribution to genome evolution.

The relatively rapid accumulation of such "macromutations"
implies that populations of individuals that are not interbreeding will diverge and then chromosome homology will be
eroded, the term homology being used here in relation to
regions of hundreds or thousands of base-pairs rather than
one or a few.

The number of "macromutations" that survive because they do
not interfere with gene expression is likely to be related
to the total DNA content. Species with large genomes, i.e.
with large proportions of non-coding, "secondary" DNA probably tolerate more "macromutations." Therefore, the genomes
of organisms with large DNA contents may diverge more rapidly
between populations than the genomes of organisms with low
DNA contents. Consequently, loss of chromosome homology due
to "macromutation" may be a more common cause of speciation for
organisms with large genomes than for organisms with small
genomes. This is a hypothesis worthy of further examination.
It emphasises that any role that "macromutation" or non-
adaptive changes play in speciation may be different in
organisms such as *Drosophila*, primates and wheat, which have
very different genomes.

REFERENCES

Anderson, D.M., Scheller, R.H., Pasakony, J.W., McAllister,
    L.B., Trabert, S.G., Beall, C., Britten, R.J. and David-
    son, E.H. (1981). Repetitive sequences of the sea urchin
    genome. Distribution of numbers of specific repetitive
    families. *J. Mol. Biol.* **145**, 5-28.
Appels, R., Driscoll, C. and Peacock, W.J. (1978). Hetero-
    chromatin and highly repeated DNA sequences in rye (*Secale
    cereale*). *Chromosoma* **70**, 67-89.
Ayonoadu, V. and Rees, H. (1973). DNA synthesis in rye chro-
    mosomes. *Heredity* **30**, 233-240.
Bedbrook, J.R., Jones, J., O'Dell, M., Thompson, R. and

Flavell, R.B. (1980a). A molecular description of telomeric heterochromatin in *Secale* species. *Cell*, 19, 545-
560.

Bedbrook, J.R., O'Dell, M. and Flavell, R.B. (1980b). Amplification of rearranged sequences in cereal plants. *Nature
(London)* **288**, 133-137.

Bedbrook, J.R., Jones, J. and Flavell, R.B. (1981). Evidence
for the involvement of recombination and amplification
events in evolution of *Secale* chromosomes. *Cold Spring
Harbor Symp. Quant. Biol.* **45**, 755-760.

Bennett, M.D. (1972). Nuclear DNA content and minimum generation time in herbaceous plants. *Proc. Roy. Soc. ser. B*
181, 109-135.

Bennett, M.D. (1973). Meiotic, gametophytic and early endo-
sperm development in Triticale. *In* "Triticale" (MacIntyre,
E. and Campbell, M., eds), pp. 137-148, International
Development Research Centre, Ottawa, Canada.

Bennett, M.D. (1977). Heterochromatin, aberrant endosperm
nuclei and grain shrivelling in wheat-rye genotypes.
Heredity **39**, 411-419.

Bennett, M.D. (1981). Nuclear instability and its manipulation in plant breeding. *Phil. Trans. Roy. Soc. ser. B*
292, 475-485.

Bennett, M.D. (1982). Nucleotypic basis of the spatial
ordering of chromosomes in eukaryotes and the implications
of the order for genome evolution and phenotypic variation.
In "Genome Evolution" (Dover, G.A. and Flavell, R.B., eds),
Academic Press, London.

Cameron, J.R., Loh, E.Y. and Davis, R.W. (1979). Evidence
for transposition of dispersed repetitive DNA families in
yeast. *Cell* 16, 739-751.

Cavalier-Smith, T. (1978). Nuclear volume control by nucleo-
skeletal DNA, selection for cell volume and cell growth
rate and the solution of the DNA C-value paradox. *J. Cell
Sci.* 34, 247-278.

Chapman, V., Miller, T.E. and Riley, R. (1976). Equivalence
of the A genome of bread wheat and that of *Triticum urartu.*
Genet. Res. (Camb.) 27, 69-76.

Davidson, E.H., Hough, B.R., Amenson, C.S. and Britten, R.J.
(1973). General interspersion of repetitive with non-
repetitive sequence elements in the DNA of *Xenopus. J. mol.
Biol.* 77, 1-23.

Dennis, E.S., Dunsmuir, P. and Peacock, W.J. (1980a). Segmental amplification in a satellite DNA: restriction enzyme analysis of the major satellite of *Macropus rufogris-
eus. Chromosoma* **79**, 179-198.

Dennis, E.S., Gerlach, W.L. and Peacock, W.J. (1980b).
Identical poly-pyrimidine-polypurine satellite DNAs in

wheat and barley. *Heredity* 44, 349–366.

Doolittle, W.F. (1982). Selfish DNA after fourteen months. *In* "Genome Evolution" (Dover, G.A. and Flavell, R.B., eds), Academic Press, London.

Dover, G.A. (1981). A role for the genome in the origin of species? *In* "Mechanisms of Speciation" (Barigozzi, C., Montalenti, G. and White, M.J.D., eds). In the press.

Dover, G.A., Brown, S.D.M., Coen, E.S., Dallas, J., Strachan, T. and Trick, M. (1982). The dynamics of genome evolution and species differentiation. *In* "Genome Evolution" (Dover, G.A. and Flavell, R.B., eds), Academic Press, London.

Dvorak, J. and McGuire, P.E. (1981). Non-structural chromosome differentiation among wheat cultivars, with special reference to differentiation of chromosomes in related species. *Genetics* 97, 391–414.

Fedak, G. (1977). Barley-Wheat hybrids. *Barley Genetics Newsletter* 7, 23–24.

Finnegan, D.J., Rubin, G.M., Young, M.W. and Hogness, D.S. (1978). Repeated gene families in *Drosophila melanogaster*. *Cold Spring Harbor Symp. Quant. Biol.* 42, 1053–1063.

Finnegan, D.J., Will, B.A., Bayer, A.A., Bowcock, A.M. and Brown, L. (1982). Transposable DNA sequences in Eukaryotes. *In* "Genome Evolution" (Dover, G.A. and Flavell, R.B., eds), Academic Press, London.

Flavell, R.B. (1980). The molecular characterisation and organisation of plant chromosomal DNA sequences. *Annu. Rev. Plant Physiol.* 31, 569–96.

Flavell, R.B. (1981). Molecular changes in chromosomal DNA organisation and origins of phenotypic variation. *Chromosomes Today* 7, 42–54.

Flavell, R.B. and Smith, D.B. (1976). Nucleotide sequence organisation in the wheat genome. *Heredity* 37, 231–252.

Flavell, R.B., Rimpau, J. and Smith, D.B. (1977). Repeated sequence DNA relationships in four cereal genomes. *Chromosoma (Berl.)* 63, 205–222.

Flavell, R.B., O'Dell, M. and Smith, D.B. (1979). Repeated sequence DNA comparisons between *Triticum* and *Aegilops* species. *Heredity* 42, 309–322.

Flavell, R.B., Bedbrook, J.R., Jones, J., O'Dell, M., Gerlach, W., Dyer, T.A. and Thompson, R.D. (1980). Molecular events in cereal genome evolution. *In* "The 4th John Innes Symposium" (Davies, D.R. and Hopwood, D.A., eds), pp. 15–30, John Innes Charity, Norwich.

Flavell, R.B., O'Dell, M. and Hutchinson, J. (1981). Nucleotide sequence organisation in plant chromosomes and evidence for sequence translocation during evolution. *Cold Spring Harbor Symp. Quant. Biol.* 45, 501–508.

Gerlach, W.L. (1977). N-banded karyotypes of wheat species.

Chromosoma **62**, 49–56.

Gerlach, W.L. and Peacock, W.J. (1980). Chromosomal locations
of highly repeated DNA sequences in wheat. *Heredity* **44**,
269–276.

Hinegardner, R. (1976). Evolution of genome size. *In* "Molecular Evolution" (Ayala, F.J., ed.), pp. 179–199, Sinauer
Associates Inc., Mass.

Holmquist, G.P. and Dancis, B. (1979). Telomere replication,
kinetochore organizers and satellite DNA evolution. *Proc.
Nat. Acad. Sci., U.S.A.* **76**, 4566–4570.

Hourcade, D., Dressler, D. and Wolfson, J. (1973). The amplification of ribosomal RNA genes involving a rolling circle
intermediate. *Proc. Nat. Acad. Sci., U.S.A.* **70**, 2926–2930.

Hutchinson, J. and Lonsdale, D.M. (1982). The chromosomal
distribution of cloned highly repetitive sequences from
hexaploid wheat. *Heredity*, in the press.

Islam, A.K.M.R., Shepherd, K.W. and Sparrow, D.H.B. (1975).
Addition of individual barley chromosomes to wheat. *Barley Genetics* 3, 260–270.

John, B. and Miklos, G.L.G. (1979). Functional aspects of
heterochromatin and satellite DNA. *Int. Rev. Cytol.* **58**,
1–114.

Jones, J.D.G. (1981). Repeated DNA sequences in rye (*Secale
cereale*) wheat (*Triticum aestivum*) and their relatives.
Ph.D thesis, University of Cambridge, England.

Jones, R.N. and Brown, L.M. (1976). Chromsome evolution and
DNA variation in *Crepis*. *Heredity* **36**, 91–104.

Jones, J.D.G. and Flavell, R.B. (1982). The structure,
amount and chromosomal localisation of defined repeated
DNA sequences in species of the genus *Secale*. *Chromosoma*,
in the press.

Macgregor, H.C. (1982). Big chromosomes and speciation
amongst amphibia. *In* "Genome Evolution" (Dover, G.A. and
Flavell, R.B., eds), Academic Press, London.

Manuelidis, L. (1982). Repeated DNA sequences and nuclear
structure. *In* "Genome Evolution" (Dover, G.A. and Flavell,
R.B., eds), Academic Press, London.

Martin, A. and Chapman, V. (1977). A hybrid between *Hordeum
chilense* and *Triticum aestivum*. *Cereal Res. Commun.* 5,
365–368.

Miller, T.E. and Chapman, V. (1976). Aneuhaploids in bread
wheat. *Genet. Res. (Camb.)* 28, 37–45.

Peacock, W.L., Dennis, E.S. and Gerlach, W.L. (1981). Satellite DNA - change and stability. *Chromsomes Today* 7, 29–
41.

Pech, M., Streeck, R.F. and Zachau, H.G. (1979). Patchwork
structure of a bovine satellite DNA. *Cell* **18**, 883–893.

Rees, H. and Hazarika, M.H. (1969). Chromosome evolution in

Lathyrus. *Chromosomes Today* **2**, 157–165.

Rees, H., Jenkins, G., Seal, A. and Hutchinson, J. (1982). Assays of the phenotypic effects of changes in DNA amounts. *In* "Genome Evolution" (Dover, G.A. and Flavell, R.B., eds), Academic Press, London.

Riley, R. (1966). The genetic regulation of meiotic behaviour in wheat and its relatives. Proc. 2nd Int. Wheat Genet. Symp. Lund. *Hereditas* (suppl.) **2**, 395–408.

Riley, R. (1981). Cytogenetic evidence on the nature of speciation in wheat and its relatives. *In* "Mechanisms of speciation" (Barigozzi, C., Montalenti, G. and White, M.J.D., eds). In the press.

Riley, R. and Chapman, V. (1967). The inheritance in wheat of crossability with rye. *Genet. Res. (Camb.)* **9**, 259–267.

Riley, R., Chapman, V. and Miller, T.E. (1973). The determination of meiotic chromosome pairing. *In* "Proc. 4th Int. Wheat Genetics Symp.", pp. 731–738, University of Missouri, Colombia.

Rimpau, J., Smith, D.B. and Flavell, R.B. (1978). Sequence organisation analysis of the wheat and rye genomes by interspecies DNA/DNA hybridisation. *J. Mol. Biol.* **123**, 327–359.

Rimpau, J., Smith, D.B. and Flavell, R.B. (1980). Sequence organisation in barley and oats chromosomes revealed by interspecies DNA/DNA hybridisation. *Heredity* **44**, 131–149.

Scherer, S. and Davis, R.W. (1980). Recombination of dispersed repeated DNA sequences in yeast. *Science* **209**, 1380–1384.

Sears, E.R. (1976). Genetic control of chromosome pairing in wheat. *Annu. Rev. Genet.* **10**, 31–51.

Smith, G.P. (1973). Unequal crossing over and the evolution of multigene families. *Cold Spring Harbor Symp. Quant. Biol.* **38**, 507–513.

Smith, G.P. (1976). Evolution of repeated DNA sequences by unequal crossover. *Science* **191**, 528–535.

Smith, D.B. and Flavell, R.B. (1974). The relatedness and evolution of repeated nucleotide sequences in the genomes of some gramineae species. *Biochem. Genet.* **12**, 243–256.

Smith, D.B. and Flavell, R.B. (1977). Nucleotide sequence organisation in the rye genome. *Biochim. Biophys. Acta* **474**, 82–97.

Thomas, J.B. and Kaltsikes, P.J. (1976). A bouquet-like attachment plate for telomeres in leptotene of rye revealed by heterochromatin staining. *Heredity* **36**, 155–162.

Thompson, W.F. and Murray, M.G. (1980). Sequence organisation in pea and mung bean DNA and a model for genome evolution. *In* "Fourth John Innes Symposium" (Davies, D.R. and Hopwood, D.A., eds), pp. 31–45, John Innes Institute, Norwich.

Ullman, J.S. and McCarthy, B.J. (1973). The relationship

between mismatched base-pairs and the thermal stability of DNA duplexes. *Biochim. Biophys. Acta (Amst.)* **294**, 405–415.

Waines, J.G. (1976). A model for the origin of diploidizing mechanisms in polyploid species. *Amer. Natur.* **110**, 415–430.

Wensink, P.C., Tabata, S. and Pachl, C. (1979). The clustered and scrambled arrangement of moderately repetitive elements in *Drosophila* DNA. *Cell* **18**, 1231–1246.

White, M.J.D. (1977). Modes of speciation. Freeman and Co., San Francisco.

Young, M.W. (1979). Middle repetitive DNA: a fluid component of the *Drosophila* genome. *Proc. Nat. Acad. Sci., U.S.A.*, **76**, 6274–6278.

Big Chromosomes and Speciation Amongst Amphibia

H.C. MACGREGOR

Department of Zoology, University of Leicester

Biologists have been looking at chromosomes for over 100 years, and it is surely true to say that no other intracellular object has proved more difficult to understand and none has offered such an endless stream of fascinating surprises. Today, the situation is no different and the problems are more challenging than ever.

There are, I think, three rather different views of a eukaryotic chromosome. In the first, we see it as a self-replicating object with a definite shape and size. And a specific pattern of behaviour during the mitotic and meiotic cycle. That is the view of the cytologist. Secondly, we may see it as an enormously long piece of DNA that goes through cycles of condensation and decondensation, local and general, and is linearly differentiated into regions that are or are not transcribed into messenger or other functional RNA molecules from time to time, according to the cellular environment in which the chromosome resides. That is the view of many who work with polytene or lampbrush chromosomes. Thirdly, a chromosome appears as an overwhelming conglomeration of lines and dots and As and Ts and Cs and Gs, and it is usually sensible and convenient to forget about most of it and concentrate on just one little bit of line, preferably without the dots. That is the view of most hard-core molecular biologists.

Whichever way we look at it, the eukaryotic chromosome is formidably large and, in general, apart from actual size differences, a chromosome from one kind of organism looks very much like that from any other. However, when we come to consider chromosome sets or karyotypes, then the situation is rather different. On the one hand, there is a useful and sometimes obvious level of variability, limited only by the need to conform to patterns of change and behaviour that are compatible with the mechanisms of meiosis. On the other hand, there is a remarkable stability and conservatism. The three chromosome sets shown in Fig. 1 are immediately

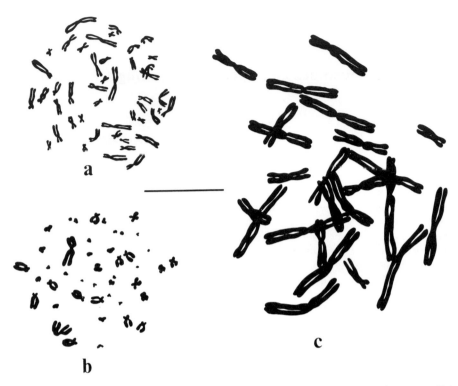

FIG. 1 *Drawings of mitotic chromosome sets from (a) man, (b)*
*(b) a reptile (*Bipes canaliculatus*) and (c) a urodele (*Triturus cristatus*). The 3 sets are reproduced on the same scale.
The bar in the middle of the Figure represents 30 μm.*

recognisable to anyone with moderate experience of animal
cytology as belonging to a mammal, a reptile and a tailed
amphibian: typical sets from whole classes of animals that
diverged more than 150 million years ago. In this paper, I
shall be concerned with the significance of this karyotypic
conservatism and the evolutionary changes that are taking
place beneath it.

  I shall concentrate mainly on chromosome growth, and behind much of what I shall say will be the notion, relatively
new to the thinking of animal cytologists and evolutionists,
that transposition of gene sequences is a major factor in
bringing about genomic and karyotypic change (Nevers and
Saedler, 1977; Shapiro, 1981). In this regard, it is important to note that there are to date just four eukaryotic systems in which transposition has been unequivocally demonstrated using techniques of genetics or experimental

molecular biology: yeast, maize, *Drosophila*, and tumour and retro-viruses. In any other case, we can at the moment only infer that transposition has happened from certain lines of circumstantial evidence. Essentially, if a sequence is repetitive, widely dispersed in different and variable locations in the karyotype, associated with other short repeated sequence elements or palindromes, then it is likely to have undergone transposition at one time or another and/or be still potentially transposable.

My main interest over the past ten years has been centred upon the very large chromosomes of tailed amphibians, and I wish now to focus in on two specific situations in newts and salamanders, one of them apparently highly stable and the other quite unstable. Both are quite well-known and have been the subject of a number of publications since 1973 (Macgregor, 1979; Macgregor and Andrews, 1977; Macgregor *et al.*, 1973, 1976, 1977, 1981; Mizuno and Macgregor, 1974; Mizuno *et al.*, 1976).

The first concerns chromosome change and speciation in the North American genus *Plethodon*, of which the commonest and best known representative is the red-backed salamander of the eastern United States, *Plethodon cinereus*. But before we look at the chromosomes of this group of organisms, there are one or two things to be said about its evolution and form. *Plethodon* is indeed a truly remarkable animal. It has been around for something like 80 million years, during which time it has radiated to form 26 good species, all of which are located on the North American continent (Highton, 1962; Wake, 1966; Highton and Larson, 1979). These species are clearly distinguishable on the basis of their geographical distribution, their adult body size, and dissimilarities in their proteins (Highton, 1962; Highton and Larson, 1979). There are also some minor skeletal differences and colour differences between species, but these are not always reliable as taxonomic characters. The most extraordinary thing about the genus *Plethodon* is that throughout the very long course of its evolution and speciation it has not changed at all in form. It is, as David Wake has described it, a morphologically monotonous genus. With regard to anatomy and osteology, all species are the same and certain species from opposite extremes of the range, species that have been separated for at least 40 million years, are almost indistinguishable to the untrained eye.

Perhaps the most astonishing feature of *Plethodon*, and the very reason why it is featuring in this article, is that notwithstanding its morphological uniformity it can be divided quite sharply into species groups on the basis of genome size. In essence, speciation has been accompanied by a dramatic

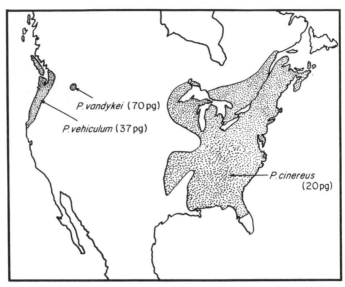

FIG. 2 *Map showing the geographic distribution and* C *values of 3 species of* Plethodon.

growth in the amount of DNA per haploid chromosome set (Mizuno and Macgregor, 1974; Macgregor *et al.*, 1976). The actual situation is outlined in Fig. 2, which shows representatives of each species group, geographical locations and *C* values. Now this would not be particularly significant were it not for a few more important facts.

First, the karyotypes of all species are identical with respect to relative lengths, centromere indices and arm ratios of corresponding chromosomes (Mizuno and Macgregor, 1974). This tells us that as the genome of *Plethodon* has expanded in the course of evolution of the western and northern species, it has done so in a balanced and uniform manner, each chromosome accumulating new DNA sequences in proportion to its existing load at any point in time, and no part of any chromosome stepping out of line and growing disproportionately.

Secondly, we see signs within the *Plethodon* genome of change and mobility. Most of the genome consists of middle repetitive DNA, and this material has diverged in sequence along with speciation, such that closely related species have as much as 90% of their middle repetitive sequences in common, whereas distantly related species with widely different genome sizes have less than 10% of their middle repetitive DNA in common. In all species there are satellite DNAs

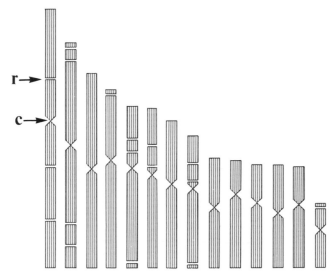

FIG. 3 *A typical idiogram of a member of the genus* Plethodon *showing all the locations at which clusters of ribosomal genes have been found. Each ribosomal cluster is indicated by a gap in the chromosome (r). Centromeres are shown as constrictions (c). Altogether, 7 species are represented in this diagram. The number of ribosomal sites for any one species varies from 1 in* P. cinereus *or* P. larselli *to 6 in* P. elongatus.

that are common to all chromosomes, and each species is quite different with regard to the numbers and locations of the major clusters of genes for ribosomal RNA (Fig. 3: Macgregor and Sherwood, 1979).

So, in *Plethodon* there has been an almost unimpeded expansion of the genome during speciation, and it may still be going on in some of the western salamanders of the group. Most of this expansion must have involved middle repetitive DNA, and it must have been accompanied by a substantial level of sequence mobility, widespread change in most DNA sequences including some biochemically important genes, but no change in karyotype and no change in the morphology of the animals.

There are just two more points worth adding about *Plethodon*. The first relates to middle repetitive sequences in two widely separate species, *P. dunni* from the Pacific northwestern region of the United States, and *P. cinereus* from the eastern United States, that have *C*-values in the ratio of 2:1. About 2.5% of the middle repetitive DNA of *P. cinereus*, or 0.75% of the genome (*C* = 20 pg) consists of sequences that are common to *P. cinereus* and *P. dunni* (Mizuno *et al.*, 1976).

The remainder of the middle repetitive DNA in these two animals is quite different. The common middle repetitive fraction does not contain ribosomal, 5 S, 4 S or histone sequences, but is quite heterogeneous. The average repetitive frequency in the families of common middle repetitive DNA is around 6000, and it is the same in both species. The common middle repetitive fraction may therefore be regarded as a group of about 250 different sequences, each between 100 and 200 nucleotides long, each repeated about 6000 times per genome, and all highly conserved with regard to both sequence and repetitive frequency over a period of 45 million years. Yet all of these sequences are really much too short to be considered as templates for biochemically significant compounds. What do they accomplish and why have they been so meticulously conserved? Are they prime examples of "selfish genes" that, in the words of Doolittle and Sapienza (1980) "... have evolved a strategy (such as transposition) which ensures (their) genomic survival", in which case, no other explanation for their existence is necessary. Whilst I regard that kind of view as quite unsatisfactory, I nevertheless think it would now be exceedingly interesting to look and see if the common middle repetitive DNA of *Plethodon* has any of the features of the transposable elements of other organisms.

Lastly, in relation to *Plethodon*, let us give just a little more thought to the matter of genome growth. Most of the genome expansion in *Plethodon* is due to accumulation, diversification and widespread scattering of middle repetitive DNA sequences. Therefore, the larger the genome becomes, the more middle repetitive DNA it will contain, the more uniform will be the distribution of this material, and the more uniform will be the pattern of genome growth. Accordingly, one would expect more similarity between chromosome sets of related species with large different-sized genomes like *P. cinereus* and *P. vandykei*, than there is between chromosome sets of species of equivalent relatedness with small different-sized genomes; and this is just what we find if we look in the right places. The genome sizes for anurans range between 1 and 17 pg as compared with urodeles, where the range is between 15 and 90 pg. In every family of the Anura there is evidence of chromosome rearrangement (Morescalchi, 1973). This is not so for the Urodela. Chromosomal variation at the population level is not uncommon in anurans. It is extremely rare in urodeles, except at the finest level. Some species of anuran seem to have experienced quite dramatic evolutionary changes in chromosome number and morphology (e.g. *Arthroleptis stenodactylus* by fusion and translocation; *Pipa parva* by fissions, and some species of *Ceratophrys* and *Xenopus* by polyploidization). There are no such instances amongst the Urodela. To be sure, there are families among the Anura that have relatively conservative karyotypes (e.g. Bufonidae and Ranidae), but there are also others that seem to be karyotypically quite volatile (e.g. Leptodactylidae) (M. Schmid, personal communication). So at least amongst the amphibia, where we do see quite wide extremes of genome size, the rule seems applicable: the larger the genome the more stable the karyotype.

The other system that I wish to examine for a moment is quite different. The species *Triturus cristatus* and *Triturus marmoratus* have become sharply separated off from all other species in the genus, partly as a consequence of a series of structural rearrangements that probably took place quite suddenly and affected only one arm of one of the longer chromosomes (number 1) of the set. The primary changes on this chromosome, which probably included a large inversion, effectively suppressed synapsis and crossing-over in the long arm. There followed a chain of events that led to the present day situation, in which males *and* females of both species are consistently heteromorphic for different forms of chromosome 1 and the heteromorphism is essential for normal development (Mancino *et al.*, 1977; Morgan, 1978; Macgregor and Horner, 1980). Individuals that have two identical chromosomes 1 die at late tail bud stage of development. Now, what is interesting about the long arm of chromosome 1 in the present context is that it is a region where there has clearly been a change in karyotype that is out of balance with events in other parts of the chromosome set and, in this sense, it is worth examining a little more closely.

First, there seems little doubt that chromosome 1 in *T. cristatus* has grown since the species has emerged. Its relative length, at 12.4 (% of the genome: Callan and Lloyd, 1960) is greater than that of any other chromosome in the genus, except for its counterpart in *T. marmoratus*. Most of this growth has happened in the long arm, since with a centromere index of 36, the chromosome is much more sub-metacentric than any of the longer chromosomes in other species of newt that have been reliably karyotyped. The long arm of chromosome 1 in *T. cristatus* represents 7.9% of the genome and the heteromorphic region of that arm comprises 5.8% of the genome (Callan and Lloyd, 1960). In absolute terms, this means that the heteromorphic region contains 1.33 pg of DNA or 1.2×10^9 base-pairs.

The heteromorphic region of chromosome 1 is disproportionately rich in highly repetitive DNA, and most of this belongs to a few families of satellite DNA sequences that together constitute between 3 and 4% of the genome (Varley *et al.*, 1980). It is therefore clear that most of the growth of

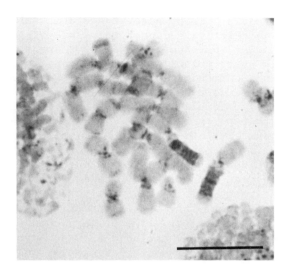

FIG. 4 *A set of chromosomes from intestinal epithelium of*
Triturus cristatus carnifex *after C-staining with Giemsa.*
The 2 heteromorphic arms of chromosomes 1 are clearly distin-
guishable at the lower right of the group, and their C- band-
ing patterns are obviously different. The bar represents
20 μm.

chromosome 1 has been a consequence of multiplication of
these short sequence elements.

Three more characteristics of chromosome 1 and its heteromorphic DNA sequences have emerged recently, and I believe
these may prove to be of considerable significance in relation to the manner in which this chromosome is behaving.
First, linear differentiation of the heteromorphic region can
be seen when the chromosome is in the lampbrush form or when
it is C-stained with Giemsa (Fig. 4) (Callan and Lloyd, 1960;
Mancino *et al.*, 1977; Morgan, 1978; Macgregor and Horner,
1980; Macgregor *et al.*, 1981). In both cases there is some
evidence for variation in the heteromorphic regions of both
chromosomes 1 between individuals within the same population,
although we do not know whether this reflects rapid change in
the heteromorphic regions within populations or the existence
of several "versions" of the chromosome that segregate and recombine at random. I might add that underlying this variation there does seem to be a basic and quite constant pattern
of C-staining and lampbrush "landmark" distribution for each
chromosome 1, suggesting that there is indeed a chromosome 1A
and a chromosome 1B, at least in the main population of *T.c.*
carnifex from which we have drawn most of our experimental

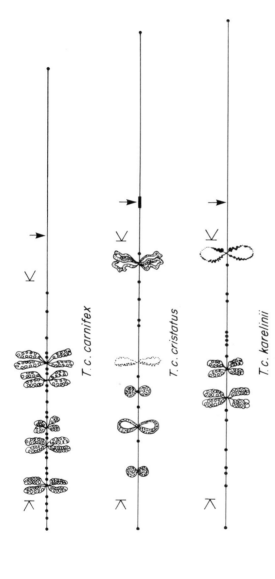

T. c. carnifex

T. c. cristatus

T. c. karelinii

FIG. 5 *Lampbrush chromosome "maps" of the chromosomes 1 from 3 sub-species of Triturus crista-tus. The distribution of "landmark" loops and other structures is clearly sub-species specific. Reproduced with the kind permission of Professor H.G. Callan.*

material. What is quite certain is that there are substantial differences in C-band patterns and lampbrush maps (Callan
and Lloyd, 1960) between sub-species of *T. cristatus* (Fig. 5).

Next, the heteromorphic regions of chromosome 1 contain
certain gene sequences that are present elsewhere in the chromosome set. There is some evidence for histone genes, perhaps
"orphons" (Childs *et al.*, 1981) – although the main clusters
of histone genes are located on chromosomes 5 and 8 (Old
et al., 1977; Gall *et al.*, 1981). The satellite sequences
found on chromosome 1 are liberally scattered around the remainder of the chromosome set, and they are present on the
chromosomes of other species of *Triturus* (Varley *et al.*,
1980). There are copies of the ribosomal genes on chromosome
1, although once again the main clusters of these are at the
nucleolus organizers on chromosome 6 and occasionally on chromosome 9 (Morgan *et al.*, 1980; Macgregor *et al.*, 1981). Moreover, the ribosomal genes on chromosome 1 are expressed in
the oocyte when others are silent, and they are transcribed
by polymerase II when the chromosomes are in the lampbrush
form (Morgan *et al.*, 1980; Macgregor *et al.*, 1981).

Lastly, when we look at the satellite sequences themselves
on chromosome 1, we find that they have some very interesting
characteristics that compare most favourably with those of
the insertion sequences found in other organisms (Shapiro,
1981). One of these satellites, designated pTcS1, is 330
base-pairs long and includes a slightly mismatched palindro-
mic sequence 14 base-pairs long (J.M. Varley, unpublished
results).

Whatever the events that led to the emergence of the
heteromorphic region of chromosome 1, it seems likely that
three factors were involved. First, a structural rearrangement that suppressed synapsis and crossing-over; secondly, a
series of events that included widespread tandem duplication
of a small number of sequences, widespread transposition of
these and other gene sequences and independent diversification of the heteromorphic regions of the two chromosomes 1;
and thirdly, the establishment of one or more homozygous
recessive lethal genes. Diversification of the two chromosomes 1 would seem to be still going on in today's population
of newts. With regard to the evolution of chromosome 1 in *T.
cristatus* and *T. marmoratus*, it would seem likely that *T.
marmoratus*, whose primary distribution was probably south of
the Pyrenees (Spurway, 1953), evolved directly from *T. cristatus*, and that the *T. marmoratus* chromosome 1 is an enlargement and peculiar diversification of the *T. cristatus* one.
The centromere index of chromosome 1 in *T. marmoratus* is 27
as compared with 36 in *T. cristatus*, and the heteromorphic
region of the former species gives the impression of being

considerably longer, although objective measurements in this regard have yet to be carried out.

What I have said so far will, I hope, serve to stress the importance of the karyotype rather than the gene. I have shown how organic form and karyotype change together, a reptile has one, a mammal another, and a newt yet another, all quite distinct and characteristic. I have shown how organic form and karyotype remain the same for long periods of evolutionary time, as in *Plethodon*. We have seen evidence of movement in the exorbitantly large genomes of tailed amphibians, implying that the exact positions of most of the DNA sequences in these genomes are of little consequence; but, on the other hand, we have seen from examples of conservatively balanced karyotypes that the *overall* distribution of DNA sequences is very important indeed. If DNA sequences were allowed complete freedom of movement and multiplication, as I think may have happened in chromosome 1 of crested newts, then I believe there would be changes in karyotype leading inevitably to changes in the relative positions of sequences that do biochemically important things, and eventually leading to asynapsis, non-disjunction, and substantially reduced reproductive fitness.

Naturally, the idea of karyotype selection is not new, but it is perhaps momentarily obscured by the vast and impressive array of gels and blots and sequences that leap out at us from every page of our journals in these times. Yet it is a very real and organismal phenomenon. The karyotype *is* a product of selection for arrangement of genes, size and number of linkage groups, and centromere position, and as such it is a means of achieving an adaptive phenotype. The situations that prove its significance are deeply rooted in the classical literature, and perhaps it is as well to be reminded of them from time to time: *balance*, which I have dealt with; *orthoselection* (White, 1978), where the fundamental number remains the same, together with the form of the chromosome arms and the kinds of change that they commonly undergo in particular taxonomic groups; and *sensitivity*, which is amply illustrated by the vulnerability of the human karyotype and phenotype to even the smallest of unorthodox changes in the relative positions of different parts of the chromosome set, and which is also seen at higher resolution in the effects of positional changes in *Drosophila*.

It is with these ideas in mind that I have come to support a kind of "neo-Goldschmidt" view of genic structure and evolution, emphasising, just as Goldschmidt did 30 or 40 years ago (Goldschmidt, 1955) the importance of genetic interdependence and hierarchical control, the importance of thinking of chromosomes and chromosome regions as functional entities.

I will, however, take care not to fall into the same trap as advocates of selfish genes, by supporting what may seem to be an inconsequential notion. To say that chromosome configuration is just another morphological character that is regulated by natural selection really gets us nowhere. The question remains: what is the genetic meaning of karyotype? The most useful answer, I think, is linkage, to which we respond: why is it so important to have the right genes in the right places? And that, of course, is exactly what most of us are looking at today when we consider our own respective little arrays of lines and dots.

Now, a few concluding remarks that I think are important in the context of this volume.

First, the matter of saltatory evolution. We know that transposition can affect gene expression. We also know that increasing differences of chromosome pattern correspond with increasing taxonomic distance. Broadly speaking, when the karyotype changes, the phenotype changes. So transposition, whether of a single sequence or a whole chromosome arm may provide a basis for an evolutionary jump. However, if we think according to the logic of selfish genes and focus back on the characteristic features of some of the very large genomes that I mentioned earlier, then another interesting idea begins to emerge. These genomes consist largely of middle repetitive DNA, which is short, unconserved and therefore unlikely to have any biochemical significance. At least three-quarters of the genome consists of this kind of material in some species. We may then ask whether the chances of effective transposition, effective in terms of development and evolution, will be greater in small genomes where sensitive targets in the form of functionally important genes will be relatively more numerous and more closely spaced?

TABLE I

Species comparisons

SPECIES	C VALUE ($g \times 10^{-12}$)	HEPATO- CYTE VOLUME	RED BLOOD CELL VOLUME	RED BLOOD CELL NUCLEUS VOLUME
Xenopus laevis	6	5.0	1.4	0.3
P. cinereus	20	14.0	5.8	1.0
P. vehiculum	37	27.8	11.5	1.4
Boletoglossa subpalmata	58	41.0	19.0	1.7

All volumes are in $\mu m^3 \times 1000$.

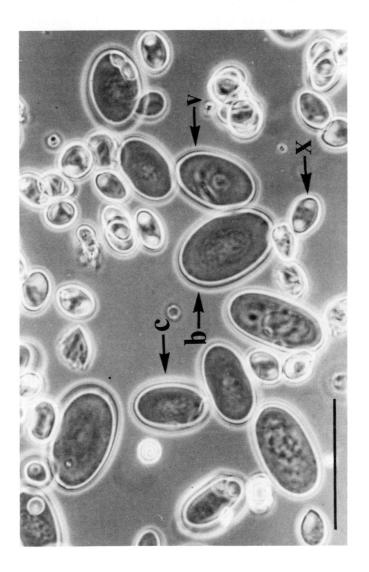

FIG. 6 *Phase contrast micrograph of a mixture of freshly isolated erythrocytes from 4 species of amphibian with widely different C-values. The smallest cells are from Xenopus laevis (X). The larger ones are from P. cinereus (c), P. vehiculum (V) and Boletoglossa subpalmata (b). The C values for these animals and the measured cell volumes are given in Table I. The bar represents 50 μm.*

Expressly, are major evolutionary saltations that are brought
about by transpositions more likely in organisms with small
genomes than in those with large genomes?

My last but one point relates to the question of genome
size and cell size. Certain aspects of this matter are fun-
damentally true. Genome size is proportional to cell volume:
big genomes, big cells (Fig. 6 and Table I). Genome size is
related to cell cycle time in comparable dividing tissues.
So, if we suppose that DNA calls the tune and sets the pace,
then more DNA leads to larger cells and slower growth. Consider for a moment the actual situation in *Plethodon*, where
P. cinereus reaches maturity with around 4×10^9 cells and *P.
vehiculum* has around 2×10^9. Since increase in cell number
and body size will proceed exponentially, it is clear that
initial development will proceed *much* more slowly in *P. vehiculum*, since its cell cycle time is twice that of *P. cinereus*,
but it will attain its final body size in just twice the time
that is taken by *P. cinereus*. It takes little persuasion for
us to accept that a difference of this kind must have selective significance. In my view there are most likely to be
quite strong selection pressures to keep genomes small in certain circumstances. But if these pressures do not exist or
are weak, then the genome will grow larger, mainly on account
of the inherent behavioural characteristics of its DNA (transposition, duplication, unequal sister chromatid exchange, and
so on) but it does not do so in a haphazard manner. The stability of the karyotype tells us that much. I would suggest
that it remains to be seen whether there is any real selective advantage to having large cells. I suspect that there is
not. But there may well be real selective advantages to having small ones.

My last point relates to the concept of useless or junk or
excess DNA. In the first place there is *no* chromosomal DNA
other than a few highly repeated satellites and perhaps some
supernumeraries that is dispensible. To be sure, there may
be local variations in amounts and distributions of certain
sequences that would seem incapable of any biochemical function, but these variations are strictly limited by the need
to preserve linkage groups and keep the chromosomes in forms
that will allow them to synapse, exchange and disjoin successfully at meiosis. In essence, chromosome shape is of the ut-
most importance. Chromosomes owe their shapes to their DNA
and its associated proteins. To be sure, some organisms have
a great deal of DNA, but so what? One of the features of DNA
seems to be that it has a propensity for multiplication.
Well, you cannot stop DNA from behaving, since it is only a
chemical substance at the mercy of its immediate environment,
but you can undoubtedly throw it overboard if it goes too far!

ACKNOWLEDGEMENTS

I am grateful to Dr Jennifer Varley and Mrs Heather Horner for allowing me to include some of their unpublished observations on satellite DNA sequences and cell volume/C value relationships in this paper.

REFERENCES

Callan, H.G. and Lloyd, L. (1960). Lampbrush chromosomes of crested newts *Triturus cristatus*. *Phil. Trans. Roy. Soc. ser. B* **243**, 135-219.

Childs, G., Maxon, R., Cohn, R.H. and Kedes, L. (1981). Orphons: dispersed genetic elements derived from tandem repetitive genes of eucaryotes. *Cell* **23**, 651-663.

Doolittle, W.F. and Sapienza, C. (1980). Selfish genes, the phenotype paradigm and genome evolution. *Nature (London)* **284**, 601-603.

Gall, J.G., Stephenson, E.C., Erba, H.P., Diaz, M.O. and Barsacchi-Pilone, G. (1981). Histone genes are located at the sphere loci of newt lampbrush chromosomes. *Chromosoma (Berlin)* **84**, 159-171.

Goldschmidt, R.B. (1955). "Theoretical Genetics", University of California Press, Berkeley and Los Angeles.

Highton, R. (1962). Revision of North American salamanders of the genus *Plethodon*. *Bull. Fla State Museum* **6**, 235-367.

Highton, R. and Larson, A. (1979). The genetic relationships of the salamanders of the genus *Plethodon*. *System. Zool.* **28**, 579-599.

Macgregor, H.C. (1979). *In situ* hybridization of highly repetitive DNA to chromosomes of *Triturus cristatus*. *Chromsoma (Berlin)* **71**, 57-64.

Macgregor, H.C. and Andrews, C. (1977). The arrangement and transcription of middle repetitive DNA sequences on lampbrush chromosomes of *Triturus*. *Chromsoma (Berlin)* **63**, 109-126.

Macgregor, H.C. and Horner, H. (1980). Heteromorphism for chromosome 1, a requirement for normal development in crested newts. *Chromosoma (Berlin)* **76**, 111-122.

Macgregor, H.C. and Sherwood, S. (1979). The nucleolus organizers of *Plethodon* and *Aneides* located by *in situ* nucleic acid hybridization with *Xenopus* ^{3}H-ribosomal RNA. *Chromosoma (Berlin)* **72**, 271-280.

Macgregor, H.C., Horner, H., Owen, C.A. and Parker, I. (1973). Observations on centromeric heterochromatin and satellite DNA in salamanders of the genus *Plethodon*. *Chromosoma (Berlin)* **43**, 329-384.

Macgregor, H.C., Mizuno, S. and Vlad, M. (1976). Chromosomes and DNA sequences in salamanders. *Chromosomes Today* 5, 331–339.

Macgregor, H.C., Vlad, M. and Barnett, L. (1977). An investigation of some problems concerning nucleolus organizers in salamanders. *Chromosoma (Berlin)* 59, 283–299.

Macgregor, H.C., Varley, J.M. and Morgan, G.T. (1981). The transcription of satellite and ribosomal DNA sequences on lampbrush chromosomes of crested newts. *In* International Cell Biology 1980–1981 (Schweuger, H.G., ed.), pp. 33–46, Springer-Verlag, Berlin and Heidelberg.

Mancino, G., Ragghianti, M. and Bucci-Innocenti, S. (1977). Cytotaxonomy and cytogenetics of European newt species. *In* The Reproductive Biology of Amphibians (Taylor, D.H. & Guttman, S.I., eds), pp. 411–447, Plenum Publishing Co., New York.

Mizuno, S. and Macgregor, H.C. (1974). Chromosomes, DNA sequences, and evolution in salamanders of the genus *Plethodon*. *Chromosoma (Berlin)* 48, 239–296.

Mizuno, S., Andrews, C. and Macgregor, H.C. (1976). Interspecific "common" repetitive DNA sequences in salamanders of the genus *Plethodon*. *Chromosoma (Berlin)* 58, 1–31.

Morescalchi, A. (1973). Amphibia. *In* Cytotaxonomy and Vertebrate Evolution (Chiarelli, A.B. & Campanna, E., eds), pp. 233–348, Academic Press, London and New York.

Morgan, G.T. (1978). Absence of chiasmata from the heteromorphic region of chromosome 1 during spermatogenesis in *Triturus cristatus carnifex*. *Chromosoma (Berlin)* 66, 269–280.

Morgan, G.T., Macgregor, H.C. and Colman, A. (1980). Multiple ribosomal gene sites revealed by *in situ* hybridization of *Xenopus* rDNA to *Triturus* lampbrush chromosomes. *Chromosoma (Berlin)* 80, 309–330.

Nevers, P. and Saedler, H. (1977). Transposable genetic elements as agents of gene instability and chromosome rearrangements. *Nature (London)* 268, 109–115.

Old, R.W., Callan, H.G. and Gross, K.W. (1977). Localization of histone gene transcripts in newt lampbrush chromosomes by *in situ* hybridization. *J. Cell Sci.* 27, 57–79.

Shapiro, J.A. (1981). Changes in gene order and gene expression. *National Cancer Institute Monograph*, in the press.

Spurway, H. (1953). Genetics of specific and sub-specific differences in European newts. *Symp. Soc. Exo. Biol.* 7, 200–237.

Varley, J.M., Macgregor, H.C., Nardi, I., Andrews, C. and Erba, H.P. (1980). Cytological evidence of transcription of highly repeated DNA sequences during the lampbrush stage in *Triturus cristatus carnifex*. *Chromosoma (Berlin)* 80, 289–307.

Wake, D.B. (1966). Comparative osteology and evolution of the lungless salamanders family Plethodontidae. *Mem. S. Calif. Acad. Sci.* **4**, 1-111.

White, M.J.D. (1978). *"Modes of Speciation"*, W.H. Freeman, San Francisco.

The Dynamics of Genome Evolution and Species Differentiation

GABRIEL DOVER, STEPHEN BROWN, ENRICO COEN,
JOHN DALLAS, TOM STRACHAN and MARTIN TRICK

Department of Genetics, University of Cambridge,
Cambridge CB2 3EH, England

"The modern theory of evolution has such enormous
facility that one could hardly imagine anything it
could not explain. Now the danger with this is that
it rules out any incentive to inquire about any other
possible mechanism that could explain the observed
facts." Peter Medawar (1974).

A THIRD MODE OF EVOLUTIONARY CHANGE

Evolutionary theory is concerned with providing mechanisms
to explain the manner in which biological variation is distributed between individuals and between species.

Until recently, natural selection and drift, alone or in
combination, have been considered the dominant forces responsible for the evolutionary progress of mutants within Mendelian populations. Both are based on the concept that mutations are, by and large, unitary and passive events. By
this we mean that they are, on average, infrequent events and
consequently rely on the processes of natural selection or
the accidents of drift to increase in abundance. There has
been some early peripheral consideration of the potential of
meiotic drive and gene conversion as processes for increasing
the frequencies of variants. These have been generally relegated to minor phenomena not considered responsible for the
observed patterns of intra- and interspecific variation.

On the basis of extensive comparative data on the patterns of variation in repetitive DNA families of genes and
non-coding sequences it is now possible to describe a third
mode of evolutionary change that is operationally different
from natural selection and drift. We call this process
molecular drive; although the name is unsatisfactory because

of its confusion with meiotic drive. They are similar only
to the extent that they give rise to patterns of non-Mendelian
segregation.

In order to understand the overall process of drive, it is
necessary to consider in detail the several molecular mechanisms that are responsible for it. Many essays in this volume
describe the range of mutational changes that have been found
in the genomes of widely disparate species. The variety and
extent of the changes at the molecular level reveals the genome as a constantly changing population of sequences. The
changes often involve the duplication and transfer of sequence
information between chromosomes and the conversion of one
variant by another in different parts of the karyotype. A
consequence of the propagation and spreading of sequences
around the karyotype is a non-Mendelian mode of inheritance
leading to the eventual fixation of variants within a population. The intragenomic mechanisms of turnover and spread can
affect the evolutionary progress of multiple-copy families,
irrespective of the copy-number, genomic distribution and
function of the family. We describe these mechanisms below.
Molecular drive would interact with natural selection and
drift depending on the family in question and on the structure
of the population. The ubiquity of the genomic mechanisms and
the widespread occurrence and numbers of multigene and non-
coding families makes the process of drive formally capable,
along with natural selection and drift, of explaining the origins of species differences.

Molecular drive suggests selfishness, in the way this has
been defined (Orgel and Crick, 1980; Doolittle and Sapienza
1980; Doolittle, 1982). However, the spread of variant
sequences throughout a population by drive can occur for both
"selfish" and "ignorant" DNA. By this we mean that in some
instances the sequence may actively influence its own propagation (selfish), and in others the sequence is passively
propagated by mechanisms that are seemingly ignorant of the
underlying DNA (Dover, 1980; Dover and Doolittle, 1980).

GENOMES AND SPECIES

Macgregor (1982), Gillespie *et al*. (1982) and Flavell (1982),
in their contributions to this volume and elsewhere, discuss
the possibility that frequent and relatively rapid changes
in components of the genome might be involved with a speciation process. The potential of fluctuations in the types and
patterns of sequences in giving rise to evolutionary novelty
was clearly forseen by Britten and Davidson (1971). Recently,
Schoph (1981) and Craig and Pounds (1981) have reiterated
these sentiments.

Is there a causal relationship between the differentiation of genomes and the differentiations of species? To answer this it is essential to provide mechanisms by which genomic mutations are able to increase in abundance. This requirement is particularly necessary for variants that are either involved directly with the reproductive isolation of species or with large-scale developmental reprogramming. In this essay, we consider widespread and evolutionary rapid processes of sequence turnover that can accidentally drive genomic mutations to fixation. It is probable that the large-scale fixation of variants throughout many multigene and non-coding families is of consequence to the phenotype. As such, it is possible that the processes of drive, involved with the accidental differentiation of populations of genomes, contribute to phenotypic differentiation and isolation of populations (see later).

The parameters that shape the dynamics of drive have recently received considerable theoretical treatment by Ohta (1980) and by Nagylaki and Petes (1982); following from the earlier studies reported by Smith (1974) and by Birky and Skavaril (1976). We examine some empirical data on the extent and rates of these processes, and assess their potential contribution to the differentiation of genomes and species.

SEQUENCE HOMOGENEITY IN MULTIPLE-COPY FAMILIES

Most higher organisms contain nuclear DNA that is several times in excess of the gene coding requirements of the organism. The degree of redundancy and of sequence composition varies extensively. A proportion of the excess DNA is composed of families of repetitive sequences in which the lengths of repeats (2 base-pairs to 5000 base-pairs) and the sizes of the families (few to several million members) are highly variable both within and between species. Each species genome can contain several hundreds of different families that are distributed in highly diverse ways relative to the chromosomes and relative to each other. Some families are confined to centromeres and to telomeres. Other families occur in complex interspersion patterns with each other and with single-copy DNA. The organisation of many such families is described in several contributions to this volumes, and have been reviewed by Davidson and Britten (1973), Flavell (1980), Dover (1980, 1981) Dover et al. (1981), Miklos and Gill (1982), John and Miklos (1979), Brutlag (1981), Peacock (1981) and Singer (1982). The organisation of multigene families such as the actin, globin and immunoglobulin genes are described in this volume by Davidson et al. (1982), Jeffreys (1982), Zachau et al. (1982) and Rabbitts et al.

(1982), and many other multigene gamilies have been reviewed
by Long and Dawid (1980), Kedes (1979) and Lewin (1980).

The existence of multiple-copy families implies a genomic
mechanism of amplification that may be either saltatory or
cumulative through successive duplications. The observation
that many families are common to a group of species suggests
that they do not often arise *de novo* during the lifespan of
a species.

For a family that is common to several species, it is possible to compare the average difference between any two members taken at random from within a species with the average
difference between any two members taken from different species. The patterns of variation within and between species
are powerful indicators of the separate modes and rates of
change occurring within the family since its inception in a
progenitor genome.

Expectations based on knowledge of point mutations in
single-copy genes and non-coding sequences predict that, for
every species, the individual members of a family would
accumulate base changes independently. On the assumption
that base changes are neutral, family heterogeneity within a
species is expected to be the same as the heterogeneity between two species. If there are selective constraints on the
rates of change in a family, however, then the patterns of
intra- and interspecific variation would reflect the degree
and mode of operation of such non-random forces. If, for
example, the selective constraints are sufficiently strong
and common to prevent any changes from accumulating, then no
variation would be observed within and between species.

In reality, family variation is observed to be distributed
within and between a group of species in a manner that could
not have been easily predicted by either the neutral or selection models in the way that they are traditionally understood.
The patterns of variation clearly indicate a higher level of
family homogeneity within a species than there is for the
same family between species. Furthermore, the same basic pattern of variation exists for many families that are shared
between a group of species.

Many families are homogeneous for sequence variants that
are diagnostic for the species. Variation is accumulating
between species, yet individual members within a species are
not evolving independently. For some families, the number of
mutational changes that have accumulated between species is
large and appears to be unconstrained by selection. Nevertheless, within each species, all members are similar with
respect to the substitutions.

Such a pattern of within-species homogeneity and between-
species heterogeneity for a family of repeated sequences is

known as concerted evolution. The processes responsible for
concerted evolution cannot be described easily in terms of
selection and drift. The observations are more easily understood in terms of an evolutionary process of fixation of
variant members within a family and within a species, as a
result of molecular mechanisms of family turnover and intra-
genomic spread of sequences.

Over the past decade, concerted evolution has been observed
in many families irrespective of their function or organisation. Many tandem and dispersed non-coding families reveal a
greater within-species than between-species homogeneity.
This has been found, for example, in plants (Rimpau *et al.*,
1978), rodents (Brown and Dover, 1981), insects (Strachan *et
al.*, 1982), echinoderms (Moore *et al.*, 1978) and primates
(Donehower and Gillespie, 1979, 1982). The concerted evolution of multigene families of ribosomal RNA, transfer RNA and
histones of several species has been reviewed by Long and
Dawid (1980), Dover and Coen (1981), Coen *et al.* (1982a),
Kedes (1979) and Arnheim *et al.* (1980). Jeffreys (1982), in
his contribution to this volume, discusses the growing literature and interest in the concerted evolution of globin genes.

A progressive homogenisation of families of sequences
seems to be the fate of large proportions of the genome.

ASSESSMENTS OF THE RATES OF HOMOGENISATION

Assessments of absolute rates of homogenisation are not possible in the absence of accurate absolute times of separation
between species. Relative rates can be assessed by comparing
the patterns of variation within and between species for a
range of families in one and the same group of species. The
choice of families makes it possible to assess the relative
constraints that might be operating on families that differ
in function and complexity.

We have chosen several families from within a tightly knit
group of seven sibling species of *Drosophila*. The families
are the ribosomal DNA, the histone genes and several non-
coding sequences. The species belong to the *melanogaster* species subgroup of the sub-genus *Sophophora*. The data are presented in full by Coen *et al.* (1982a) and by Strachan *et al.*
(1982). Here, we present the data in a stylised form to
highlight features of concerted evolution that are relevant
to the points we raise in the general introduction.

The patterns of biological relationships between the seven
species are well-understood from a variety of criteria, from
which a reliable consensus phylogeny can be extracted (see
Fig. 6). These studies include polytene chromosome inversion
patterns (Lemeunier and Ashburner, 1976), inter-sterilities

(Lemeunier, 1979), mating songs (Cowling and Burnet, 1981), protein allelomorphs (Eisses *et al.*, 1979), heat-shock genes (Leigh Brown and Ish-Horowicz, 1981), larval serum proteins (Brock and Roberts, personal communication), the alcohol dehydrogenase gene (M. Bodmer, personal communication), mobile elements (Young and Schwartz, 1980), mitochondrial genomes (Fauron and Wolstenholme, 1980) and the ribosomal RNA, histone and non-coding families (Tartof, 1979; Barnes *et al.*, 1978; Coen *et al.*, 1982; Strachan *et al.*, 1982). The relationship of the sub-group to the rest of the African Drosopholids has been described by Tsacas and his colleagues (1981), and a summary account based on the extensive work of Tsacas *et al.* on the fertility and ecological relationships between the species is given by Dover (1981).

The family of sequences that comprise the ribosomal RNA genes in *Drosophila* consists of a regular interspersion of the two major genes (18 S RNA and 28 S RNA genes) with two spacers (Fig. 1). The large spacer (NTS) is, for the most part, not transcribed, and the smaller spacer (ITS) is transcribed and eventually spliced from the functional RNAs. In *D. melanogaster* there are approximately 200 copies of such a repeating unit on each of the X and Y chromosomes. About 60% of the 200 units on the X chromosome contain an insertion sequence (type I - INS) within the 28 S gene, and about 15% of the X and Y units contain a different insertion (type II-INS). There is a region of short repeating sequences within the non-transcribed longer spacer (NTS).

We have used standard techniques of restriction mapping and Southern hybridisation (Southern, 1975a) to examine rDNA variation. The results show that each region of the compound rDNA unit has evolved in concert. Most regions contain characteristic restriction sites that are diagnostic for a species throughout most, if not all, of its copies.

The pattern of distribution of variants in the NTS of three of the species is illustrated by Fig. 2.

FIG. 1 *Repeating unit of rDNA in* D. melanogaster, *(see Long and David 1980 for details). 18 S and 28 S are the coding regions. ETS and ITS are the external and internal transcribed spacers. NTS is a non-transcribed spacer. INS is an intron appearing in some rDNA units. Small inverted triangles represent a region of internal repetition with a 250 base-pair periodicity recognised by* AluI.

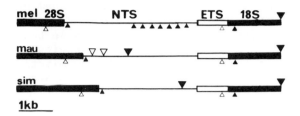

FIG. 2 *Restriction maps of the NTSs of the rDNA units of* D. melanogaster *(mel) (cf. Fig. 1);* D. mauritiana *(mau) and* D. simulans *(sim). Restriction sites are indicated by symbols* Δ, Hae*III;* ▲ Alu*I;* ▼ EcoRI; ∇, EcoRI *sites probably not present in all the repeats. Other sites are present in most if not all the repeats of a species.*

Each spacer can be identified unequivocally as to its species of origin by the presence of one or more restriction sites. The NTSs throughout the rDNA arrays of all seven species are identifiable in this manner. The region of repetition within the NTSs of *D. melanogaster* can be identified by *Alu*I-sensitive sites, which are spaced at 250 base-pair intervals. The NTSs of the other six species do not contain *Alu*I sites. Heteroduplex electron microscopy between the NTSs of some of the species have shown that there is full heteroduplex matching throughout the NTS (Tartof, 1979). From this, we conclude that the internal region of repetition within *D. melanogaster* has not resulted from the insertion of extraneous sequences into the NTSs of this species, but that the NTSs of the other species contain internal repetitions that are homologous to this region. Repetitions at 250 base-pair intervals have been found in the NTSs of the more distantly related species, *D. hydei* and *D. virilis* (Renkawitz-Pohl *et al.*, 1980; Rae *et al.*, 1980). In addition, the separate examination of the X and Y rDNA arrays in *D. melanogaster* indicates that each array contains NTS internal repeats that are *Alu*I sensitive (Coen *et al.*, 1982b). From the extensive studies made on rDNA in populations of *D. melanogaster* from around the world, it is clear that the *Alu*I sensitivities of these internal repeats are widespread.

The data suggest that there is some mechanism able to progressively fix an *Alu*I-sensitive variant, appearing sometime in the lineage of *D. melanogaster*, so that it has replaced pre-existing sequences at different levels. First, at the level of the 250 base-pair repeats within an NTS; secondly, for all these regions throughout the NTSs; and thirdly, for each of the X and Y chromosome arrays.

The mechanism most likely to be responsible for

homogenisation within rDNA arrays is unequal chromatid
exchange (Tartof, 1974; Smith, 1974). Regions of internal
repetition in other RNA gene clusters have been implicated as
the basis for unequal exchange (for a review, see Federof,
1979), and experimental verification of mitotic and meiotic
unequal exchanges in the rDNA of yeast is available (Szostak
and Wu, 1980; Petes, 1980).

A process of unequal exchange is expected to vary the copy
number of a tandem series of repeats. Unequal exchange within
the internal region of the NTS spacer would alter the number
of 250 base-pair repeats and hence lead to variation in
length of the total NTS by such intervals. This is illustrated in the top half of Fig. 3. Our measurements of NTS length
variability within and between the species show that they are
based on multiples of 250 base-pairs. In addition, the analysis of single X and Y chromosome rDNA arrays in *D. melanogas-*
ter shows that within each array, the copy number of a particular length of NTS varies considerable. In the lower part of
Fig. 3 we explain how this might arise as the result of
unequal exchange at the longer periodicity of the total rDNA
unit. A process of exchange between total rDNA units also
explains the observation that several regions within the
compound repeating unit have been replaced by variant sequences at equivalent rates. We discuss this later.

Smith (1974) and Kimura and Ohta (1979) describe the parameters involved with unequal exchange that influence the rate
at which sequence identity is achieved between members in an
array.

Some indication of the rate of unequal exchange in rDNA is

Unequal exchange within rDNA units

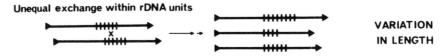

VARIATION
IN LENGTH

Unequal exchange between rDNA units

x 50   VARIATION
x 20   IN LENGTH
x 130   COPY-NUMBER

FIG. 3 (Top) *Unequal exchange at the periodicity of an*
internal region of repetition within a larger repeating unit,
(cf. rDNA: Fig. 1). This generates differences in length of
the longer repeat by multiples of the internal repeat length.
(Bottom) *Unequal exchange at the periodicity of the longer*
repeat unit (e.g. total rDNA unit) generates variation in
copy-number of the different lengths of the unit.

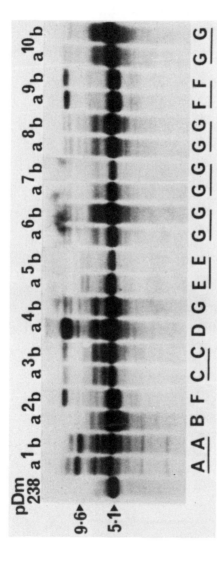

Fig. 4 Hybridisation patterns of the Hae III restricted DNA of XO males derived from 10 iso-female lines established from a wild population of D. melanogaster (see text). Each line was divided into a and b sublines at the F-1 stage. Probe used is a clone of the NTS (see Fig. 1). Patterns classified into 7 X rDNA types A-G. Lines in which sublines are indistinguishable are underlined. Left-most track is the control. Sizes in kilo bases.

derived from studies of differences in the organisation of
single X and Y rDNA arrays of individual flies.   Seven X
chromosome rDNA types and ten Y chromosome types of organisation can be recognised with respect to length and copy number
of individual lengths from amongst ten lines established from
single pregnant females hovering around a water melon in Okla-
homa (Coen *et al.*, 1982b), (Fig. 4).   After 230 generations,
the data show that the different rDNA types are inherited
stably, and that they are apparently selectively neutral.   The
rate of unequal exchange is sufficiently high to produce extensive polymorphism in the original population but not high
enough to produce new types during the period of laboratory
culture.   We have been able to assess the rate of generation
of new lengths of rDNA, at a copy number at which they can be
detected, at about 1 in 10^3 to 10^4 generations.

Within a species, there is considerable heterogeneity for
lengths and copy number of the several lengths of rDNA (Coen
et al., 1982b; Endow and Glover, 1979; Indik and Tartof,
1980), (see Fig. 4).   However, there is extensive homogeneity
throughout all the repeats, no matter what their length and
abundance, for the sequence variants that are diagnostic for
a species.   These features are summarised in Fig. 5 (see the
legend).

We discuss later the extent to which the rate of sister

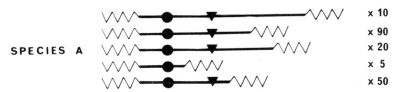

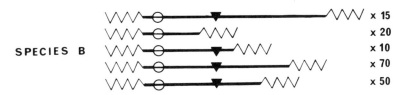

FIG. 5 *Symbolic comparison of a family shared by 2 species A
and B.   Within each species there is heterogeneity of length
and copy-number of lengths of the repeat unit.   These are
generated as in Fig. 3.   Species A can be distinguished by a
sequence variant ● that is diagnostic and present throughout
all the repeats.   Similarly ○ is diagnostic for species B. ▼
is present in both species.*

chromatid unequal exchange in rDNA can account for the extensive distribution of sequence variants within a species. We
also raise the possibility that the sequence identity between
the X and Y rDNA arrays with respect to the NTSs requires an
additional mechanism for the transfer of information between
chromosomes. Mechanisms of interchromosome transfer have the
potential to accelerate the process of fixation of variants.

COMPARATIVE RATES OF HOMOGENISATION IN RIBOSOMAL, HISTONE AND NON-CODING SEQUENCE FAMILIES

We have extended the discussion on the rDNA family in order to
illustrate the basic process of turnover and homogenisation of
new variants relative to the rate of production of new variants. Other tandem arrays of histone genes and non-coding
families are subject to the same process. For example, each
species can be characterised by diagnostic variants that are
present throughout the spacers and coding regions of the
histone repeating unit (Coen *et al.*, 1982a).

Concerted evolution in the more abundant families of non-
coding sequences has been detected with respect to three features of these families (Strachan *et al.*, 1982). These
include restriction sites, patterns of organisation and overall sequence divergence. There are several different major
density components in each of the species (Barnes *et al.*,
1978; Dover, 1981; for a review, see Brutlag, 1981), which
vary in copy number between the species. Many of these belong
to two ancient families of sequences that have evolved independently within each species since the time of separation.
We have called these the 500 and 360 families, from the
lengths in base-pairs of the most common basic repeating unit
within each species.

The changes within the families have led both to new patterns of organisation and to substantial sequence divergence.
The data show that the rates of turnover in these abundant
families, whose copy number varies from 3000 to 20 000, are
sufficient to lead to the progressive homogenisation of novel
patterns and sequence variants within the arrays. In the
majority of pairwise comparisons between the species there is
a greater degree of sequence and pattern homogeneity within a
species than there is between two species. However, the data
indicate that no major difference in sequence or pattern have
become fixed in the families between the two most closely
related species, *D. mauritiana* and *D. simulans*. This is in
contrast to the smaller families of rDNA and histone genes,
in which novel sequence variants have become fixed in all
seven species.

In order to illustrate these differences, we have

constructed a phylogeny that is based on the extents to which each region of a family can be used to discriminate between the species (Fig. 6). For example, regions such as the coding sequences of the rDNA, which are homogeneous for the same sequences in all seven species, are placed at level F. There are coding regions of the histone genes that are equally invariant, but these are not shown. At the other extreme, level A represents the within-species polymorphism for rDNA NTS length and copy-number of the different lengths (see Figs 3 and 4). An intermediate level, for example C, is illustrated by the length variation of the rDNA ITS, which cannot discriminate between *D. mauritiana* and *D. simulans*.

In this phylogeny, we are able to place the minimum rates at which fixation has proceeded for the rDNA, histone, 500 and 360 units. These are indicated at level B for the rDNA and histone genes, and at level C for the 500 and 360 families. In order to explain how we arrived at these minimal rates, we make the following points.

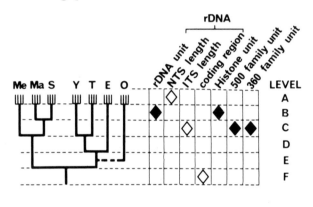

♦ = RATE OF FIXATION

◇ = RATE OF CHANGE

ᵚ = INTRASPECIFIC VARIATION

FIG. 6 *Phylogeny of the 7 sibling species of the* melanogaster *species subgroup.* Me, *D.* melanogaster; Ma, *D.* mauritiana; S, *D.* simulans; Y, *D.* yakuba; T, *D.* teissieri; E, *D.* erecta *and* O, *D.* orena. *The phylogeny describes the extent to which variation in different families is able to discriminate between species. It also shows the minimum rates of fixation and change for different repeating units and regions within them (see the text for explanation). Intraspecific variation is as shown in Figs 3 and 4.*

The rate of evolution of each region of a family can be partitioned into its separate components. The rate of evolution in a multiple-copy family can be considered to be an outcome of the rate of change and the rate of fixation. The rate of change is understood as the rate of production of effectively neutral mutations that are able to enter the homogenisation process and become fixed. The rate of production of effectively neutral mutations is clearly different in regions that are under different selective constraints. The rate of fixation of a neutral variant unit in a population would depend on:

(a) the stochastic rate of homogenisation of a variant unit in an array on a chromosome;
(b) the stochastic rate of fixation of the chromosomal array in a population;
(c) the rate of "transfer" of the variant units between chromosomes.

The dynamics of (a) and (b) have been examined extensively by Kimura and Ohta (1979) and by Ohta (1980,1982). The probability of fixation at both levels is dependent on statistical accidents of drift arising from fluctuations in copy-number of arrays and populations. The fluctuations in copy-number within an array arise from the process of unequal sister chromatid exchange (see Fig. 3). The rate of unequal exchange within the chromosome would determine the rate of homogenisation within an array. In addition, the rate of intrachromosomal gene conversion has been assessed as to its influence on the rate of homogenisation within an array (Birky and Skavaril, 1976; Ohta, 1977; Nagylaki and Petes, 1982). Interchromosomal transfer of sequence information (c), is a process that becomes increasingly important, in addition to (a) and (b), for the fixation of variants in large families that are distributed widely in the genome. We discuss this process later. We are able to illustrate these points by reference to Fig. 6.

From our previous discussion on the process of turnover and homogenisation in the rDNA, we concluded that unequal exchange is occurring at two different levels in the unit (see Fig. 3). Unequal exchange between total rDNA units ensures the co-homogenisation of all regions in the rDNA unit. For example, the coding regions (level F) and the ITSs (level C) are equally homogeneous in the species as the NTSs. The polymorphism at level A is a reflection of the change in copy-number and length generated by the unequal exchange process in rDNA (see Fig. 3). This indicates that the rate of unequal exchange is a recurrent and rapid process on an evolutionary time-scale. The rapidity of this process contributes to a rate of fixation that is sufficiently fast to fix species-specific variants of the NTS in all seven species (level B). Accordingly, a minimal rate of fixation of the total rDNA unit can be placed at level B. Similarly, the minimal rate of fixation of the histone unit can be placed at level B. If the rates of fixation within these two families were very much faster, then we would expect to observe the fixation of variants that are diagnostic for different populations within each species.

There are other regions in the rDNA unit in which the pattern of variation cannot discriminate between all seven species. For example, there is no ITS length variation between *D. mauritiana* and *D. simulans* (level C); also, there is no inter-specific variation in coding regions of the rDNA unit (level F). However, all regions within any one unit are being co-homogenised (see Fig. 3) and, accordingly, the rate of fixation is the same. It is clear, therefore, that the rates of change within different regions of the rDNA unit are different. This might be due to differences in the rate of mutation, or to selective constraints on the number and types of mutation that are permitted to proceed and enter the homogenisation process. Given the biological function of the coding region, it is reasonable to suggest that the latter is the case for this region. The placing of ITS length at level C suggests that either the rate of change of ITS length is slower than that of the NTS (level A) (possibly due to the absence of an internal region of repetition) or that there are selective constraints on changes in the length of the ITS transcribed spacer.

It is important to clarify the role of selection in the fixation process. We have shown that the rate of change in some regions of the rDNA and histone units are lower than others. By this we mean that the rate of production of neutral variants is constrained by selection in some regions, and hence fewer variants can enter the homogenisation processes and become fixed. Clearly, there could be an interaction between selection and homogenisation. For example, the detrimental effects of one variant member of a multigene family might go unrecognised until is had increased to a given number of repeats within the array. In like manner, selection might go unrecognised until it had increased to a given num- once it had reached a given number.

The rates of fixation within the 500 and 360 families are slower than those within the rDNA and histone families, in that no species-specific variants have become fixed in these two families during the time of separation of *D. simulans* and *D. mauritiana* (level C). The rate of production of neutral mutations is probably maximal in these non-coding families. The absence of variation between the two most closely related species is therefore due to a lower rate of fixation. This might be due to a lower rate of homogenisation (unequal exchange within an array) or to a lower rate by which sequence variants are transferred between chromosomes. For families such as the 500, which are dispersed in the karyotype, the rate at which interchromosome homogenisation takes place influences the overall rate of fixation. The mechanisms responsible for interchromosome homogenisation have the potential to drive a variant to fixation in a population.

MOLECULAR DRIVE AND THE RATE OF FIXATION

In order to determine whether the stochastic forces of fixation (a) and (b) are sufficiently rapid to ensure the fixation of different rDNA variants in the two species $D.$ $melanogaster$ and $D.$ $simulans$, we have made the following calculations based on the considerations of Tartof (1974), Smith (1974) and Ohta (1980).

Assuming that only unequal sister chromatid exchange occurs, i.e. within a single chromosomal array, and that the number of repeats is constrained to lie within 10% of their mean, then the time required for the spread of a neutral variant unit to 200 copies of an array will be of the order of 10^3 to 10^4 crossover generations. Given a rate of unequal exchange of approximately 3×10^{-4} per generation (see above; and Coen et $al.$, 1981b; Frankham et $al.$, 1980), the time required for homogenisation within an array is of the order of 10^7 generations. The average time taken for a chromosome, carrying an array fixed for a neutral variant, to spread throughout the population is $4\bar{N}_e$ generations, where $\bar{N}_e$ is the mean effective population size. Thus, if $\bar{N}_e$ is 10^7 or less, the time taken for fixation would be limited by the rate of homogenisation and hence be of the order of 10^7 generations or about 10^6 years, assuming ten generations per year.

Estimates of the time of divergence of $D.$ $melanogaster$ and $D.$ $simulans$ vary within an order of magnitude. If the most recent estimate of 10^6 generations is to be believed (Leigh Brown and Ish-Horowicz, 1981), then the time is insufficient for fixation by the stochastic processes (a) and (b). The lack of time becomes more critical in considering the fixation of species-diagnostic variants of the rDNA and histone families in the two most closely related species, $D.$ $simulans$ and $D.$ $mauritiana.$

The rate of fixation can be increased either by positive selection on each variant as it arises or by mechanisms with the potential to drive a variant through a population. The role of selection, although formally possible, is hard to envisage. Selection would need to recognise the effect on fitness of variants that initially comprise only a very small proportion of the family. Secondly, it requires a belief in the adaptive necessity of every point mutational change within non-coding families that differ widely in sequence composition, periodicity and interspersion. The high number of different types of families showing concerted evolution suggests a degree of neutrality for the finer points of variation in each member. This is not to say that a family, as a collective of repeats that have evolved in concert, is of necessity with zero effect on the phenotype. There might be an effect on phenotype but there would not necessarily be differences in phenotype between individuals upon which natural selection could act. This is because the processes responsible for achieving family uniformity lead inevitably to the uniformity taking shape concomitantly through a population of individuals. The implications of this are developed later.

The necessity for involving molecular drive in the process of fixation of variants of families in populations is strengthened by the observations of concerted evolution in very large non-coding families that are finely distributed in the genome. The possibility of drive becomes formal given the activities of mechanisms that are able to transfer sequence information from one chromosome to another in a non-reciprocal and biased manner (see below). There are specific features of the variation in large families that indicate that interchromosome sequence transfer is intimately involved with the processes of fixation. To illustrate these, we refer to the patterns of interspecific variation of abundant tandem and interspersed families in species of rodents.

One particular family of about 20 000 members is interspersed in the genomes of several rodents. This has been designated MIF-1 (Mouse Interspersed Family-1) (Brown and Dover, 1981). Figure 7 illustrates some features of these families that are indicative of concerted evolution in the four rodent species.

From Fig. 7 it can be seen that restriction sites are distributed to different extents between the species. For example, there are two sites that are invariant and present throughout most copies of the family in each species. At the other extreme, there are sites that are diagnostic for a species, and which have become fixed throughout the family. Interestingly, the family is divided into four mainly non-overlapping sub-populations of repeats, each of which can be distinguished by the presence of a diagnostic site. For example, the subscripts 1, 2, 3 and 4 in Fig. 7 represent the positions of *Hin*dIII sites in the family of *Mus musculus*. The majority of repeats of any one sub-population contain only one of these four sites. Sub-populations or "segments" of

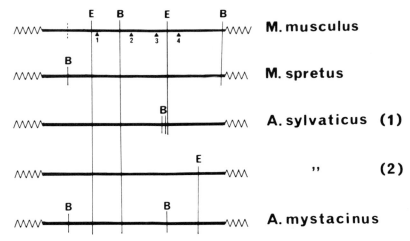

FIG. 7 *Restriction maps of an approximately 3 kilobase repeating unit of an abundant (approx. 20 000 copies) interspersed repetitive family in 4 rodent genomes. Family is designated MIF-1 (see Brown & Dover 1981); previously called the EcoRI 1.3 kb family. E, EcoRI site; B, BglII site. Four triangles numbered 1, 2, 3 and 4 are the positions of HindIII sites, each of which occurs separately in a subpopulation of repeats (see the text). The family in* Apodemus sylvaticus *consists of 2 related subpopulations distinguishable by variation at the right-hand EcoRI and BglII sites.*

families have been extensively characterised in large tandem arrays of sequences (Hörz and Zachau, 1977; Southern, 1975b; Brown and Dover, 1980a; for a review, see Dover *et al.*, 1981). It is possible that they reflect a slower process of homogenisation, relative to the rate of production of new variants in high copy-number families.

It has been suggested that sub-populations of tandem arrays are localised to single chromosomes and reflect the separate evolution of the chromosome sequences. A widespread karyotype distribution of many sub-populations of the satellite DNA of *M. musculus* indicate that to a large degree the chromosomes are not evolving independently in this respect (Brown and Dover, 1980a). This suggests that there is a continual process of interchromosomal exchange of variant sequences.

The elucidation of sub-populations in MIF-1, together with the fact of the fine interspersion of the members with many other sequences, suggests that the process of homogenisation is able to transgress chromosomes. This could not be achieved by unequal exchange, which is an homogenisation mechanism within tandem arrays only. The homogenisation of non-tandem families is probably the result of gene conversion.

Scherer and Davis (1980) have shown that gene conversion events in yeast take place irrespective of the non-homologous chromosome distribution of the two sequences involved. In the case of MIF-1, the process needs to achieve homogeneity for approximately 20 000 dispersed copies in each species. If there is no bias in the rates of the forward and backward directions of conversion (i.e. $A \to a = a \to A$), then it is unlikely that extensive homogeneity could be achieved in a stochastic manner. There is a requirement for a degree of directionality, which could be achieved either by selection or by drive. The arguments against selection are the same as those we raise above. If unbiased gene conversion is occurring continuously and with equal probability throughout the karyotype, then it would be difficult for selection to promote any one variant or chromosome through the generations. Any bias, however, in the direction of conversion (i.e. $A \to a \neq a \to A$) would lead to the progressive fixation of a favoured variant.

Molecular drive is a process of fixation that arises from biases in the non-reciprocal spread of information between chromosomes.

INTERCHROMOSOME SPREAD AND MOLECULAR DRIVE

There are two mechanisms that are capable of generating biases in the direction of transfer between chromosomes: sequence transposition and gene conversion.

Transposition involves the duplication and movement of a sequence from one position to another. A sequence with a high propensity to duplicate and move to another chromosome increases its probability of spreading in a karyotype, and hence through a population. The molecular basis of mobility is discussed in this volume by Finnegan and his colleagues (1982), and the possibility that interspersed families are the result of the repeated insertion of transposable elements, reverse-transcribed from RNA, has been discussed by Jagadeeswaran *et al.* (1981).

Gene conversion is a non-reciprocal transfer of information that can operate in both directions. It is considered to result from the mismatch repair of heteroduplexes, formed during recombination, giving rise to a homoduplex. The direction of repair can be biased by the sequences of one of the strands involved in the exchange, in a way that is not understood. A preferential repair of errors in a non-methylated strand during DNA replication has been shown to occur in prokaryotes (reviewed by Whitehouse, 1982). A delayed premeiotic replication and methylation in certain regions of meiotic chromosomes might influence the direction of meiotic conversion.

A bias in the direction of conversion can be observed as a departure from parity in the ratios of the expected products. Nagylaki and Petes (1982) and Whitehouse (1982) have reviewed the fungal data on the extent to which bias is observed in the conversion of different mutant sites. In yeast, approximately half the mutants depart from parity; and in some cases the departure can give rise to a 2 to 1 ratio in favour of the conversion of one allele by another. For other sites, in other fungi, the deviations from parity are as high as 100%.

A consistent bias in the direction of conversion within a chromosome, even if small, can have a dramatic effect on the probability of homogenising a favoured variant within a family on one chromosome (Nagylaki and Petes, 1982). It is clear that precisely the same dynamics ensue regarding the fixation of a favoured variant in a dispersed family by biased conversions between chromosomes.

The outcome of biased conversion and transposition is a persistent non-Mendelian segregation of one sequence. In this respect, both mechanisms are similar to meiotic drive. They differ, however, in two important respects. Meiotic drive is a non-Mendelian segregation resulting from the biased propagation of one homologue over another, or one haplotype over another: often the result of gametic natural selection. In favouring one element over another, there is a reduction in the total number of elements. There is no reduction in the total number of elements resulting from gene conversion or transposition.

Mechanisms of non-reciprocal transfer by gene conversion or transposition are proving to be widespread and involved with the population genetics of many multiple-copy families (see also Jeffreys, 1982). They are less of an evolutionary side-line than meiotic drive.

We depict the pathways of interchromosome spread in Fig. 8.

A new variant sequence arising in chromosome 1 can be fixed throughout its family on the chromosome of origin either by stochastic forces of unequal exchange and unbiased conversion, or be driven to fixation by biased intrachromosome conversion. There is evidence that the rate of intrachromosome spread (a) is faster to a limited degree than the rate of spread between homologous (b) or non-homologous (c) chromosomes (for data and discussion, see Dover *et al.*, 1981; Brown and Dover, 1980a; Coen *et al.*, 1982a; Strachan *et al.*, 1982). There is evidence from studies of some mobile elements in *Drosophila* (Ising and Block, 1981) that the rate of (b) is faster than the rate of (a). Sequences transfer may also occur between the regular (A) chromosomes and the supernumerary (B) chromosomes (Amos and Dover, 1981). The rates at which the different pathways of transfer proceed

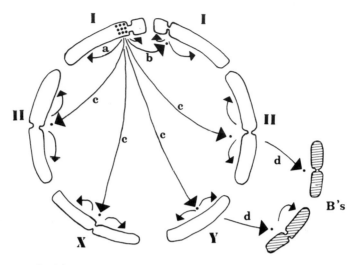

FIG. 8 *Symbolic representation of an intragenomic spread of a sequence from the chromosome of origin 1 to the homologous and non-homologous chromosomes. This could be due either to the transposition of an element or to a "biased" conversion of wild-type members of a dispersed family by a variant member (see the text). The relative rates of pathways a to d are described in the text.*

also differ between mitotic and meiotic cells. For example, there is a 400-fold difference in the rate of intrachromosome conversion in yeast between meiosis (Klein and Petes, 1981) and mitosis (Jackson and Fink, 1981).

The difference in rates between intra- and interchromosome spread is critical to the degree to which molecular drive alone is responsible for the fixation of a variant in a population. If the rate of (a) is relatively high, then a sufficiently large and rapid difference might accumulate between homologous chromosomes for natural selection to be able to recognise the difference. If there is no transfer between the homologues, then the fixation in a population of a variant array on a single chromosome would depend on natural selection, and the statistics of drift in the way this has been modelled (Ohta, 1980; Nagylaki and Petes, 1982). If (b) and (c) are relatively rapid, then the influence of drive on the fixation process may become substantial. In this circumstance, the net effect would be a progressive change in a family of sequences; not, however, in one chromosome or a set of chromosomes of one individual, but in a population of chromosomes that are shared by many sexual individuals. The biological effects of coincidental changes within a population on the differentiation of species need to be considered.

MOLECULAR DRIVE:
AN ACCIDENTAL MODE OF SPECIES DIFFERENTIATION

We discussed in the Introduction the obligation on models that
are concerned with the inception of evolutionary novelty and
new species to provide a mechanism by which the required mutations may spread in a population. If the mechanism is considered to be natural selection operating on the phenotypic
differences between individuals, then the mutational effects
on phenotype need to be relatively small. Natural selection
has no difficulty in increasing the frequencies of micromuta-
tions, leading to a gradual, small-step adaptation of a population. Mutations with larger phenotypic effects (macromutations) cannot enjoy the services of natural selection. This
is a point that is well-understood by population geneticists,
but which is curiously underestimated in the studies of
several macromutationists. The reason might well be histori-
cal, in that Goldschmidt himself was self-confessedly unaware
of the need to explain how a novel monster phenotype would
survive in the population of origin (see Mayr, 1980). A Gold-
schmidtian blindspot afflicts proponents of evolutionary and
speciation models to different extents. It is difficult to
understand from the studies reported by Stanley (1979), Gould
and Eldridge (1977), Schopf (1981) and White (1978) precisely
how a novel type of mutation effecting developmental reprogramming or interspecific sterilities is considered to survive
in a population, without invoking very special circumstances.
The problem exists for all types of macromutation whether
chromosomal, genomic or genic. The history of this issue is
discussed elsewhere (Dover, 1982).

There are mathematical solutions to this problem that man-
ipulate the traditional parameters affecting natural selection in both large and isolated Mendelian populations (for a
review, see Charlesworth *et al*., 1982). These, however, have
the effect of converting macromutations back to micromuta-
tions; which is begging the question.

Molecular drive is a mechanism with the potential to
increase the frequency of mutations with zero, small or large
effects on the ontogeny or reproductive compatibility of organisms. This is achieved, not by increasing the progeny of
one set of phenotypes at the expense of another, but by
inducing concerted phenotypic changes in a population of
individuals.

Let us consider a family of genes or non-coding sequences
that affect allometric growth rates of an organism. The
family could be considered a set of regulatory elements. If
a mutational change were to arise in one repeat and spread
rapidly in the family within one individual, then there might

be an immediate reduction in fitness of the individual or its
progeny, resulting from a disharmony in growth patterns. In
this circumstance, we can consider the mutant family as a
whole to be similar to a single mutant gene with macroeffects,
and hence suffering the same selectional disadvantages. It is
unlikely, however, given the experimental assessments of the
rates of family homogenisation, that the family would become
fixed for a variant within the lifetime of an individual. A
large family dispersed over several chromosomes would take
time to become homogeneous for a variant, by which time fixation would have occurred in a population of individuals. In
a sexual population that is continuously exchanging chromosomes, there could be a gradual simultaneous homogenisation
of the family on all the chromosomes in a population. The co-
efficient of identity between any two chromosomes with a
population with respect to the numbers of variant repeats it
has acquired would depend on the rates of intra- and interchromosomal homogenisation, in relation to the rates at which
chromosomes within the population are being effectively randomised. This latter process clearly depends on the reproductive biology and generation times of a population. The
respective rates of these parameters could ensure that at any
given time there is, on average, the same ratio of new to old
variants of a large dispersed family in all individuals.

In essence this means that the genomes of any two individuals taken at random from within a population would be homozygous for an average number of new and old variants of a
family. In addition, the same degree of homozygosity would
ensue for many families evolving by the same processes.
Accordingly, if such families affect the phenotype then the
degree of biological incompatibility between phenotypes
within a population is kept to a minimum. This is in marked
contrast to the maximum biological incompatibility that
might result from high levels of family heterozygosity existing between two individuals from two populations that are
becoming fixed and homogeneous for different variants.

In view of this, we are not unduly concerned with the magnitude of the phenotypic effects of a family that is differ-
entiating by drive. Clearly, the fixation of variants in
multigene families such as histones, rDNA, globins, etc.
might affect the phenotype. There would, however, be an
effect on phenotype that does not necessarily affect differential fitnesses between individuals, and hence would go
unrecognised by natural selection. This is not because the
effect is small, but because the effect is appearing in concert in a population of individuals as a family of sequences
evolves in concert. A creeping concerted change in the phenotypes of a population of individuals would considerably

reduce the probability of fitness differences between individuals. A monster syndrome is only seen to be monstrous in relation to the wild-type population. If there is a gradual simultaneous increase in individual monstrosity throughout a population, then no individual is more or less fit.

If the gradual increase in number of a new variant were to affect the viability of an organism, then there would be an average effect on the whole population. The population stands and falls as a unit.

The process is analogous to directional selection on a polygenic system with additivity between all loci. A gradual shift in the mean number of variants in a multiple-copy family occurs simultaneously within all individuals as a result of drive. The variance between individuals does not increase.

Molecular drive has the potential to impart cohesiveness to a population. At the same time, it would accentuate the biological discontinuity between two distinct but internally cohesive populations. The extent to which a discontinuity between the phenotypes of any two randomly chosen individuals from two such populations becomes a major problem of sexual incompatibility (speciation), depends on the rates and nature of the families undergoing concerted homogenisation in each population. If future studies reveal that many non-coding families affect the biology of chromosomes, either with respect to the production and processing of transcripts or with respect to chromosome behaviour during division, then the process of generating population cohesion and distinction would be of considerable evolutionary significance.

The molecular drive of a variant in a population depends on the mechanisms of non-reciprocal interchromosome transfer, as described above. There might be circumstances in which the occasional migration of individuals from one population to another would greatly increase the population size to which a variant would spread. The numbers of boundaries (biological discontinuities) that eventually take shape would reflect the breeding structures and dispersal ranges of different organisms. The observation that concerted evolution begins and ends at the species boundaries suggests that the biological discontinuities, produced by the molecular processes that impart separate cohesiveness on populations, might be involved with the erection of such boundaries (for further details, see Dover, 1982).

Sequence turnover and homogenisation are occurring continuously. The degree to which they lead to the concerted differentiation of a family depends on the rate of production of new variants. The process is analogous to the continual turnover of new for old banknotes. The process goes largely unrecorded without the introduction of a new design. There might be in

concerted evolution an explanation for the long periods of
stasis punctuated by more rapid periods of change that is
observed in the fossil record (Gould and Eldridge, 1977;
Stanley 1979; Williamson, 1981). Stasis could be a reflection
of a slow mutation rate; and punctuation a reflection of the
more rapid rates of turnover and fixation in the genome. In
this respect, molecular drive, which appears to be the fate
of many families of sequences, would contribute to the tempo
of the evolutionary process responsible for species discontinuities in biological organisation.

Molecular drive is a third evolutionary mode of change that
is independent of natural selection and drift, but which would
interact with them in a continuum of permutations arising from
a multitude of real situations. Evolution is probably an outcome of adaptive, accidental and driven processes of change.

ACKNOWLEDGEMENTS

These studies have been supported by grant nos GR/A/64379
and GR/B 6810.7 from the Science Research Council (G.B.).

REFERENCES

Amos, C.A. and Dover, G.A. (1981). The distribution of repet-
    itive DNAs between regular and supernumerary chromosomes in
    species of *Glossina*: a two-step process in the origin of
    supernumeraries. *Chromosoma* 81, 673-690.
Arnheim, N., Krystal, M., Schmickel, R., Wilson, G., Ryder,
    O. and Zimmer, E. (1980). Molecular evidence for genetic
    exchanges among ribosomal genes on nonhomologous chromo-
    somes in man and apes. *Proc. Nat. Acad. Sci., U.S.A.* 77,
    7323-7327.
Barnes, S.R., Webb, D.A. and Dover, G.A. (1978). The distri-
    butions of satellite and main-band DNA components in the
    melanogaster species subgroup of *Drosophila*. *Chromosoma*
    67, 341-363.
Birky, C.W. Jr and Skavaril, R.V. (1976). Maintenance of
    genetic homogeneity in systems with multiple genomes.
    Genet. Res. (Camb.) 27, 249-265.
Britten, R.J. and Davidson, E.A. (1971). Repetitive and non-
    repetitive DNA sequences and a speculation on the origins
    of evolutionary novelty. *Quart. Rev. Biol.* 46, 111-138.
Brown, S.D.M. and Dover, G.A. (1980a). The specific organisa-
    tion of satellite DNA sequences on the X-chromosome of
    Mus musculus: partial independence of chromosome evolu-
    tion. *Nucl. Acids Res.* 8, 781-792.
Brown, S.D.M. and Dover, G.A. (1980b). Conservation of seg-
    mental variants of satellite DNA of *Mus musculus* in a

related species: *Mus spretus*. *Nature (London)* **285**, 47–49.

Brown, S.D.M. and Dover, G.A. (1981). The organisation and evolutionary progress of a dispersed repetitive family in widely separated rodent genomes. *J. Mol. Biol.* **150**, 441–466.

Brutlag, D.L. (1981). Molecular arrangements and evolution of heterochromatic DNA. *Annu. Rev. Genet.* **14**, 121–144.

Charlesworth, B., Lande, R. and Slatkin, M. (1982). A defense of neo-Darwinism. *Evolution*, in the press.

Coen, E.S., Strachan, T. and Dover, G.A. (1982a). Rates of concerted evolution in regions of rDNA and histone gene families of *Drosophila*. *J. Mol. Biol.* (Submitted).

Coen, E.S., Thoday, J.M. and Dover, G.A. (1982b). Extensive, yet stably inherited, structural polymorphism in the rDNA multigene family of *D. Melanogaster*. *Nature (London)*, in the press.

Cowling, D.E. and Burnet, B. (1981). Courtship songs and genetic control of their acoustic characteristics in sibling species of the *Drosophila melanogaster* subgroup. *Anim. Behav.* **29**, 924–935.

Craig, S.D. and Pounds, J.A. (1981). The re-organisation of repetitive DNA as a possible factor in the differentiation of small isolated populations. (manuscript: personal communication).

Davidson, E.H. and Britten, R.J. (1973). Organisation, transcription and regulation in the animal genome. *Quart. Rev. Biol.* **48**, 565–613.

Davidson, E.H., Thomas, T.L., Scheller, R.H. and Britten, R.J. (1982). The sea urchin actin genes, and a speculation on the evolutionary significance of small gene families. *In* "Genome Evolution" (Dover, G.A. and Flavell, R.B., eds), Academic Press, London.

Donehower, L. and Gillespie, D. (1979). Restriction site periodicities in highly repetitive DNA of primates. *J. Mol. Biol.* **134**, 805–834.

Doolittle, W.F. and Sapienza, C. (1980). Selfish genes, the phenotype paradigm and genome evolution. *Nature (London)* **284**, 601–603.

Dover, G.A. (1980). Ignorant DNA? *Nature (London)* **285**, 618–620.

Dover, G.A. (1981). The evolution of DNA sequences common to closely related insect genomes. *In* "Insect Cytogenetics" (Blackman, R.L., Hewitt, G.M. and Ashburner, M., eds), *Symp. Roy. Ent. Soc. Lond.* vol. 10, pp. 13–35, Blackwell, London.

Dover, G.A. (1982). A role for the genome in the origin of species? *In* "Mechanisms of Speciation" (Barigozzi, C., Montalenti, G. and White, M.J.D., eds), Alan Lis, New York. In the press.

Dover, G.A. and Coen, E.S. (1981). Springcleaning ribosomal
    DNA: a model for multigene evolution? *Nature (London)*
    290, 731-732.
Dover, G.A. and Doolittle, W.F. (1980). Modes of genome evo-
    lution. *Nature (London)* **288**, 645-647.
Dover, G.A., Strachan, T. and Brown, S.D.M. (1981). The evo-
    lution of genomes in closely related species. *In* "Evolu-
    tion Today" (Scudder, G.G.E. and Reveal, J.L., eds), *Proc.
    2nd Int. Congr. System and Evol. Biol.* pp. 337-349, Hunt
    Institute, Pittsburgh.
Eisses, K.T., Dijk, H.V. and Delden, W.V. (1979). Genetic
    differentiation within the *melanogaster* group of the genus
    Drosophila (Sophophora). *Evolution* **33**, 1063-1068.
Endow, S.A. and Glover, D.M. (1979). Differential replication
    of ribosomal gene repeats in polytene nuclei of *Drosphila.*
    Cell **17**, 597-605.
Fauron, C.M.-R. and Wolstenholme, D.R. (1980). Extensive di-
    versity among *Drosophila* species with respect to nucleotide
    sequences within the adenine and thymine-rich region of
    mitochrondial DNA molecules. *Nucl. Acids Res.* **11**, 2439-
    2452.
Federof, N. (1979). On spacers. *Cell* **16**, 697-670.
Finnegan, D.J., Will, B.M., Bayev, A.A., Bowcock, A.M. and
    Brown, L. (1982). Transposable DNA sequences in eukary-
    otes. *In* "Genome Evolution" (Dover, G.A. and Flavell,
    R.B., eds), Academic Press, London.
Flavell, R.B. (1980). The molecular characterization and
    organisation of plant chromosomal DNA sequences. *Annu.
    Rev. Plant Physiol.* **31**, 569-596.
Flavell, R.B. (1982). Sequence amplification, deletion and
    rearrangement: major sources of variation during genome
    evolution. *In* "Genome Evolution" (Dover, G.A. and Flavell,
    R.B., eds), Academic Press, London.
Frankham, R., Briscoe, D.A. and Nurthen, R.K. (1980). Unequal
    crossing over at the rRNA tandon as a source of quantita-
    tive genetic variation in *Drosophila.* *Genetics* **95**, 727-
    742.
Gillespie, D., Donehower, L. and Strayer, D. (1982). Evolu-
    tion of primate DNA organisation. *In* "Genome Evolution"
    (Dover, G.A. and Flavell, R.B., eds), Academic Press,
    London.
Gould, S.J. and Eldridge, N. (1977). Punctuated equilibria:
    the tempo and mode of evolution reconsidered. *Palaeobio-
    logy* **3**, 23-40.
Hörz, W. and Zachau, H.G. (1977). Characterization of dis-
    tinct segments in mouse satellite DNA by restriction
    nucleases. *Eur. J. Biochem.* **73**, 383-392.
Indik, Z.K. and Tartof, K.D. (1980). Long spacers among

ribosomal genes of *Drosophila melanogaster*. *Nature (London)* **284**, 477-479.

Ising, G. and Block, K. (1981). Derivation-dependent distribution of insertion sites for a *Drosophila* transposon. *Cold Spring Harbor Symp. Quant. Biol.* **45**, 527-544.

Jackson, J.A. and Fink, G. (1981). Gene conversion between duplicate genetic elements in yeast. *Nature (London)* **292**, 306-307.

Jagadeeswaran, P., Forget, B.G. and Weissman, S.M. (1981). Short interspersed repetitive DNA elements in eukaryotes: transposable DNA elements generated by reverse transcription of RNA Pol III transcripts? *Cell* **26**, 141-143.

Jeffreys, A.J. (1982). Evolution of globin genes. *In* "Genome Evolution" (Dover, G.A. and Flavell, R.B., eds), Academic Press, London.

John, B. and Miklos, G.L.G. (1979). Functional aspects of satellite DNA and heterochromatin. *Int. Rev. Cytol.* **58**, 1-114.

Kedes, L.H. (1979). Histone Genes and histone messengers. *Annu. Rev. Biochem.* **48**, 837-870.

Kimura, M. and Ohta, T. (1979). Population genetics of multigene family with special reference to decrease of genetic correlation with distance between gene numbers on a chromosome. *Proc. Nat. Acad. Sci., U.S.A.* **76**, 4001-4005.

Klein, H.L. and Petes, T. (1981). Intrachromosomal gene conversion in yeast: a new type of genetic exchange. *Nature (London)* **289**, 144-148.

Leigh Brown, A.J. and Ish-Horowicz, D. (1978). Evolution of the 87A and 87C heat-shock loci in *Drosophila*. *Nature (London)* **290**, 677-682.

Lemeunier, F. (1979). Ph.D. thesis. University of Paris VI, Paris.

Lemeunier, F. and Ashburner, M. (1976). Relationships within the *melanogaster* species subgroup of the genus *Drosophila* (Sophophora). *Proc. Roy. Soc. ser. Lond. B.* **193**, 275-294.

Lewin, B. (1980). "Gene Expression: vol. 2, Eucaryotic Chromosomes". 2nd edit., John Wiley, New York.

Long, E.H. and David, I.B. (1980). Repeated genes in eukaryotes. *Annu. Rev. Biochem.* **49**, 727-764.

Macgregor, H.C. (1982). Big chromosomes and speciation amongst amphibia. *In* "Genome Evolution" (Dover, G.A. and Flavell, R.B., eds), Academic Press, London.

Mayr, E. (1980). *In* "The Evolutionary Synthesis" (Mayr, E. and Provine, W.B., eds), Harvard University Press, Cambridge.

Medawar, P. (1974). Discussion following: On chance and necessity (J. Monod). *In* "Studies in the Philosophy of Biology" (Ayala, J., ed.), p. 357, University of California Press, Berkeley.

Miklos, G.L.G. (1982). Sequencing and manipulating highly
    repeated DNA. *In* "Genome Evolution" (Dover, G.A. and
    Flavell, R.B., eds), Academic Press, London.

Miklos. G.L.G. and Gill, A.C. (1982). Nucleotide sequences
    of highly repeated DNAs: compilation and comments. *Genet.
    Res*. In the press.

Moore, G.P., Scheller, R.H., Davidson, E.H. and Britten, R.J.
    (1978). Evolutionary change in the repetition frequency
    of sea-urchin DNA sequences. *Cell* 15, 649-660.

Nagylaki, T. and Petes, T.D. (1982). Intrachromosomal gene
    conversion and the maintenance of sequence homogeneity
    among repeated genes. *Genetics*, in the press.

Ohta, T. (1977). On the gene conversion model as a mechanism
    for maintenance of homogeneity in systems with multiple
    genomes. *Genet. Res. (Camb.)* 30, 89-91.

Ohta, T. (1980). "Evolution and Variation of Multigene Fami-
    lies" (Lecture Notes in Biomathematics), p. 37, Springer,
    New York.

Ohta, T. (1982). Further study on genetic correlation between
    members of a multigene family. *Genetics*, in the press.

Orgel, L.E. and Crick, F.H.C. (1980). Selfish DNA: the ulti-
    mate parasite. *Nature (London)* 284, 604-607.

Peacock, J. (1981). Satellite DNA - change and stability.
    In "Chromosomes Today" (Bennet, M.D., Bobrow, M. and
    Hewitt, G.M., eds), p. 30, Allen, London.

Petes, T.D. (1980). Unequal meiotic recombination within
    tandem arrays of yeast ribosomal DNA genes. *Cell* 19, 765-
    774.

Rabbitts, T.H., Bentley, D.L., Forster, A., Milstein, C.P.
    and Mattyhssens, G. (1982). Human antibody genes: evolu-
    tionary comparisons as a guide to function and the mech-
    anisms of DNA rearrangement. *In* "Genome Evolution"
    (Dover, G.A. and Flavell, R.B., eds), Academic Press,
    London.

Rae, P.M.M., Kohorn, B.D. and Wade, R.P. (1980). The 10 kb
    Drosophila virilis 28S rDNA intervening sequence is
    flanked by a direct repeat of 14 base-pairs of coding
    sequence. *Nucl. Acids Res.* 8, 3491-3504.

Renkawitz - Pohl, R., Glatzer, K.H. and Kunz, W. (1980).
    Characterisation of cloned ribosomal DNA from *Drosophila
    hydei*. *Nucl. Acids Res*. 8, 4593-4609.

Rimpau, J., Smith, D.B. and Flavell, R.B. (1978). Sequence
    organisation analysis of the wheat and rye genomes by
    interspecies DNA/DNA hybridisation. *J. Mol. Biol.* 123,
    327-359.

Scherer, S. and Davis, R.W. (1980). Recombination of dis-
    persed repeated DNA sequences in yeast. *Science* 209, 1380-
    1384.

Schopf, T.J.M. (1981). Evidence from findings of molecular biology with regard to the rapidity of genomic change: implications for species durations. *In* "Palaeobotany, Paleoecology and Evolution" (Banks, H.P. and Niklas, K.J., eds), pp. 91-142 Praeger, New York.

Singer, M.F. (1982). Highly repeated sequences in mammalian genomes. *Int. Rev. Cytol.* In the press.

Smith, G.P. (1974). Unequal crossover and the evolution of multigene families. *Cold Spring Harbor Symp. Quant. Biol.* **38**, 507-513.

Southern, E.M. (1975a). Detection of specific sequences among DNA fragments separated by gel electrophoresis. *J. Mol. Biol.* 98, 503-517.

Southern, E.M. (1975b). Long range periodicities in mouse satellite DNA. *J. Mol. Biol.* **94**, 52-69.

Stanley, S.M. (1979). "Macroevolution". Freeman, San Francisco.

Strachan, T., Coen, E., Webb, D.A. and Dover, G.A. (1982). Modes and rates of change of abundant DNA families of *Drosophila*. *J. Mol. Biol.* submitted.

Szostak, J.W. and Wu, R. (1980). Unequal crossing over in the ribosomal DNA of *Saccharomyces cerevisiae*. *Nature (London)* **284**, 426-430.

Tartof, K. (1974). Unequal mitotic sister chromatid exchange and disproportionate replication as mechanisms regulating ribosomal RNA gene redundancies. *Cold Spring Harbor Symp. Quant. Biol.* **38**, 491-500.

Tartof, K. (1979). Evolution of transcribed and spacer sequences in the ribosomal RNA genes of *Drosophila*. *Cell* **17**, 607-614.

Tsacas, L., Lachaise, D. and David, J.R. (1981). Composition and biogeography of the Afrotropical Drosophilid fauna. *In* "The Genetics and Biology of Drosophila 3a" (Ashburner, M., Carson, H.L. and Thompson, J.N. Jr, eds), pp. 197-255, Academic Press, London.

White, M.J.D. (1978). "Modes of Speciation". Freeman, San Francisco.

Whitehouse, H.L.K. (1982). "Genetic Recombination - Understanding the Mechanisms". John Wiley, New York. (in preparation).

Williamson, P.G. (1981). Paleontolological documentation of speciation in Cenozoic molluscs from Turkana Basin. *Nature (London)* **293**, 437-

Young, M.W. and Schwartz, H.E. (1980). Nomadic gene families in *Drosophila*. *Cold Spring Harbor Symp. Quant. Biol.* **45**, 629-640.

Zachau, H.G., Höchtl, J., Neumaier, P.S., Pech, M. and Schnell, H. (1982). On the generation of antibody diversity and on computer aided analysis of V kappa gene sequences. *In* "Genome Evolution" (Dover, G.A. and Flavell, R.B., eds), Academic Press, London.

PART V
SUMMARY

Overview — Unsolved Evolutionary Problems

J. MAYNARD SMITH

School of Biological Sciences, University of Sussex, England

In asking for an "overview", the organisers of this conference
cannot have expected a summary of the proceedings; if so, they
would presumably have invited someone who works in the field
of genome evolution. I shall therefore interpret the word to
mean "a view over the fence"; that is, a view of the present
state of the subject from someone who works in the neighbouring field of population genetics. I was also asked to talk
about "unsolved problems". That will present no difficulty.
At this conference, few problems have been solved; we are
still at the stage of trying to ask sensible questions.

The evolution of the large-scale features of the genome, at
a level above that of the single gene, is one of the most difficult, perhaps *the* most difficult, question in evolutionary
biology. There are two reasons for this. The first is that
we do not understand the relation, if any, between the structure of the genome and the processes of development. The second is that it is always difficult to understand the evolution
of an organ whose most important effects are, not on the individual possessing that organ, but on the population of individuals. I will discuss these two difficulties in turn.

Consider first the relation between the structure of the
genome and development. Two extreme views can be held on this,
together with a spectrum of opinions lying between them. The
first, which I will call the "null hypothesis", was supported
during our discussions by Bodmer and by Miklos. It is that
there is no structure on a scale larger than the gene family
(i.e. a group of closely linked genes with related functions,
together with sequences controlling them) that is developmentally relevant. Genes and gene families are tied together on
chromosomes as a means of ensuring proper disjunction during
cell division, but their large-scale arrangement has no developmental relevance. Supporters of this view would accept that
differences in the large-scale arrangement of the two chromosome sets of a single individual can lead to infertility, by

interfering with disjunction during meiosis.

The alternative view, which, following Macgregor*, I will call "neo-Goldschmidtian", is that the large-scale structure of the genome is significant for morphological development, and hence that significant morphological evolution requires genomic reorganisation, and morphological conservatism will be accompanied by conservatism of the genome. This view was supported at the conference by Bennett, Davidson and Macgregor, and, at least to some extent, by several other speakers.

Before commenting on some of the evidence bearing on the issue, I should perhaps declare my own position. I have a bias, but not an unshakeable prejudice, in favour of the null hypothesis. There are two reasons for this. One is that it seems sensible to hold that the world is simple until one is forced by the evidence to accept that it is complicated. The other, more personal, reason is that I worked for many years with *Drosophila subobscura*, a species in which all natural populations are polymorphic for large paracentric inversions affecting all autosomal arms, none of which has any morphological effect whatever. More generally, I know of no evidence that forces one to abandon the null hypothesis. It may prove to be false, but it does not seem unreasonably obstinate to accept it.

I have not time to discuss all the evidence relevant to this question that was presented at the conference, but I want briefly to comment on the papers of Bennett, Wilson and Macgregor. Bennett's data suggest that the shapes of chromosomes influence the ways in which they are arranged in metaphase. While his evidence on this seems convincing, it would be well to remember the data presented by Miklos and by Rees, showing that cell division and chromosome disjunction can proceed successfully despite rather large difference in DNA amount between homologous chromosomes. However, the open question is whether chromosome shape influences morphogenesis, not whether it influences the mechanics of chromosome pairing and disjunction. Both Bennett and Manuelidis argue that the way in which chromosomes are arranged in interphase, including the proximity in space of specific regions of different chromosomes, has important consequences. This is an interesting possibility but, even if the proximity of specific regions *is* important, the morphological irrelevance of *Drosophila* inversions, and the viability of translocation heterozygotes in many organisms, suggests that the necessary proximity can be achieved despite large changes in the arrangement of genes

---

*All references in this overview are to papers read at this conference.

on chromosomes.

Wilson's data are more directly relevant to the problem. In earlier papers, he has pointed out that, if one compares mammals and anurans, both morphological evolution and change in genome structure have been more rapid in the former than the latter (although molecular evolution has proceeded at very similar rates), and has suggested that the explanation of this association is that structural changes in chromosomes cause changes in the regulation of development, and hence changes in morphology. In his paper at this meeting he has reported that if one analyses narrower taxonomic groups, there is no longer a correlation between genomic and morphological change. This suggests that no direct causal connection exists between genomic and morphological change, and requires that we seek for some other explanations of the rapidity of both these processes in mammals. My own view is that any attempt to explain the rapidity of morphological evolution that ignores the intensity of directional natural selection is doomed to failure.

Macgregor argued for a neo-Goldschmidtian view from the fact that the plethodontid salamanders have been remarkably conservative both in morphology and in the relative shapes of their chromosomes, although large (approximately fourfold) differences in the amount of DNA in the genomes of congeneric species do exist. These data are certainly consistent with the neo-Goldschmidtian view. Unfortunately, they are also precisely what one would expect on an extreme "selfish DNA" hypothesis. Thus, if there are transposable elements with relatively little influence on adult phenotype, which insert into regions of non-essential DNA with a probability proportional to the quantity of DNA already present, the result would be a general increase in chromosome size, without change of shape. This illustrates the difficulty of interpreting karyotypic data.

I now turn to the difficulties that arise because of the different levels at which selection can act. Studies of the evolution of the genome, particularly of the plant genome, have never fully recovered from the confusion generated by Darlington's "Evolution of Genetic Systems". Changes in genome structure have effects on the viability and fertility of individuals with the changed structure and their immediate progeny, and also on the evolutionary properties of the populations of which those individuals are members. Unless we distinguish between these two kinds of effect, confusion is unavoidable.

This difficulty surfaced on the first day of the meeting, when we found ourselves arguing about the meaning of the word function. Some people clearly felt that to argue about words was fruitless. However, when people from different disiplines meet, these is a real danger that they will misunderstand one another because they differ in their use of some apparently innocuous word. Therefore, at the risk of seeming Jesuitical, I want to discuss the use of the word function. This word has a very precise meaning in some areas of evolutionary biology (e.g. in the evolution of behaviour). If we say that the function of the heart is to pump blood round the body, we do not mean merely that the heart does, as a matter of fact, pump blood. We mean that the heart evolved because it pumped blood; that is, those animals whose hearts were better pumps survived and left more descendents. Hence, the effects or consequences of an organ can be different from its functions. Thus, horses have very stiff backbones, and a consequence of this is that people can ride them. However, we would not say that the function of a horse's backbone is to enable people to ride horses, because we do not think that horses' backbones evolved as they did so as to enable people, in the future, to ride.

When, during our discussion, Miklos remarked (of the effect of satellite DNA in reducing recombination in its neighbourhood) "That's an effect, not a function", he meant that, although satellite DNA may indeed reduce recombination, it did not evolve for that reason.

Usually, the distinction between an effect and a function is fairly clear, but difficulties arise when the effect is not on the individual, but on the species. Thus Finnegan pointed out that transposable elements are important mutagenic agents in prokaryotes, and probably in eukaryotes. However, he clearly thought that this is an effect, but not a function in the sense in which I have used the word above, and he is almost certainly right. To say that the function of transposable elements is to cause mutation would mean something of the following kind: "Those species that have many transposable elements have high mutation rates, and consequently have much genetic variability. Hence, such species compete successfully with species with fewer transposable elements. Thus, differential survival and extinction of species with and without transposable elements explains the prevalence of such elements in existing species". This explanation is not logically impossible, but I think it very implausible. Inter-species selection of this kind is a very weak force, basically because the extinction of species is a rare event compared to the death of individuals. In the case of transposable elements, there are probably much stronger selective forces acting at the level of the individual organisms and of the elements themselves.

This brings me to the point that transposons introduce

a third level of selection into an already complex situation.
In the classical models of population genetics, it is assumed
that if an allele A is inherited from only one parent, it will
be transmitted (on average) to exactly half the offspring.
If so, the only way in which allele A can increase in frequency in the population is *via* its effects on the fitness of
individual organisms (I have not forgotten drift, hitch-hiking,
or even the "elevator effect"; I just prefer to discuss one
complication at a time). But an element such as a transposon,
which can multiply horizontally in the individual, may be inherited from only one parent, yet be transmitted to more than
half the offspring. If intra-genomic multiplication of genetic elements occurs, and if elements with different sequences
have different probabilities of multiplication, it follows
necessarily that the evolution of "selfish" (or, equivalently,
"parasitic") DNA will take place. The conditions for evolution by natural selection exist.

It is important to understand that the proposal that selection is operating at the DNA level, as well as at the organism
and the population levels, is not an additional hypothesis
introduced to explain otherwise anomalous observations. It
is a necessary consequence of multiplication and variation at
the DNA level. In the same way, Hamilton's earlier proposal
of kin selection was not an additional *ad hoc* hypothesis, but
a necessary consequence of the fact that organisms interact
with their relatives. Those many biologists who are accus-
tomed to thinking of organisms as the essential units in biology, and of the parts of organisms as having functions in the
the survival and reproduction of organisms, should remember
that the same logic that explains why hearts pump blood will
predict that transposable elements will have sequences ensuring their own transposition, even if they do not contribute
to the fitness of the organism. It should be emphasized that
it is not essential to the selfish DNA hypothesis that the
presence of such DNA have no effects on the fitness of organisms in which it is present, any more than it is essential
for the existence of a parasite that it have no effect on its
host.

The questions that have to be answered are; first, are
there sequence differences between DNA elements that affect
their success in multiplication; and second, how many sequences, and which kinds, are properly thought of as selfish? On
the former question, the similarities described by Finnegan
between *Drosophila* transposable elements and bacterial transposons on the one hand and DNA proviruses of RNA retroviruses
on the other, is very striking. On the latter question,
clearly not all reiterated sequences are without important
functions for the organism. We already know of some sequences

that are involved in the control of gene action, and doubtless more will be discovered. As Davidson emphasized during
the discussion, we know far too little about gene regulation
to assume that a sequence is functionless merely because we
have not yet discovered its function.

Some evidence for an organismic function of repeated elements was reported at the meeting. One hint was Jeffreys'
observation that the non-coding sequences lying between genes
in the haemoglobin region are rather conservative in evolution. Jones provided more direct evidence of a sequence that
is highly conserved in evolution, and which seems to have some
function in sex determination. However, the difficulty of
interpreting data in this field was brought out when Charlesworth and Cavalier-Smith pointed out during the discussion
that the data were equally compatible with the relevant sequence being a parasitic DNA adapted for life in non-recombining
regions of chromosomes.

As a population geneticist, I wonder whether the time has
not arrived for more formal mathematical models of these processes. In fact, there already existed, in the interstices
of the conference, a small group of population geneticists
exchanging pieces of paper covered in algebra. Further, there
is considerable formal similarity between the spread of a
transposable element and meiotic drive, and there is a substantial body of mathematics on the latter topic. It might
seem that there are as yet so many unknowns that mathematical
models would be premature. However, such models would confer
two advantages. First, they would get us away from awkward
words like "selfish" and "function", because the assumptions
are made explicit by the model. Second, as soon as one attempts to formulate a model, it becomes more apparent what it
is one needs to know or to measure in order to understand the
process.

I want to mention only one aspect that is brought into the
open by an attempt to model transposition. This is the question of how the number of copies per genome of a transposable
element such as *copia* is regulated. I can think of at least
four possibilities, but doubtless there are more.

(1) There is a limited number of possible sites (this is
made unlikely by the fact that an element is found at different sites in different organisms).

(2) There is a balance between multiplication of the element, and selection operating against organisms with large
numbers of copies.

(3) The element itself carries a gene that limits its own
replication, as in the case of some bacterial transposons.
It is not clear to me whether, or in what circumstances, such
a gene could be favoured by selection in a transposon in the

genome of a diploid sexual organism.

(4) The element is limited by organismic genes (i.e. by genes that themselves obey Mendel's laws). This possibility is at least hinted at by the phenomenon of hybrid dysgenesis.

I have spent perhaps too long on the topic of selfish DNA. My excuse is that (as illustrated by some of the quotations in Doolittle's paper) there is still much conceptual confusion, which happens to be logically similar to confusions in other fields of biology, notably in ethology, with which I have been struggling for many years. There are two other topics, gene families and speciation, on which I will touch briefly.

The session on the evolution of gene families was the one in which one could most readily feel that problems were being solved, rather than that questions were being asked. The reason for this, I think, is that neither of the two difficulties I mentioned at the beginning of this overview are relevant here. We have a reasonably good idea of the function of haemoglobins, immunoglobins and actins (although it is far from obvious why sea urchins need 11 actin genes; here we may be getting into the difficult area of gene regulation), and in each case the word function is properly used, since the molecules do contribute in a well-understood way to the survival of individual organisms. The progress reported in this session was exciting, but it is not progress to which I can usefully contribute, except by making generally encouraging noises.

The role of repeated DNA sequences in speciation is inevitably more controversial. The argument put forward by Dover and by Flavell is essentially as follows. Processes are going on in the genome that can lead both to rather rapid changes in the total amount of reiterated DNA, and to concerted changes in the sequences present. Consequently, populations separated from one another for an appreciable time will come to differ in the amount and kind of reiterated DNA. These differences can be expected to interfere with the viability and fertility of hybrids, and hence lead to speciation. The evidence is equivocal. Flavell shows that repeated sequences do have a dysgenic effect in wheat-rye hybrids, which are more viable when these sequences are lost. Rees and Miklos, however, reported cases, in grasses and *Drosophila*, respectively, in which large quantitative differences have little effect on the hybrids. The suggested mechanism may well play a part in the origin of isolating mechanisms between species. Those cases in plants in which the diploid hybrid is sterile but the allotetraploid is fertile are candidates for such an explanation.

It is important, however, for molecular biologists to

appreciate that a failure of chromosome pairing, however it
arises, is certainly not always the primary cause of isolation between species. We know this because there are numerous cases in which two related species of animals are isolated by behavioural differences in courtship but, if interspecific mating is brought about in captivity, the resulting
hybrids are perfectly fertile. This commonly happens, for
example, in birds, grasshoppers and *Drosophila*. This illustrates the point that, although at the molecular level a
single process often turns out to have a single cause, unitary explanations are less commonly true at the organismic
level. There are general principles in evolutionary biology,
notably the principle of natural selection, but the details
of the process are irreducibly complex.

   I conclude with two remarks, one addressed to old-
fashioned evolutionary biologists like myself, and the other
to molecular biologists who have recently become interested
in evolutionary problems. Those on my side of the fence are
going to have to take seriously the mass of new data and
ideas that have been discussed at this conference. Concept-
ually, the most important fact that we shall have to digest
is the widespread occurrence of genetic elements capable of
intra-genomic multiplication. To molecular biologists, I
will only remark that people have been thinking about evolution for a long time, and that some of their conclusions may
be worthy of attention.